SPRINGER
LAB MANUAL

Springer-Verlag Berlin Heidelberg GmbH

Melody S. Clark (Ed.)

Plant Molecular Biology – A Laboratory Manual

With 37 Figures, Including a Color Plate

Springer

Dr. Melody S. Clark

Molecular Genetics Unit
University of Cambridge
School of Clinical Medicine
Department of Medicine, Level 5
Addenbrooke's Hospital
Hills Road
Cambridge CB2 2QQ
UK

ISBN 978-3-642-49144-3 ISBN 978-3-642-87873-2 (eBook)
DOI 10.1007/978-3-642-87873-2

Library of Congress Cataloging-in-Publication Data. Plant molecular biology: a laboratory manual/ Melody S. Clark (ed.). p. cm. - (Springer lab manual) Includes biliographical references and index. ISBN 3-540-58405-6 (alk. paper) 1. Plant molecular biology--Laboratory manuals. 2. Plant molecular genetics--Laboratory manuals. I. Clark, Melody. II. Series. QK728, P536 1996 581.8'8-dc20 96-23895

Originally published by Springer-Verlag Berlin Heidelberg New York in 1997.

Cover design: Design & Production GmbH, Heidelberg

Typesetting: Scientific Publishing Services (P) Ltd, Madras

SPIN: 10469549 39/3137/SPS - 5 4 3 2 1 0 - Printed on acid-free paper

Preface

The aim of this manual is to encompass a broad range of the latest plant molecular biology techniques. However, it is acknowledged that any manual will be read (and hopefully) used by a wide range of people with different levels of experience. Hence the remit of the manual was widened to include a full range of basic molecular techniques, so that novices do not have to consult several texts to enable the execution of each major experiment.

The manual is divided into three main parts:

Part I: Basic Molecular Techniques

The raison d'être behind this part is to provide a background knowledge of molecular techniques, but also to reduce duplication in later chapters (this is particularly true of the methods contained in Chap. 1). All authors provided very detailed methods and often forgot that some of these would be covered earlier. A particular favourite was DNA extraction methods, where everyone managed to provide a slightly different variant! My view was that it is far less confusing for the reader to be presented with one standard protocol and accompanying troubleshooting tips, than to read a different version in each chapter. In this way the basic techniques are addressed more in depth (and my apologies to all authors for judicious use of the delete key!).

RNA methodology is covered in Chapter 3. This proceeds from the fundamentals of extraction, northern blotting etc., to cDNA libraries. The more complex selective cloning techniques of differential display reverse transcriptase PCR and subtractive hybridization are fully described to aid cloning of specific RNA populations, an increasingly important consideration in the search for novel genes and developmentally expressed sequences.

DNA cloning techniques are provided alongside characterization of clones (Chaps. 2 and 4). Whilst not uniquely plant-based techniques, they are essential for the furtherance and analysis of more complex biotechnology. Their inclusion also excludes the need for further consultation with yet more text books and hopefully provides a more even flow through with the planning and execution of experiments.

Part II: Characterization of Plant DNA

This deals with the analysis of the plant genome as a whole, at both the organelle and genomic levels. This is particularly important, not only for the mapping and isolation of specific gene sequences , but also for the broad genomic characterization of a plant species prior to DNA manipulation.

Various mapping approaches and techniques (RFLPs and RAPDs) are discussed in depth. Emphasis is placed on the initial choice of starting material and potential constraints. Methods of analysis and statistical parameters are extensively covered, in particular for RFLPs, which are still by far the most popular method of mapping plant genomes.

Part III: Genetic Engineering Methodology and Analysis

The final part deals with what is "classically" called plant biotechnology or gene manipulation.

Chapter 8 covers gene transfer at both a single-gene level using *Agrobacterium*-mediated transfer, electroporation and particle bombardment, and also the more "random" fusion technology, the use of which depends on the crop, i.e. how developed the tissue culture and regeneration systems are, what gene sequences are available and whether breeding barriers between wild and domesticated family members exist.

Subsequent chapters deal with the molecular characterization of transformed plants. The approaches differ depending on whether single genes and reporter systems were used or somatic fusion. The former utilizes Southern analysis, GUS and LUC assays, whilst the latter requires a more genome-wide approach, consistent with the massive genomic disruption induced by fusion of whole or partial donors to recipient cells.

The inclusion in the final chapters of in situ hybridization techniques is an acknowledgement that molecular biology, these days, cannot stand alone. It is becoming increasingly integrated with other fields. For example, whilst it is important to clone a gene, in order to gain a fuller understanding of its action and immediate environmental influences, physical mapping is required. Additional emphasis is placed on the manipulation of small chromosomes, in acknowledgement of the fact that, in the vast majority of crop plants (apart from the cereals), the commercial usefulness often seems to be inversely proportional to the size of the chromosomes! Whilst prometaphase analysis is possible in these crops, developments in the field of human genetics such as hybridization to extended chromatin fibres may soon be extended to plants. However, the latter technique is probably far too specialized for the average plant molecular lab, and is outside the scope of this manual.

Best of luck!

Cambridge, January 1996 MELODY S. CLARK

Contents

List of Contributors

BENNETT, M.D.
Royal Botanic Gardens, Kew
Richmond
Surrey TW9 3DS
UK

BERNARD, M.
INRA
Station d'Amélioration
des Plantes
63039 Clermont-Ferrand Cedex 2
France

BRADY, G.
School of Biological Sciences
University of Birmingham
Edgbaston
Birmingham B15 2TT
UK

CAISSARD, J.C.
INRA
Station d'Amélioration
des Plantes
63039 Clermont-Ferrand Cedex 2
France

CHASALOW, S.D.
Biomathematics
and Statistics Scotland
Scottish Crop Research Institute
Invergowrie, Dundee
Scotland DD2 5DA
UK

CLARK, J.R.
Animal Health Discovery
Pfizer Central Research
Sandwich, Kent CT13 5NJ
UK

CLARK, M.S.
Molecular Genetics Unit
University of Cambridge
School of Clinical Medicine
Department of Medicine, Level 5
Addenbrooke's Hospital
Hills Road
Cambridge CB2 2QQ
UK

DECROOCQ, V.
CSIRO, Division of Plant Industry
PO Box 1600
2601 ACT Canberra
Australia

ELGAR, G.
Molecular Genetics Unit
University of Cambridge
School of Clinical Medicine
Department of Medicine, Level 5
Addenbrooke's Hospital
Hills Road
Cambridge CB2 2QQ
UK

Foote, H.C.C.
Department of Biology
UCLA
405 Hilgard Avenue
Los Angeles, CA 90024-1606
USA

Franklin, C.H.
School of Biological Sciences
University of Birmingham
Edgbaston
Birmingham B15 2TT
UK

Glimelius, K.
Department of Plant Breeding
Research, Swedish University
of Agricultural Sciences
Box 7003
75007 Uppsala
Sweden

Guevara-García, A.
Departmento de Ingeniería
Genética de Plantas
Centro de Investigación
y de Estudios Avanzados del IPN
Unidad Irapuato
Apartado Postal 629
36500 Irapuato, Guanajuato
Mexico

Guzmán, P.A.
Departmento de Ingeniería
Genética de Plantas
Centro de Investigación
y de Estudios Avanzados del IPN
Unidad Irapuato
Apartado Postal 629
36500 Irapuato, Guanjuato
Mexico

Herrera-Estrella, L.
Departmento de Ingeniería
Genética de Plantas
Centro de Investigación
y de Estudios Avanzados del IPN
Unidad Irapuato
Apartado Postal 629
36500 Irapuato, Guanajuato
Mexico

Jones, M.G.K.
The State Agricultural
Biotechnology Centre
Murdoch University
Perth, Western Australia 6150

Kenton, A.Y.
Royal Botanic Gardens, Kew
Richmond
Surrey TW9 3DS
UK

Landgren, M.
Department of Plant Breeding
Research, Swedish University
of Agricultural Sciences
Box 7003
75007 Uppsala
Sweden

Lapitan, N.
Department of Agronomy
Colorado State University
Fort Collins
CO 80523
USA

Leitch, I.J.
Royal Botanic Gardens, Kew
Richmond
Surrey TW9 3DS
UK

Leroy, P.
INRA
Station d'Amélioration
des Plantes
63039 Clermont-Ferrand Cedex 2
France

Lessard, P.
Institut de Biotechnologie
des Plantes
URA CNRS 1128
Laboratoire de Biologie
du Développment des Plantes
Bâtiment 630
Université de Paris Sud
91405 Orsay Cedex
France

Lindsey, K.
Department of Biological
Sciences
University of Durham
South Road
Durham DH1 3LE
UK

Liu, C.-N.
Department of Agronomy
Purdue University
West Lafayette, IN 47907-1150
USA

Lu, Y.H.
INRA
Station d'Amélioration
des Plantes
63039 Clermont-Ferrand Cedex 2
France

Matallana, E.
Departmento de Bioquímica
y Biología Molecular
Facultad de Ciencias
Universitat de Valencia
46100 Burjassot, Valencia
Spain

Merlino, M.
INRA
Station d'Amélioration
des Plantes
63039 Clermont-Ferrand Cedex 2
France

Nègre, S.
INRA
Station d'Amélioration
des Plantes
63039 Clermont-Ferrand Cedex 2
France

Olmedo-Alvarez, G.
Departmento de Ingeniería
Genética de Plantas
Centro de Investigación
y de Estudios Avanzados del IPN
Unidad Irapuato
Apartado Postal 629
36500 Irapuato, Guanajuato
Mexico

Parokonny, A.S.
Royal Botanic Gardens, Kew
Richmond
Surrey TW9 3DS
UK

PEHU, E.
Department of Plant Production
Box 27
University of Helsinki
00014 Helsinki
Finland

PEHU, T.
Department of Plant Production
Box 27
University of Helsinki
00014 Helsinki
Finland

POTTER, R.H.
The State Agricultural
Biotechnology Centre
Murdoch University
Perth, Western Australia 6150

SIMPSON, J.
Departmento de Ingeniería
Genética de Plantas
Centro de Investigación
y de Estudios Avanzados
del IPN
Unidad Irapuato
Apartado Postal 629
36500 Irapuato, Guanajuato
Mexico

SOURDILLE, P.
INRA
Station d'Amélioration
des Plantes
63039 Clermont-Ferrand Cedex 2
France

THOMAS, M.
Institut de Biotechnologie
des Plantes
URA CNRS 1128
Laboratoire de Biologie
du Dévéloppement des Plantes
Bâtiment 630
Université de Paris Sud
91405 Orsay Cedex
France

THORLBY, G.J.
School of Biological Sciences
University of Birmingham
Edgbaston
Birmingham B15 2TT
UK

TOPPING, J.F.
Department of Biological Sciences
University of Durham
South Road
Durham DH1 3LE
UK

VAN DEN BERG, J.H.
Nunhems Zaden BV
PO Box 4005
6080 AA Haelen
The Netherlands

WAUGH, R.
Department of Cellular
and Molecular Genetics
Scottish Crop Research Institute
Invergowrie, Dundee
Scotland DD2 5DA
UK

WILKIE, S.
Department of Molecular
Genetics
Institute of Opthalmology
University College London
Bath Street
London EC1V 9EL
UK

XU, Y.-S.
Department of Anatomy
and Cell Biology
Temple University
School of Medicine
3400 N. Broad Street
Philadelphia, PA 19140
USA

List of Abbreviations

The following is a complete list of abbreviations used throughout the book.

4-MU	4-Methyl Umbelliferone
ANOVA	ANalysis Of VAriance
AP	Alkaline Phosphatase
AP-PCR	Arbitrarily Primed Polymerase Chain Reaction
ATP	Adenosine Triphosphate
BC	BackCross
BCIP	5-Bromo-4-Chloro-3-Indolyl-Phosphate, also known as X-phosphate
bp	base pairs
BSA	Bovine Serum Albumen
cDNA	complementary DeoxyriboNucleic Acid
CIAP	Calf Intestinal Alkaline Phosphatase
cM	centiMorgan
CMS	Cytoplasmic Male Sterility
cos	cohesive end sites
cpDNA	chloroplast DNA
CTAB	Cetyl TriethylAmmonium Bromide
DAB	DiAminoBenzidine
DAPI	4′-6-DiAmidino-2-PhenylIndole
dATP	deoxyAdenosine TriPhosphate
dCTP	deoxyCytidine TriPhosphate
DEPC	DiEthylPyroCarbonate
dGTP	deoxyGuanidine TriPhosphate
DH	Doubled Haploids
DIG	DIGoxigenin
DMSO	DiMethyl SulphOxide
DNA	DeoxyriboNucleic Acid
DNase	DeoxyriboNuclease
dNTP	deoxyNucleotide TriPhosphates (usually a mix of dATP/dTTP/dCTP/dGTP)
DTT	DiThioThreitol

dTTP	deoxyThymidine TriPhosphate
dUTP	deoxyUridine TriPhosphate
EDTA	ethyleneDiamine Tetracetic Acid
EGTA	Ethylene Glycol Tetraacetic Acid
EPPS	(N-[hydroxyEthyl]Piperazine-N-[3-Propane-Sulphonic acid];HEPPS)
EtOH/EthOH	ethanol
FITC	Fluoroscene IsoThioCyanate
gDNA	genomic DeoxyriboNucleic Acid
GISH	Genomic In Situ Hybridization
GUS	β-GlUcoronidaSe
HAP	HydroxyAPatite
IPTG	IsoPropyl β ThioGalactopyranoside
ISH	In Situ Hybridization
KAc	potassium acetate
kb	kilobase pairs
LB	Luria-Bertani media
LOD	Log Of Odds
LUC	Firefly LUciferase
Lumigen PPD	4 methoxy-4-(3-phosphatephenyl)spiro-(1,2-dioxetane-3,2′-adamantaneh)
MAS	Marker-Assisted Selection
MLE	Maximum Likelihood Estimator
mRNA	messenger RNA
mtDNA	mitochondrial DNA
MUG	4-Methyl Umbelliferone Glucuronide
NaAc/NaOAc	sodium Acetate
NBT	4-Nitro Blue Tetrazolium chloride
NIL	Near Isogenic Lines
O/N	OverNight
ORF	Open Reading Frame
PAH	PhenAnthroline Hydrate
PCR	Polymerase Chain Reaction
PEG	PolyEthylene Glycol
PFGE	Pulsed Field Gel Electrophoresis
PI	Propidium Iodide
PVP	PolyVinylPyrrilodone
QTL	Quantitative Trait Loci
RAPD	Randomly Amplified Polymorphic DNAs
RFLP	Restriction Fragment Length Polymorphism
Ri	*Agrobacterium rhizogenes*

RI	Recombinant Inbred lines
RNA	RiboNucleic Acid
RT-PCR	Reverse Transcriptase Polymerase Chain Reaction
SDS	Sodium Dodecyl Sulphate
SDS-PAGE	SDS-PolyAcrylamide Gel Electrophoresis
SDW	Sterile Distilled Water
SSC	Standard Saline Citrate
T-DNA	region of plasmid DNA transferred and integrated into plant with *Agrobacterium* transformation
TB	Terrific Broth
TCA	TriChloroacetic Acid
TEMED	N,N,N′,N′-TEtraMethylEneDiamine
Ti	*Agrobacterium tumifaciens*
UV	Ultraviolet
X-gal	5-bromo-4-chloro-3-indolyl β D galactoside
X-gluc	5-bromo-4-chloro-3-indolyl glucuronide
YAC	Yeast Artificial Chromosome

Part I

Basic Molecular Techniques

Chapter 1

Genomic DNA Isolation, Southern Blotting and Hybridization

1.1 Isolation of Total Genomic DNA

S. Wilkie

1.1.1 Introduction

The isolation of plant nucleic acids is a fundamental requirement for most genome characterization and mapping procedures involving the use of genetic markers, and for the identification and isolation of plant genes for genetic engineering. The degree of purity and quality required of the DNA isolated varies from application to application. On the one hand, high molecular weight, high purity DNA is required for the production of genomic DNA libraries, which are screened for plant gene sequences and for other genetic markers such as RFLPs. For genetic analysis, on the other hand, the degree of purity required may be lower, but other factors, such as the yield of DNA, may be more important.

In general, a good extraction procedure for the isolation of DNA should fulfil three major criteria.

- It should yield DNA of reasonable purity for the subsequent manipulations to which it is to be subjected. In the case of RFLP analysis this means that it must be pure enough to be cut completely by restriction endonucleases and to be transferred successfully to membranes for Southern analysis. For analysis involving use of the polymerase chain reaction (PCR), the DNA extract must not contain contaminants which interfere with the requirements of PCR.

- The DNA should be sufficiently intact to give accurate and reproducible migration patterns following gel electrophoresis.

- The yield of DNA must be adequate. For example, in order to study the segregation of a range of markers in a population of individual plants, this means that the yield of DNA from a single plant must be sufficient for perhaps a hundred different analyses. On the other hand, for screening of a large number of plants for a single marker, the yield of DNA from a micro-scale extraction procedure carried out in a single Eppendorf tube may be more relevant.

  For routine application in large-scale plant screening programmes, further criteria may be added.

- The procedure should be quick, simple and cheap and, if possible, avoid the use of dangerous chemicals.

1.1.2
Theory Underlying Extraction Procedures

Extraction procedures for plant DNA must accomplish the following. Firstly, the cell walls must be broken (or digested away) in order to release the cellular constituents. Unfortunately, many techniques for breaking open cells also shear DNA and thus any method must be a compromise between DNA length and yield. For the isolation of whole genomic DNA, the usual procedure involves dehydration of the plant tissue and then grinding it to a fine powder. Alternatively, fresh plant tissue may be fast frozen in dry ice or liquid nitrogen and, while still frozen, ground to a powder using a pestle and mortar. For the isolation of nuclear or organellar DNA, the procedures used must be gentler so as to avoid premature damage to internal membranes. In this case, homogenization at 4 °C in a buffer containing an osmoticum is generally used.

Secondly, the cell membranes must be disrupted so that the DNA is released into the extraction buffer. This is accomplished by using a detergent, for example sodium dodecyl sulphate (SDS) or cetyl trimethylammonium bromide (CTAB). The DNA must be protected from endogenous nucleases. The detergents serve this purpose, and EDTA is often included in the extraction buffer to chelate magnesium ions, a necessary co-factor for most nucleases.

Finally, shearing of the DNA once released must be minimized. High molecular weight DNA in solution can be broken by exposure to turbulence, for example by vortexing or by being drawn quickly through a small orifice. Typically, DNA of molecular weight 50–100 kb can be obtained with reasonable care.

The extraction of high molecular weight DNA is, however, only one part of the task, as initial DNA extracts often contain large amounts of RNA, proteins, polysaccharides, tannins and pigments which, in some cases, are difficult to separate from the DNA. Most proteins can be removed by denaturation and precipitation from the extract, for example using chloroform and/or phenol. Treatment of the extract with heat-treated RNAse A will remove most of the RNA. Polysaccharide-like contaminants, however, are often more difficult to remove and, when present in high concentration, may cause the DNA extract to assume an almost gel-like constitution. More importantly, they may interfere in subsequent procedures even at relatively low concentrations. They can inhibit the activity of certain DNA-modifying enzymes, including restriction enzymes, which can lead to difficulties when performing Southern analysis of DNA or when attempting genomic cloning. They may also interfere in the quantification of nucleic acids by spectrophotometric methods.

A few commonly used procedures for the extraction of plant DNA for different purposes are described in this chapter. Two basic extraction protocols, the CTAB method (Sects. 1.1.3 and 1.1.4) and the method of Dellaporta et al. (Sect. 1.1.5), are presented which, in the past, have been used successfully with a wide range of species to give a good yield of reasonably high molecular weight DNA, albeit of variable purity. Both these methods can be readily scaled down to yield DNA from as little as 20 mg of dehydrated or 0.5 g of fresh plant material. An alternative protocol is then presented for use with species where the basic protocols fail to give DNA of sufficient purity (Sect. 1.1.6). This relies on performing an initial isolation of intact nuclei, which are then lysed to release their DNA. Finally, a method of purifying DNA from crude DNA extracts by differential density centrifugation on a caesium chloride gradient is given (Sect. 1.1.7).

1.1.3
Extraction of Whole Genomic DNA by the CTAB Method

This method, a modification of the methods described by Murray and Thompson (1980) and Saghai-Maroof et al. (1984), is relatively simple, and has been used successfully with a wide range of monocot and dicot species. The method may be used with either fresh or dehydrated plant material (Note 1). Although the DNA obtained is not of high purity, it is generally of sufficient quality for restriction analysis or for amplification by PCR. Yields are typically of the order of 100–200 μg/g fresh weight of tissue.

Reagents

- Liquid nitrogen
- 1x CTAB extraction buffer: 50 mM Tris-HCl pH 8, 0.7 M NaCl, 10 mM EDTA pH 8, 1% CTAB, 20 mM 2-mercaptoethanol (added immediately prior to use)
- Chloroform/isoamyl alcohol (24:1)
- RNAse A: 2000 U/ml, 10 mg/ml in 0.15 M NaCl, 0.015 M trisodium citrate; heat-treat at 100 °C for 15 min prior to use to remove residual DNAse activity
- 2-Propanol
- 76% aqueous ethanol, 0.2 M sodium acetate
- 70% aqueous ethanol
- TE buffer: 10 mM Tris-HCl, 1mM EDTA pH 8

Equipment

- Balance
- Pestle and mortar
- 50 ml polypropylene centrifuge tubes
- 37, 65 and 90 °C water baths
- Wide-bore Pasteur pipettes
- Centrifuge (capable of achieving at least 5000 *g*)
- 4 °C refrigerator

Procedure

1. Add 20 ml 1x CTAB extraction buffer, pre-heated to 65 °C, to 700–800 mg dried and ground plant tissue (Note 1) [preferably de-starched (Note 2)] in a 50 ml polypropylene centrifuge tube. Swirl gently to ensure even dispersal of the plant material in the buffer, and then incubate at 65 °C for 90 min with occasional swirling.

   Alternatively, if using fresh plant material, quick-freeze 10 g de-starched (Note 2) plant tissue in liquid nitrogen and grind to a fine powder using a pestle and mortar. Transfer the powder, while still frozen, to a 50-ml polypropylene centrifuge tube and add 20 ml 2x CTAB extraction buffer, pre-heated to 90 °C (Note 3). Swirl gently to mix and incubate at 65 °C for 90 min as above.

2. Cool the mixture to room temperature and then add an equal volume of chloroform/isoamyl alcohol. Invert the tube repeatedly but

gently to form an emulsion. Centrifuge at 5000 *g* for 10 min at room temperature to separate the phases.

3. Remove the upper (aqueous) phase to a clean tube and add 1/100 vol RNAse A stock solution. Mix by inversion and incubate at 37 °C for 30 min.

4. Add 0.7 vol 2-propanol, mix carefully and stand at room temperature for 15 min to allow the DNA to precipitate out.

5. Lift out the DNA with a glass hook (Notes 4,5) and transfer to a clean tube containing 5 ml 76% aqueous ethanol, 0.2 M sodium acetate. Allow to stand on ice for 20 min, then transfer to a tube containing 5 ml 70% aqueous ethanol, again on ice, and leave for 5 min.

6. Remove the DNA to a clean tube and allow to air-dry for 15 min before dissolving in the minimum volume of TE buffer (up to 5 ml, depending on yield and purity). To aid the solution process and to remove any residual DNAse activity, the tube may be heated in a 65 °C water bath for 10 min. Store at 4 °C.

Notes

- *Note 1.* Dehydration is a convenient way of storing plant material before DNA isolation. It is most commonly achieved by lyophilization (freeze drying). Alternatively, Tai and Tanksley (1990) have reported success using a commercial food dehydrator in which the tissue is dried at 45 to 55 °C by passing over warm, dry air for a period of a least 12 h. The dried tissue may be disrupted very efficiently by dry milling or grinding to yield a fine powder, which can be stored long term at room temperature.
- *Note 2.* De-starching is achieved by keeping the plant in the dark for 24 h immediately prior to harvesting.
- *Note 3.* CTAB solutions will precipitate out at temperatures below 15 °C, so it is important to pre-heat the 2x CTAB buffer before adding it to the frozen plant material.
- *Note 4.* In the optimum situation the DNA-CTAB precipitate is white and fibrous and may be easily hooked out of the solution in a single wad. However, with some species, the DNA may contain contaminants, in particular polysaccharides, which change its constitution to a more flocculant or gelatinous mass. Isolation of the

DNA-CTAB precipitate by gentle centrifugation may be required. However, care should be taken to avoid over-compaction of the pellet, as such pellets may prove extremely difficult to redissolve.

- *Note 5.* Discoloration of the DNA may occur with certain species as a result of the presence of brown polyphenolic compounds. Often the presence of such discoloration does not affect subsequent procedures. However, where there is interference, the inclusion of 2% (w/v) polyvinylpyrrolidone (M_r 10 000) in the extraction buffer may help to remove polyphenolic contaminants (Couch and Fritz 1990).

1.1.4
Micro-Isolation of DNA by the CTAB Method

Reagents

- CTAB extraction buffer (–CTAB): 50 mM Tris-HCl pH 8, 0.7 M NaCl, 10 mM EDTA pH 8
- CTAB extraction buffer (+CTAB): as above plus 2% CTAB and 40 mM 2-mercaptoethanol
- Other reagents as for the large-scale CTAB method

Equipment

- Eppendorf tubes
- Glass rod with a rounded end ground to fit the bottom of an Eppendorf tube
- Microfuge
- 65 °C water bath
- Stretched glass Pasteur pipettes
- 4 °C refrigerator
- Micropipettes and tips

Procedure

1. Homogenize approximately 1–2 cm^2 of fresh young leaf in 400 µl ice-cold CTAB extraction buffer (–CTAB) in an Eppendorf tube using the ground glass rod.

2. Add 500 µl CTAB buffer (+CTAB), pre-heated to 65 °C, mix and incubate at 65 °C for 90 min with occasional inversion.

3. Cool to room temperature before adding 450 µl chloroform/isoamyl alcohol. Emulsify by repeated inversion – do not vortex. Spin in a microfuge for 2 min to separate the layers.

4. Remove the aqueous phase to a clean Eppendorf tube. Add 5 µl RNAse A stock and incubate at room temperature for 30 min.

5. Add 600 μl 2-propanol and mix by inversion.

6. Pellet the precipitated DNA by spinning in a microfuge for 10 min. Pour off the supernatant and rinse the pellet first with 800 μl 76% aqueous ethanol, 0.2 M sodium acetate for 5 min and then with 100 μl 70% aqueous ethanol for 5 min.

7. Spin the pellet in the microfuge for 5 min and then remove all traces of the ethanol rinses with a stretched glass Pasteur pipette.

8. Resuspend the DNA pellet in 100 μl TE buffer and store at 4 °C.

1.1.5
Extraction of Whole Genomic DNA by the Method of Dellaporta, Wood and Hicks (1983)

This alternative procedure again relies on an initial extraction with hot detergent (SDS). High molarity potassium acetate is then added to the extract at 0 °C and the mixture is centrifuged to remove insoluble materials. This is purported to remove contaminating proteins and polysaccharides.

Reagents

- Liquid nitrogen
- Extraction buffer: 500 mM NaCl, 100 mM Tris-HCl pH 8, 50 mM EDTA pH 8, 10 mM 2-mercaptoethanol (added immediately prior to use)
- 20% SDS pH 7.2
- 5 M potassium acetate
- Miracloth
- 50 TE buffer: 50 mM Tris-HCl, 10 mM EDTA, pH 8
- RNAse A: 2000 U/ml (heat-treated)
- 3 M sodium acetate
- 2-propanol
- TE buffer: 10 mM Tris-HCl, 1 mM EDTA, pH 8

Equipment

- Balance
- Pestle and mortar
- 50 ml centrifuge tubes
- 37 and 65 °C water baths
- Ice bucket
- Refrigerated centrifuge (capable of achieving at least 25 000 *g*)
- –20 °C freezer

- Eppendorf tubes
- Micropipettes and tips
- Microcentrifuge
- Glass Pasteur pipettes
- 4 °C refrigerator

Procedure

1. Quick freeze 2 g fresh, de-starched plant tissue in liquid nitrogen and grind to a fine powder using a pestle and mortar.
2. Transfer the powder, while still frozen, to a 50-ml centrifuge tube and add 15 ml extraction buffer. Swirl to disperse the tissue.
3. Add 1 ml 20% SDS, mix and incubate at 65 °C for 10 min with occasional swirling to mix (Note 1).
4. Add 5 ml 5 M potassium acetate, mix and place on ice for 20–30 min (Note 2).
5. Centrifuge at 25 000 *g* for 20 min at 4 °C.
6. Pour the supernatant through a filter comprising two layers of Miracloth into a clean 50-ml centrifuge tube. Add 10 ml 2-propanol, mix gently by inversion and leave at −20 °C for 40 min to allow the nucleic acids to precipitate.
7. Centrifuge at 20 000 *g* for 20 min at 4 °C. Pour off the supernatant and allow the pellet to drain for 5 min.
8. Resuspend the pellet in 700 µl 50 TE buffer and transfer to an Eppendorf tube (Note 3).
9. Add 7 µl RNAse A stock solution and incubate at 37 °C for 1 h.
10. Add 75 µl 3 M sodium acetate, mix by flicking and spin in a microfuge for 15 min to pellet insoluble debris.
11. Transfer the supernatant to a clean tube containing 500 µl 2-propanol, mix gently and allow to stand at ambient temperature for 5 min.
12. Pellet the DNA in a microfuge for 15 min. Pour off the supernatant and rinse the pellet with 500 µl 70% aqueous ethanol. Re-pellet in the microfuge for 5 min and remove the ethanol with a stretched glass Pasteur pipette.
13. Resuspend the pellet in 200 µl TE buffer and store at 4 °C.

Notes

- *Note 1.* The SDS may precipitate when added to the thawed extract. Ensure that it redissolves and that the extract is well mixed during the incubation at 65 °C.
- *Note 2.* In the presence of high potassium ion concentration (> 1 M) sodium dodecyl sulphate (SDS) forms insoluble potassium dodecyl sulphate complexes with proteins and polysaccharides. These are often flocculant and have to be removed by centrifugation at a relatively high *g* force and filtration through Miracloth.
- *Note 3.* If the pellet is difficult to resuspend, the tube may be heated at 65 °C for 10 min.

1.1.6
Extraction of DNA from Isolated Nuclei

With "difficult" plants, for example plants high in polyphenolics or polysaccharides, the basic protocols described above may fail to give DNA extracts of a purity sufficient for subsequent manipulations. One way round this problem is to attempt an initial concentration of the nuclei away from most of the cytoplasmic components prior to lysis.

The method for isolating intact nuclei described here (Henfrey and Slater 1988) requires the use of fresh plant material. It involves disrupting the cells under gentle conditions designed to minimize damage to the nuclear membranes, filtering off the general cell debris and collecting a crude nuclei-enriched pellet by gentle centrifugation of the filtrate. The DNA extracted from isolated nuclei is generally of high quality and purity. Although the extract will contain a predominance of nuclear DNA, some organellar DNA, in particular chloroplast DNA, is also likely to be present. A major drawback of this approach, however, is that overall yields tend to be low, typically of the order of 10–20 μg/g fresh weight of tissue.

Reagents and Equipment

In addition to the equipment and reagents required for the CTAB method the following are required.

- Nuclei isolation buffer (NIB): 0.44 M sucrose, 2.5% Ficoll (M_r 400 000), 5% Dextran 40, 25 mM Tris-HCl pH 7.6, 10 mM magnesium chloride, 10 mM 2-mercaptoethanol (added immediately prior to use)
- Nuclei lysis buffer: 10 mM Tris-HCl, 10 mM EDTA, 0.5% Triton X-100, pH 7.2
- Miracloth

Procedure

All solutions and equipment used in steps **1–4** must be pre-chilled to 4 °C in order to reduce the activity of any endogenous proteases and nucleases present.

1. Cut 10 g fresh, de-starched leaf tissue into small pieces and homogenize in 10 cm^3 NIB using a pestle and mortar.

2. Dilute the homogenate with 2 vol NIB and filter through two layers of Miracloth (pre-wetted with NIB). Wash the residue with a further 5 vol NIB.

3. Divide the filtrate between two 50-ml polypropylene centrifuge tubes and centrifuge at 500 *g* for 10 min at 4 °C. Discard the supernatants.

4. Resuspend the pellets in a total of 20 ml NIB, combine and centrifuge again at 500 *g* for 10 min at 4 °C. Discard the supernatant.

5. Resuspend the crude nuclear pellet in 6 ml nuclei lysis buffer. Add 6 ml 2x CTAB extraction buffer, pre-heated to 65 °C, swirl to mix and incubate at 65 °C for 1 h with occasional swirling.

6. Follow the procedure for the basic CTAB method from step 2, Section 1.1.3.

1.1.7
Purification of DNA Extracts by Differential Density Centrifugation on a Caesium Chloride Gradient

This technique will remove most protein, RNA, polysaccharide and other contaminants from crude DNA extracts produced by any of the foregoing methods. It is useful, therefore, when a really pure sample of DNA is required, for example to prepare a DNA library. The technique, however, involves the use of both toxic chemicals (ethidium bromide and caesium chloride) and sophisticated equipment (an ultracentrifuge). It is unsuitable, therefore, for routine application. Furthermore, it does not lend itself to multi-sample purifications.

Reagents

- Caesium chloride (Molecular Biology grade)
- Ethidium bromide, 5 mg/ml
- Caesium chloride – saturated isoamyl alcohol or 2-propanol
- 3 M sodium acetate
- Absolute alcohol

- 70% aqueous ethanol
- TE buffer: 10 mM Tris-HCl, 1 mM EDTA, pH 8

Equipment

- Ultracentrifuge (capable of achieving at least 55 000 rpm)
- Vertical rotor (for example a Sorvall TV-865)
- Polyallomer centrifuge tubes
- Long wave UV (320 nm) lamp
- 1-ml syringe fitted with a wide-gauge needle
- Eppendorf tubes
- Micropipettes and tips
- Microcentrifuge
- –70 °C freezer
- Stretched glass Pasteur pipettes

Procedure

1. Centrifuge the crude DNA extract briefly to pellet any insoluble debris and remove 3.6 ml to a clean tube. Add 4 g caesium chloride and allow to dissolve by heating in a warm water bath. When completely dissolved, add 0.4 ml ethidium bromide solution and mix.
2. Transfer the solution to a 5 ml polyallomer centrifuge tube and, if necessary, fill to the top with 1 g/ml caesium chloride solution.
3. Centrifuge at 55 000 rpm in a Sorvall TV-865 rotor or its equivalent for at least 12 h. A longer time may be required if a vertical rotor is not used.
4. Visualize the DNA band under long wave UV (320 nm) illumination and remove it to an Eppendorf tube by side puncture of the centrifuge tube using a 1-ml syringe fitted with a wide-gauge needle.
5. Remove the ethidium bromide by repeated extraction with caesium chloride–saturated isoamyl alcohol or 2-propanol until the organic phase is completely colourless.
6. Dilute the aqueous solution four times with TE buffer.
7. Precipitate the DNA by adding 0.1 vol 3 M sodium acetate and 2.5 vol ice-cold absolute ethanol, mixing and allowing to stand at –70 °C for at least 30 min. Pellet by spinning in a microfuge for 10 min. Remove the supernatant and rinse the pellet with 500 ml 70% aqueous ethanol. Spin in the microfuge for 5 min and remove the ethanol wash solution with a stretched glass Pasteur pipette.
8. Redissolve the DNA pellet in 100 ml TE buffer. Store at 4 °C.

1.1.8
Estimation of DNA Concentration, Purity and Quality

Quantification of reasonably pure DNA solutions may be achieved by UV absorbance spectrophotometry over the wavelength range 200–300 nm. The DNA should show a clear absorbance peak at 260 nm. The A_{260} value provides a measure of concentration (1.0 A_{260} = 50 µg/ml in a 1-cm light path) whilst the A_{260}/A_{280} ratio provides as estimate of purity (a pure DNA solution has an A_{260}/A_{280} ratio of 1.8 ± 0.1).

Where the DNA extract is less pure, data obtained from UV spectrophotometry may be misleading because of interference by RNA or non-nucleic acid contaminants. In this case, quantification of the DNA in the extract may be achieved by analysis on an agarose electrophoresis gel with ethidium bromide staining. Aliquots of the extract containing approximately 50–100 ng DNA are analyzed on a low percentage agarose gel (0.65% agarose) alongside a range of uncut lambda DNA standards containing from 25 to 200 ng DNA. High molecular weight DNA (MW > 50 kb) appears as a well-resolved band alongside the lambda DNA bands, whilst smearing below the band indicates mechanical and/or chemical degredation. An additional smeared band towards the bottom of the gel is indicative of the presence of RNA in the extract.

Quantification of the DNA is achieved by comparing the intensity of the DNA bands from the extract with the intensity of the bands from the lambda DNA standards. A rough estimate of DNA content (±10 ng) may be obtained by comparing band intensities by eye. However, greater precision may be gained with the aid of commercial gel imaging equipment and gel analysis software (for example UVP (Cambridge) software package SW2000). This allows band intensities to be assigned numerical values and from a calibration plot of band intensity vs. DNA loading for the lambda standards, a more precise estimate of DNA content in the unknown extract may be made.

References

Couch JA, Fritz PJ (1990) Isolation of DNA from plants high in polyphenolics. Plant Mol Biol Rep 8(1):8–12

Dellaporta SL, Wood J, Hicks JB (1983) A plant DNA minipreparation:version II. Plant Mol Biol Rep 1(4):19–21

Henfrey RD, Slater RJ (1988) Isolation of plant nuclei. In: Walker JM (ed) Methods in molecular biology, vol 4. Humana Press, Totowa, New Jersey, pp 447–452
Murray MG, Thompson WF (1980) Rapid isolation of high molecular weight plant DNA. Nucleic Acids Res 8:4321–4325
Saghai-Maroof MA, Soliman KM, Jorgensen RA, Allard RW (1984) Ribosomal DNA spacer-length polymorphisms in barley: Mendelian inheritance, chromosomal location and population dynamics. Proc Natl Acad Sci USA 81:8014–8018
Tai TH, Tanksley SD (1990) A rapid and inexpensive method for isolation of total DNA from dehydrated plant tissue. Plant Mol Biol Rep 8(4):297–303

1.2 Southern Blotting

M.S. Clark

1.2.1 Introduction

Southern blotting is the transfer of DNA fragments from agarose gels to nitrocellulose or nylon membranes. This chapter will describe the preparation of a Southern blot from digested total genomic DNA; however, the same transfer technique can be used for PCR products, digests of plasmids, phage, cosmids etc. The technique is essentially unaltered since it was first developed by Southern in 1975. The only significant change has been from the use of nitrocellulose to nylon membranes. The latter are easier to handle, virtually indestructable and more efficient at binding DNA, particularly with low ionic strength buffers. They also have the advantage that the DNA can be bound to the membrane using UV irradiation; this takes only a few minutes compared to the use of an 80 °C oven which requires 2 h (Khandijian 1987). Southern blotting is one of the fundamental techniques underpinning many plant molecular biology techniques, such as RFLP analysis, characterization of clones, cultivar identification etc.

Preparation of the blot itself involves several stages:

- Isolation of genomic DNA
- Restriction enzyme digestion of genomic DNA and check mini-gel

- Agarose gel electrophoresis and transfer of DNA to nylon membrane

1.2.2
Isolation of Total Genomic DNA

This has been thoroughly dealt with in the previous section (1.1). The method of choice really depends on the problems encountered with the plant under study.

DNA for use in a Southern blot must be full length, with no shearing or degradation. Samples should be treated with RNase A to remove all RNA.

1.2.3
Restriction Enzyme Digestion of Genomic DNA and Check Mini-Gel

Reagents

- Restriction enzymes plus digestion buffer (available commercially)
- Sterile distilled water (SDW)
- 40 mM Spermidine
- TE buffer: 10 mM Tris/1 mM EDTA, pH 8
- DNA molecular weight marker, e.g. Lambda HindIII
- Glycerol loading buffer (GLB) 0.1 M EDTA pH 8/8% saturated bromophenol blue in SDW/50% glycerol
- 0.8% agarose gel (50 ml volume) in 1x TBE buffer
- 1x TBE running buffer (0.045 M Tris-borate/0.001 M EDTA)
- Ethidium bromide (10 mg/ml stock)

Equipment

- 37 °C waterbath
- 65 °C waterbath
- Mini-gel tank and power supply
- Apparatus for viewing and photographing gel
- Micropipettes and tips
- Eppendorf tubes

Procedure

1. Set up the following reaction mix:
   11 μg genomic DNA in TE buffer or SDW
   1x restriction enzyme digestion buffer
   4 mM spermidine
   40U enzyme

Make up the solution to an appropriate volume using SDW (this depends on the initial volume of the DNA). Do not vortex to mix, as this will shear the DNA.

2. Incubate overnight at 37 °C.

3. Stop reaction by incubation at 65 °C for 10 min.

4. Remove a sample from the digestion reaction equivalent to 1 μg (the volume of which should be approximately 10 μl, if not ethanol precipitate and resuspend in 10 μl TE buffer) add to 3 μl GLB and run on 0.8% agarose mini-gel with DNA size marker. To speed up the mini-gel stage it is possible to add ethidium bromide directly to the gel at a concentration of 0.5 μl stock to 10 ml gel (final concentration of 500 ng/ml). Store the remaining digest reaction at –20 °C until required.

5. If ethidium bromide was not added directly to gel, stain and view. If digestion was successful, proceed onto following section: agarose gel electrophoresis and transfer of DNA to nylon membrane.
The DNA in the lanes should appear as an even smear (see Fig. 1.2.1). Often there are brighter bands within the smear, these are repetitious sequences, not artefacts.

Notes

- The digest does not need to be left overnight, but a longer digestion time means that less enzyme can be used. A higher concentration of enzyme is generally used for genomic DNA digests when compared to those of vector/insert characterization etc.

- The DNA molecular weight marker should encompass a wide size range, e.g. lambda HindIII, which covers 23 kb to 500 bp. Most commercial companies supply a range of different size markers, which cover every possible requirement.

- The addition of a final concentration of 4 mM spermidine enhances the digestion reaction. However, it should be noted that higher concentrations of spermidine will inhibit the reaction.

- Which enzyme to use is a matter of personal choice. Often relevant literature may supply the answer, but if very little work has been carried out on the plant under study, try the cheapest enzymes recognizing six base pair sequences first, e.g. EcoRI, HindIII, BamHI. It is best to steer clear of methylation-sensitive enzymes (Kochert

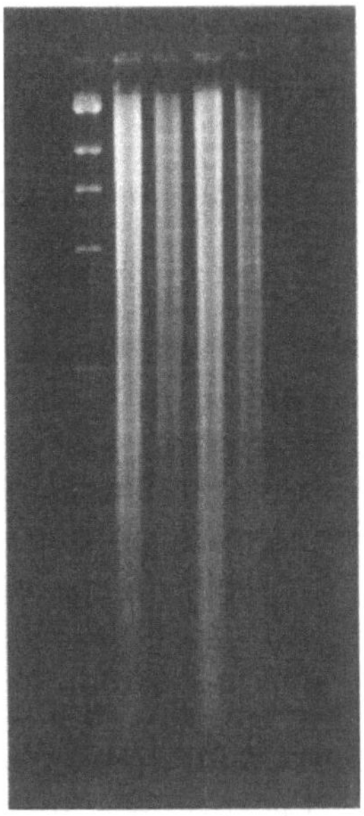

Fig. 1.2.1. Digest of total genomic DNA of date palm (*Phoenix dactylifera* L.). *Lane 1* Lambda HindIII marker; *lanes 2 and 3* 10 μg and 5 μg of male DNA; *lanes 4 and 5* 10 μg and 5 μg of female DNA

et al. 1991) and 4 base-pair cutters (Landry et al. 1987; McCouch et al. 1988). It has been found that the average size of genomic fragments generated by different enzymes is positively correlated with the amount of polymorphism generated in RFLP analysis (McCouch et al. 1988; Miller and Tanksley 1990). In experiments with tomato, the average size generated by cutting with DraI, EcoRI or HindIII was 2.8; 6.6 and 6.3 kb respectively. EcoRV and XbaI produced average fragments sizes of 8.6 and 8.9 kb respectively, with most polymorphism generated on blots of the latter two. This phenomenon is explained if polymorphism is generated by insertions/deletions/rearrangements and the bigger the fragment generated by restriction enzyme digestion, the more likely these events are to have occurred.

- Whilst certain factors can be maximized, such as careful choice of restriction enzymes for digestion of total genomic DNA, ultimately it is a hit-and-miss affair as to whether any given probe will produce high levels of polymorphism for any given plant. What is certain is that levels of polymorphism are greater between wild species when compared to cultivated and inbred species (Kochert et al. 1991). Therefore researchers should be prepared to undertake more extensive investigations when studying cultivated species.

Troubleshooting

- If the DNA in the lane is much brighter at the top than at the bottom, incomplete digestion has occurred. This could be due to several reasons:
  - Digestion was not left to proceed for long enough (this does not apply if the reaction was left overnight).
  - Not enough enzyme was added or enzyme was only partially active; the latter can happen if, for example, the enzyme was taken from the end of an old batch.
  - If the original DNA sample was dissolved in TE buffer, and this comprised a large proportion of the restriction enzyme digest reaction, the EDTA in the TE buffer could have inhibited the digestion reaction. Ethanol-precipitate the original sample and either dissolve in a smaller quantity of TE buffer or use SDW.
  - The DNA is highly methylated and the enzyme used was cytosine methylation sensitive, e.g. PstI, SalI, ClaI, XhoI, etc. If possible, set up the reaction again using a different enzyme. Chemical companies selling restriction enzymes usually list any methylation sensitivity in the product data.
  - The DNA could have been overloaded for the comb size. Reduce the concentration and re-run the gel.

- If the DNA in the lane is "streaky", this is due to either:
  - the DNA being only partially dissolved in the original buffer,
  - some ethanol remaining in the DNA sample after an ethanol precipitation.

  Ethanol-precipitate the original DNA sample and resuspend again. Check on a mini-gel that the DNA has completely dissolved (i.e. single discrete band produced). This problem also causes incomplete digestion. Genomic DNA should always be checked on a gel before digestion to ensure that is has fully dissolved in the buffer and also that it has not degraded.

- The DNA digests should show the same brightness in all the lanes. If not, this is because different concentrations of DNA have been digested. DNA concentrations are often difficult to determine accurately. This can be corrected by adjusting the amount of sample loaded onto the final gel.

- The DNA is much brighter at the bottom of the lane, i.e. very short fragments and diffuse. This could be residual RNA, but if RNase treatment has been carried out and the DNA checked afterwards, it is more likely to be degradation of DNA. The cause is usually to be found in the DNA isolation procedure, not the restriction enzyme digestion reaction. The DNA has to be re-isolated, perhaps using a different method.

- If problems do occur, it is usually best to set up a fresh reaction, as too much manipulation of full-length DNA will lead to its degradation and decrease the possibility of obtaining a good result with the Southern blot.

1.2.4
Agarose Gel Electrophoresis and Transfer of DNA to Nylon Membrane

If the digest and check mini-gel are satisfactory, the DNA samples can then be loaded onto a larger gel. Often, ethanol precipitation is required to concentrate the samples so that they can be loaded on the gel. Re-dissolve these samples in TE buffer.

This example describes the preparation of a Southern blot for total genomic DNA, but the same blotting procedures apply for plasmid, lambda, RAPD blots etc. The protocol for treatment of the gel is based on Reed and Mann (1985). Alternative transfer buffers include high salt (Southern 1975) and 1 M ammonium acetate (Kreike et al. 1990).

Reagents

- Digested DNA samples
- DNA molecular weight marker, e.g. lambda HindIII
- TE buffer: 10 mM Tris/1 mM EDTA, pH 8.0
- Glycerol loading buffer (GLB)
- 0.8% agarose gel (200 ml vol) in 1x TBE
- 1x TBE running buffer
- Ethidium bromide (10 mg/ml)
- Sterile distilled water (SDW)
- Nylon membrane, e.g. GeneScreen Plus (NEN, Dupont)
- 0.25 M HCl
- Denaturation solution: 0.6 M NaCl/0.4 M NaOH
- Neutralization solution: 1 M NaCl/0.5 M Tris pH 7.2

Equipment

- Large gel (maxi, approximately 200–250-ml capacity gel tray) tank and power supply
- Staining tray

- Apparatus for viewing and photographing gel
- Plastic trays (e.g. sandwich boxes) for treating gel
- Shaking platform
- Micropipettes and tips

Procedure

1. Load samples (DNA plus 5 µl GLB) onto gel. Also include a DNA size marker, such as lambda HindIII, on one of the outside lanes. Run overnight at 25 V.
2. The following morning, stain the gel in ethidium bromide (use 5 µl of stock solution to 100 ml of SDW) for approximately half an hour. View the gel under UV. The DNA should have run 2/3 to 3/4 of the length of the gel.
3. If the gel has not run far enough, re-run for longer. When ready, take a photograph, with ruler alongside the marker lane, making sure that the ruler markings are legible on the photograph. Trim gel by cutting off excess agarose (un-used lanes, for example, and agarose above wells). Cut a small piece diagonally off the bottom left-hand corner (this helps to orientate the gel in the following procedures). Do not cut directly on the UV apparatus, as the quartz plate scratches very easily. Treat the gel prior to blotting as follows:
4. Incubate twice in 500 ml 0.25 M HCl for 15 min each time and then rinse in SDW.
5. Set up blot as described below (see Fig. 1.2.2). Always use gloves when handling the nylon membrane.

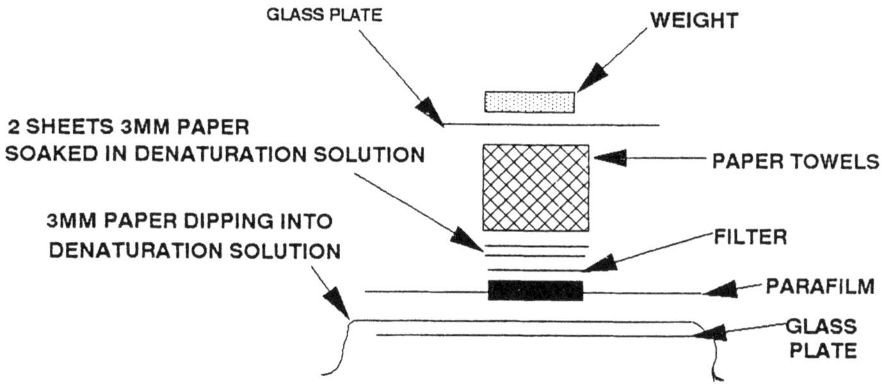

Fig. 1.2.2. Set up for Southern blot

a) Place a glass plate across a sandwich box. It should be longer than the sandwich box, but narrower, such that 3-MM paper can be draped across and dip down into the box to act as a wick.
b) Cut a piece of 3-MM paper, the same width as the sandwich box, but about double the length. Place the paper on the glass plate, so that the long ends dip into the box.
c) Soak the paper in denaturation solution and add a further 250 ml denaturation solution to the box.
d) Place the gel in the centre of the plate and make sure that there are no air bubbles present between the gel and the 3 MM paper (use a glass pipette as a "rolling pin"). Surround the gel with parafilm.
e) Cut a piece of nylon membrane to the exact size of the gel, not forgetting to snip off the bottom left-hand corner. Place the membrane over the gel, removing all air bubbbles (mark the well lanes with pencil, if required).
f) Cut two sheets of 3 MM paper to the exact size of the gel, soak in denaturation solution and place on top of the nylon membrane.
g) Next place a stack of paper towels on top of the 3 MM paper, followed by a glass plate and a weight. Leave overnight, but make sure that the reservoir of denaturation solution in the sandwich box does not dry up and that there are enough towels to ensure a constant pull of water through the system.

6. Leave blot overnight. The following morning dismantle the blotting "apparatus". Carefully remove the nylon membrane and incubate in neutralization solution for 30 min. Take the gel, re-stain and view under UV to check that transfer was complete.

7. Air-dry the nylon membrane and fix DNA to the filter. This can be done either by baking for 2 h at 80 °C or, more commonly, by exposing to UV irradiation (254 nm) either using a purpose-built cross-linker or a standard UV transilluminator (the length of time varies according to the type of transilluminator, but is generally about 4 min). This blot can now be stored sealed either in a plastic bag or Saran wrap at room temperature until required.

Once a Southern blot has been fixed, it can be hybridized with either radioactive or non-isotopic probes.

Notes

- How much DNA should be loaded onto the gel to ensure that a signal will be obtained when the blot is hybridized? This largely depends on the size and specific activity of the probe, the proportion of the genome complementary to the probe and the amount of DNA on the filter. The latter depends on the genome size of the plant. Small genomes such as tomato (0.74 pg) require only 3 μg per lane (Bernatzky and Tanksley 1986), which is the equivalent of 4×10^6 genomes, to obtain signals from single-copy sequences after an overnight exposure.

However, more DNA, e.g. 10 μg, is required from plants with large genomes such as wheat (15.7 pg). This only works out at 6×10^5 genome equivalents and therefore the proportion of the genome hybridizing to the probe would be much less than, for example, with tomato, and signal will develop much more slowly. In fact, exposures of 2 to 3 weeks are not unusual for single-copy sequences. However, if much more than 10 μg is loaded onto a gel, overloading will occur, which will distort migration rates and produce misleading results. For comprehensive lists of DNA amounts in plants consult Bennett et al. (1982) and Bennett and Smith (1976, 1991).

- There are various types of membrane on the market made from nylon and nitrocellulose. Nylon is more popular, due to its durability. Nitrocellulose tends to chip if knocked or bent, which limits its lifespan. The treatment of the gel prior to blotting may vary, depending on the brand of, membrane used, so it is always best to check the manufacturer's instructions before use. The choice of membrane may be influenced by the method of probing the blot, i.e. use of either radioactivity or non-isotopic methods. Certain types of membrane are not recommended for the latter procedure, but these are detailed by the manufacturers of the detection system. Whatever the type of membrane, pre-treatment of the gel should be kept to a minimum to maintain resolution; excessive treatment results in diffuse bands.

- For genomic Southern blots, it is best to leave the transfer set-up overnight, as transfer of large fragments around 15 kb in length take about 15 h (Olszewska 1988). However, if smaller fragments are to be transferred, e.g. plasmid digests, the process can be speeded up by periodically replacing the wet towels with dry ones and completed within a few hours.

- Any exposure of the gel to UV irradiation should be kept to a minimum, as the ability of DNA to form hybrids is compromized by such exposure (Reed and Mann 1985).

- Resist the temptation to put really heavy weights on top of the Southern blot set-up. Transfer works by capillary action and excessive weights may result in a squashed gel and diffuse bands on probing.

- There are commercial vacuum blotters available, which will blot gels within 30 min. The "old" method of using capillary action is presented here as it requires the least amount of sophisticated equipment. The procedure for vacuum blotting varies according to make, and so manufacturer's instructions should be consulted before use. Advocates of vacuum blotting claim, along with faster transfer, improved resolution of bands and more efficient transfer of DNA (Medveczky et al. 1987; Olszeska and Jones 1988).

- To be able to determine the approximate size of any hybridization signals produced when the Southern blot is subjected to probing, it is necessary to photograph the gel with a ruler alongside the marker lane. The size of each marker band (in base pairs) is then read off against the ruler, taking the length of migration in mm. These results are then transferred to log/linear graph paper. Across the bottom (linear scale) mark migration lengths in mm, on the vertical axis mark off in base pairs. Plot the results of the measurements for each marker band. The results should give a reasonably straight line (large deviations indicate that the gel has not run very well and any calculations will not be accurate). After probing, any signals can be measured on the autoradiograph, taking the distance from the well line to the signal. The length (mm) is then plotted onto the marker graph. By reading up to the line drawn for the markers and across at the intercept you can obtain an approximate size for the band.

Troubleshooting

- Providing the mini-gel check was satisfactory, there should be no problems with this stage.

- If some DNA remained in the gel after blotting, i.e. some luminescence from the ethidium bromide remained in the gel, this could be due to:
  - The blotting was not carried out for long enough (not applicable if left overnight).
  - The depurination treatment was not 100% sucessful.
  - Not enough towels were added to maintain the capillary action and draw the DNA up into the filter. This is seen in the morning as completely soaked towels.

- The parafilm border around the gel was not complete and water was drawn up directly into the towels, by-passing the gel.

If only a small amount of DNA was left, carry on with the blot.

- If the lanes of DNA are distorted in any way, this is due to the agarose not being dissolved properly or partially solidifying before being poured. Any distortion will give an inaccurate picture of band size, and so gel should be discarded.

References

Bennett MD, Smith JB (1976) Nuclear DNA amounts in angiosperms. Phil Trans R Soc Lond B 274:227–274

Bennett MD, Smith JB (1991) Nuclear DNA amounts in angiosperms. Philos Trans R Soc Lond B 334:309–345

Bennett MD, Smith JB, Heslop-Harrison JS (1982) Nuclear DNA amounts in angiosperms. Philos Trans R Soc Lond B 216:179–199

Bernatzky R, Tanksley SD (1986) Methods for detection of single or low copy sequences in tomato on Southern blots. Plant Mol Biol Rep 4:37–41

Khandijian EW (1987) Optimised hybridization of DNA blotted and fixed to nitrocellulose and nylon membranes. Bio/Technology 5:165–167

Kochert G, Halward T, Branch WD, Simpson CE (1991) RFLP variability in peanut (*Arachis hypogaea* L.) cultivars and wild species. Theor Appl Genet 81:565–570

Kreike CM, de Koning JRA, Krens FA (1990) Non-radioactive detection of single copy DNA-DNA hybrids. Plant Mol Biol Rep 8:172–179

Landry BS, Kesseli R, Hei Leung, Michelmore RW (1987) Comparison of restriction endonucleases and sources of probes for their efficiency in detecting restriction fragment polymorphisms in lettuce (*Lactuca sativa* L.). Theor Appl Genet 74:646–653

McCouch SR, Kochert G, Yu ZH, Wang ZY, Khush GS, Coffman WR, Tanksley SD (1988) Molecular mapping of rice chromosomes. Theor Appl Genet 76:815–829

Medveczky P, Chang CW, Oste C, Mulder C (1987) Rapid vacuum-driven transfer of DNA and RNA from gels to solid supports. Biotechniques 5:242–246

Miller JC, Tanksley SD (1990) Effect of different restriction enzymes, probe source, and probe length on detecting restriction fragment length polymorphism in tomato. Theor Appl Genet 80:385–389

Olszewska E, Jones K (1988) Vacuum blotting enhances nucleic acid transfer. TIG 4:92–94

Reed KC, Mann DA (1985) Rapid transfer of DNA from agarose gels to nylon membranes. Nucleic Acids Res 13:7207–7221

Southern EM (1975) Detection of specific sequences among DNA fragments separated by gel electrophoresis. J Mol Biol 98:503–517

1.3 Hybridization with Radioactive Probes

M.S. Clark

1.3.1 Introduction

The general procedure for hybridization of radioactive probes to Southern blots has changed very little since it was first described (Southern 1975), as it is essentially a kinetics-based technique (Meinkoth and Wahl 1984). Modifications have since been introduced, such as the use of the more durable nylon filters instead of nitrocellulose, and more efficient labelling techniques, such that the detection of single-copy sequences has been routine for many years. Considerable effort recently has been put into hardware development, with the production of special hybridization ovens, which are much more convenient and safer to use than previous home-made designs such as bags and sandwich boxes (although this is not to say that these methods are not still in heavy use!). However, the biggest change in recent years has to be the development of non-isotopic detection methods. The most popular available within the UK, in kit form, include Digoxigenin (Boehringer) and ECL (Amersham). A protocol for non-radioactive labelling and detection (Sect. 1.4) follows this section.

Whilst non-radioactive methods have been available for several years, there is still debate as to whether they are as sensitive as ^{32}P, and the answer depends on which laboratory is canvassed for their opinion. Likewise, the choice of whether to use them or remain with radioactivity generally lies with the experience of particular laboratories and individuals, some of which have made the non-isotopic techniques work very efficiently, whilst others have either failed completely or given up trying. There is no doubt that with a minimum of training (largely concerned with learning safety procedures) anyone can perform radioactive hybridizations and obtain satisfactory results; non-isotopic methods require a certain period of time for development and optimization when a laboratory first starts to use them.

There are undoubtedly other pros and cons concerning the use of radioactive labelling and detection. Disadvantages include the effects of exposure to isotopes and the requirement for protective equipment and monitoring procedures. ^{32}P has a short half-life (2 weeks) and consequently probes have a very short lifespan and can usually only be used once or twice with any efficiency.

On the plus side is the fact that the technique is well established, highly sensitive and reproducible. Radiation levels can be monitored throughout the procedure and protocols changed accordingly. This can be of particular importance with heterologous probes where the effect of washing stringency has to be closely monitored. Radiation is constantly being emitted from hybridized sequences, albeit at decreasing levels, but this does mean that weak signals can be left for several weeks to develop and results still obtained, whereas signals from non-isotopic methods have a much shorter lifespan.

Radioactive hybridization involves several stages:

- Probe labelling
- Probe purification
- Hybridization
- Filter stripping

1.3.2 Probe Labelling

In this section, the two most popular methods, PCR and oligolabelling will be described.

1.3.2.1 *PCR Labelling of Probes*

This technique is used to amplify and label the sequence of interest when

- the sequence is in a vector for which specific primers exist around the multiple cloning site, e.g. M13 primers;
- specific primers exist. These can be either primers to amplify directly from genomic DNA or cDNA;
- the sequence can be efficiently amplified by PCR, i.e. generally up to 2–3 kb.

Reagents

- 10x dNTP mix (2 mM each of dATP, dGTP, dTTP, dCTP)
- 10x hot dNTP mix (2 mM each of dATP, dGTP, dTTP and 0.2 mM dCTP)
- Taq DNA polymerase
- 10x reaction buffer (usually supplied commercially with the Taq polymerase)
- Specific primers for the sequence of interest or vector (10 μM stock solution)
- Sterile distilled water
- Mineral oil
- 0.8% mini-gel in 1x TBE
- 1xTBE running buffer

Equipment

- PCR machine
- Minigel apparatus and equipment for viewing and photographing gel
- Micropipettes and tips
- 0.5 ml Eppendorf tubes

Procedure

1. The following reaction (50 μl final volume) is set up in a 0.5-ml microcentrifuge tube:
   1x reaction buffer
   1x hot dNTP mix
   1 μM of primer 1
   1 μM primer 2
   10 ng DNA
   1 U Taq polymerase
   20 μCi α^{32}P dCTP
   A duplicate reaction is set up, as a positive control, without the ^{32}P dCTP, and replacing the hot dNTP mix with the 10x dNTP mix.

2. The reactions are overlaid with 70 μl mineral oil and subjected to the following PCR cycle:
   95 °C 2 min
   95 °C 15 s
   55 °C 30 s
   72 °C 1 min
   Repeated 25 times
   95 °C 30 s
   55 °C 30 s
   72 °C 3 min

3. The positive control is run out on a mini-gel, to check that an amplified product was produced and also that a single band was produced with no contamination.

Notes

- The temperature of annealing (in this example 55 °C) depends on the primers used. This can be roughly calculated by counting the number of bases in the primer: with +4 °C for each G or C and +2 °C for each A or T.
- The length of the extension time depends on the length of sequence to be amplified. The theoretical maximum efficiency of DNA Taq polymerase is 3.6 kb per min. Erring on the side of caution, use half this level, i.e. a 30-s extension time will successfully amplify 1 kb.
- When amplifying directly from genomic DNA, use 30–35 cycles.

1.3.2.2
Oligolabelling

An alternative to PCR amplification is oligolabelling with ^{32}P dCTP (Feinberg and Vogelstein 1983, 1984). This method is generally used when

- the sequence of interest is too big to be amplified efficiently using PCR,
- total genomic DNA is required as a probe, a band from a digest is excised and labelled directly without going through a subcloning stage,
- amplifying the insert via PCR would include some vector sequence, which would cross-hybridize unfavourably with other sequences on the blot, creating high background labelling.

With this method, the Klenow fragment of DNA polymerase I utilizes random oligonucleotides as primers for copying single-stranded templates incorporating radioactivity at high efficiency. It requires only small amounts of DNA, e.g. 25 ng, and is also fairly tolerant of impurities in the DNA sample.

The easiest way to label a probe is with a commercial kit, of which there are many available, hence the procedure will not be described here. This is the best way if you are just starting out, as all the kit components are optimized, also a standard DNA sample is included which is useful for troubleshooting purposes.

1.3.2.3
Isolation of Fragments from Gels for Labelling

A DNA fragment can be isolated from a gel for labelling directly by

- *Using low melting point agarose.* Run sample on a 0.8% low melting point agarose gel using 1x TAE buffer containing 0.5 μg/ml of ethidium bromide. View gel under UV and cut out the fragment in the smallest piece of agarose.

  Place in an Eppendorf tube and heat to 70 °C for 15 min in order to melt gel slice. The gel will remain liquid at 37 °C, so a small portion of melted gel can be directly added to the labelling reaction (Feinberg and Vogelstein 1984).

- *Freeze-squeeze (Tautz and Renz 1983).* Run a 0.8% agarose gel and excise gel slice in minimal volume. Add the slice to 1 ml disposable sterile syringe, replace plunger and store at –20 °C for approximately 1 h until the gel slice is frozen solid. Take out of freezer and slowly depress the plunger. The gel slice slowly melts and the liquid emerging contains the fragment. Do not squeeze so much that the agarose also comes out of the syringe.

  The solution containing the fragment can then either be used directly in the labelling reaction or purified with phenol chloroform first. Whilst the latter procedure purifies the fragment and therefore labelling will probably be more efficient, the yield of DNA is decreased.

- *Commercial preparations.* Kits are sold commercially for the extraction of DNA from gels, e.g. GeneClean II (Stratech. Scientific Ltd) and Qiaex gel extraction kit (Qiagen). The resulting products are much purer than obtained by the above methods, and DNA recovery more efficient, but they do work out much more expensive. Methods vary according to manufacturer's instructions.

1.3.3 Purification of Probe

Some manufacturers of labelling kits state that it is not necessary to purify the probe before use, i.e. remove the unincorporated nucleotides. This short cut can be taken with repetitive probes, but it is still preferable to purify when the filters have to be exposed for a long time, as background levels may be significantly lower. It does also provide a check that the probe has actually labelled efficiently, for example when there is no positive control, such as in oligolabelling. There are two methods described here: Sepharose CL6B spin columns and Sephadex G-50, the former of which requires the provision of a bench top centrifuge in the radioactive work area.

1.3.3.1 *CL6B Spin Columns*

Reagents

- Glass beads
- CL6B Sepharose resin, washed and resuspended in TE buffer
- TEN buffer (10 mM Tris/1 mM EDTA, 100 mM NaCl pH 8)

Equipment

- Small-bore steel syringe needle
- 50-ml centrifuge tubes
- Bench top (non-refrigerated) centrifuge capable of at least 1500 rpm
- Micropipettes and tips
- 1.5-ml Eppendorf tubes, flip top and screw cap

Procedure

1. Cut the lid off a 1.5-ml flip top microcentrifuge tube. Pierce the bottom of the tube with a small-bore needle. Add a few glass beads to the bottom of the tube, followed by 1 ml CL6B resin
2. Place this tube on top of another 1.5-ml microcentrifuge tube, put both inside a 50-ml centrifuge tube and spin at 1500 rpm for 2 min.
3. Check that the surface of the resin is dry, if not, briefly re-spin. Change the lower tube for a screw-cap tube. Re-assemble in 50-ml centrifuge tube. Add the probe on top of the resin. Add 100 μl of TEN buffer. Spin for 2 min at 1500 rpm.
4. Monitor the liquid in the lower tube; this should be the radioactively labelled probe.

1.3.3.2
Sephadex G-50 Column

Reagents

- Sephadex G-50 soln: 1 g Sephadex G-50 beads in 50 ml TE buffer. The beads always settle out, so shake and resuspend before use.
- TE buffer: 10 mM Tris/1 mM EDTA pH 8
- Blue dextran dye: 0.05 g blue dextran/0.012 g Orange G in 10 ml SDW

Apparatus

- 1 ml disposable sterile syringe
- Glass fibre filter paper, e.g. Whatman GFC
- Boss, clamp and stand
- Sterile glass Pasteur pipettes
- Eppendorf tube rack
- Micropipettes and tips
- 1.5-ml Eppendorf tubes

Procedure

1. Using the plunger, put a small square of glass fibre filter paper at the bottom of the syringe. Use a Pasteur pipette to fill syringe with G-50 Sephadex solution up to 1-ml mark. Clamp to a stand above a tray of Eppendorf tubes. The TE buffer will drain out, leaving a packed column. More Sephadex will need to be added to make final column volume of 1 ml. Add 400 μl of TE buffer to wash the column through.

2. Add 50 μl blue dextran dye to the labelling reaction and carefully apply to the top of column. Add 100 μl aliquots of TE buffer to run probe through the column.

3. The dye separates into two colours, the blue dye runs with the labelled probe, whilst the yellow runs slower with the unincorporated nucleotides. Collect the column fractions 100 μl before the blue dye starts to come off, all of the fractions containing the dye, and 100 μl after the dye finishes (approximately 400 μl). Collect the fractions in one screw-capped Eppendorf tube (these are safer when denaturing the probe).

Notes

- Recommended radioactivity levels for probes vary between 1–10 x 10^8dpm/μg. Obviously, the more radioactive the probe, the more quickly the result is obtained.

- Precise levels of labelling can be measured in a scintillation counter by spotting 1 µl onto a small piece of filter paper (^{32}P can be detected on dry filters and does not need to be added to a toluene-based scintillation cocktail). For a more precise protocol on how to measure the specific activity of a probe, consult Sambrook et al. (1989), although most institutes have their own isotope department, with their own particular methods and machines.
- Many people do not bother to check incorporation or radioactivity levels accurately, other than quickly passing the probe past a Geiger counter (useful particularly if the probe is required immediately or there are many samples). If the Geiger counter bleeps and goes off the scale, i.e. greater than 2000 cps, labelling has been very successful; if not, counts showing around 500 cps are still usable for low-copy number sequences, but the filters will have to be exposed for much longer. Any lower than this, and something has gone wrong with the labelling reaction.

Troubleshooting

- *Probe is not very radioactive.* There are several potential reasons for this.
  - Contaminants in the original DNA sample are inhibiting the action of the labelling enzyme in the oligolabelling kit. Test-label the standard DNA supplied with the kit. If the standard DNA labels well, the original DNA sample is at fault (re-isolate probe or phenol chloroform treat to purify).
  - The labelling enzyme is losing some of its activity. Test-label the standard DNA supplied with the kit. If the standard DNA does not label, the enzyme is at fault. Buy new enzyme.
  - The label is old ($\alpha^{32}P$ has a half-life of 2 weeks). Always check reference date of label before use. If detection of low or single-copy sequences is required, the probes should be labelled and used as soon as possible after the reference date of the isotope.
  - The PCR reaction has not worked. Check annealing temperatures of primers (and that correct primers have been used). Replace dNTP mix with fresh and also check that enzyme is effective.

1.3.4 Hybridization Protocol

After probe labelling it is best to use the probe as quickly as possible. With a half-life of only 2 weeks, storage for a day or two can significantly reduce the activity of a probe, and also the radioactive emissions will damage the labelled DNA.

This protocol (hybridization temperature and washes) assumes that a homologous probe is used for probing a genomic Southern blot. Different conditions apply for heterologous probes and library screening (see notes for details).

Reagents

- Labelled probe
- Prepared Southern blot filters
- 2x SSC: 0.3 M NaCl/0.03 M sodium citrate
- Hybridization solution: 3x SSC/0.02% Ficoll/0.02% PVP/1% SDS/ 50 μg/ml salmon sperm carrier DNA
- 2x SSC/0.1% SDS
- 0.1x SSC/0.1% SDS

Equipment

- Heating block set at 95 °C
- Ice
- Hybridization oven and tubes
- Nylon mesh (usually supplied commercially with hybridization oven/tubes)
- Waterbath at 65 °C
- Saran wrap or cling film
- Autoradiographic cassettes with intensifying screens
- Autoradiographic film
- –80 °C freezer

Procedure

1. Wet the Southern blot in 2x SSC. Lay on top of a square of nylon mesh and roll the two up together. Add to hybridization oven tubes. Add 50 ml 2x SSC and unroll the filter and mesh so that they lie flat on the side of the tube. Remove the 2x SSC.
2. Add 15 ml of hybridization solution to each tube. Incubate in the hybridization oven (with rotation) for at least 1 h at 65 °C to prehybridize.
3. Denature the probe in the heating block for 5 min and then quench on ice.

4. Add the denatured probe to the hybridization solution and hybridize overnight at 65 °C (with rotation).

5. The following morning, remove the radioactive hybridization solution and wash the filters as follows:
2x SSC/0.1% SDS, 15 mins, 65 °C.
0.1x SSC/0.1% SDS, 15 min, 65 °C. Repeat once.

6. Remove the filter and nylon mesh from the tube. Drain excess moisture from filter and wrap in cling film/Saran wrap. Check the number of counts remaining on the filter with a Geiger counter. Put into an autoradiographic cassette with intensifying screens and add film. Store cassette at –80 °C until ready to develop.

Notes on parameters

- *Hybridization temperature.* Hybridization should be carried out at 20–25 °C below the melting temperature (T_m) of the hybrid. This can be calculated using the formula below:
  T_m = 81.5 °C – 16.6 [log10(Na+)] + 0.41 (%G + C) – 0.63 (%formamide) - (600/l), where l = length of the hybrid in base pairs.
  For heterologous probes 1% mismatch in homology reduces the Tm by 1.5 °C, or 1.5 °C below the T_M allows for 1% mismatch.
  A more complete explanation is given in Meinkoth and Wahl (1984) and Sambrook et al. (1989). As a general rule, however, homologous probes are hybridized at 65 °C with lower temperatures required for heterologous sequences. If a heterologous probe is used for the first time, it is more efficient to hybridize at a relatively low temperature, e.g. 50 °C, with increased stringency being applied via the washes if necessary.

- *Use of formamide.* Fifty percent formamide and hybridization at 42 °C is often advocated for use with heterologous probes with a high percentage of mismatch. Hybridization progresses at a slower rate in formamide, and this may result in more efficient hybridization and detection. However, formamide is a toxic substance and is best avoided.

- *Stringency of washes.* The higher the salt content, the less stringent the wash. These should be as stringent as possible to minimize levels of background. However, these stringencies can also be calculated and 12–20 °C below the T_m is recommended.

If a probe is used for the first time, it is best to monitor the levels of radioactivity during the washes and alter as appropriate, i.e. the stringency increased if levels of radioactivity remain high. It is also possible to re-wash the filter after it has been exposed to an autoradiographic film, providing the filter has not dried out and the exposure was only for a day or two.

For heterologous probes it is best to start with washes at room temperature and monitor every 5 min. It is surprising how quickly these sequences can wash down. If in doubt, expose the filter and wash more stringently after seeing the result on the autoradiograph.

For library filters, e.g. phage or cosmid, start with room temperature washes, even with homologous probes. Cosmid filters, in particular wash down very quickly.

- *Exposure time*. This will depend on several factors, the proportion of DNA on the filter homologous to the probe, the amount of the DNA on the filter and the size and specific activity of the probe. There are no hard and fast rules and this is usually learnt by experience! Filters showing a Geiger counter reading of 10–20 cps can be developed after a few hours, but readings around background level should be left for 1 to 2 weeks. If unsure, err on the side of caution.

  It is also possible to sandwich a filter between two sheets of X-ray film, developing the first after a short time and leaving the second for longer, if necessary. NB/anchor the filter to the second film to prevent movement during removal of the first film, otherwise double images may occur.

- *Analysis of signals*. The size of hybridization signals (in kb) can be worked out by measuring the distance from the well line to the signal and then referring to the graph on log/linear paper prepared earlier from the Southern gel with the ruler alongside the DNA molecular weight markers. For a fuller explanation, see the section on Southern blots (Sect. 1.2). For further information on the sequence of interest, for example how to determine copy number, see details given by Cullis et al. (1984) and Rivin (1986), or whether the sequence is translated Chapter 3 on RNA.

General notes Hybridization ovens are now the most popular method for the radioactive hybridization of filters. The volume of liquid added to the tubes varies according to the size of the tube, and this information will be available in the manufacturer's instructions. The advantage of these ovens is that small volumes of liquid are used, thus making the probe more concentrated. However, hybridization efficiency decreases dramatically if too many filters are added to the bottles at the same time (I would suggest no more than five laid on top of each other).

When using a hybridization oven, always check the filter after it has been rotating for 10 min. If the glass tubes are inserted in the wrong way, the filter and nylon mesh gradually roll into an ever tighter circle. Putting the tube in the other way will untangle them and ensure that they both stay pressed against the side of the tube.

Alternative methods

Alternative methods of performing the hybridizations which are still popular include:

- *In sealed bags*. Filters are placed in plastic bags with the hybridization solution, all air bubbles removed, and the bag sealed. These are then placed in sandwich boxes and incubated overnight. The problem with these is that the bags often leak, which is a radioactive hazard. Also all the air bubbles have to be removed before hybridizing, and although there are commercial pumps which help with the process, this can also result in problems with radioactive spillage.

- *In a sandwich box*. Filter and hybridization solution are added directly to plastic sandwich boxes. The problem with this method is that comparatively large volumes (50 ml) of hybridization solution have to be used, which results in a less concentrated probe and less efficient hybridization. This method is frequently used when screening genomic libraries (these usually consist of many large filters which are difficult to fit into hybridization bottles efficiently.

Both of these methods expose the user to greater levels of radioactivity than the thick glass tubes used in hybridization ovens.

Using the two methods above, it is not necessary to agitate the filters overnight whilst hybridizing. However, if probing multiple filters, nylon mesh should be put between each filter to prevent them sticking together; gentle agitation may also help.

For best results, the washes should be pre-heated and kept in a heated waterbath until required. It is possible to cheat and give the washes a quick heat in a microwave oven, but make sure not to overheat.

In this protocol, sheared, denatured salmon sperm DNA, in conjunction with 1% SDS, is used to block non-specific binding of the probe. Alternatives include Denhardt's solution, usually used at 10x concentration (stock solution of 100x: 10% polyvinylpyrrolidone (PVP), 10% bovine serum albumin (BSA) and 10% Ficoll 400 in water) and non-fat dried milk. All seem to work equally well, and hence the choice is a question of personal preference.

The blocking agents may obliterate any very weak signals and, particularly if using low stringency conditions and a heterologous probe, more success may be achieved by hybridizing in Church-Gilbert solution (hybridization: 1 mM EDTA, 0.5 M $NaHPO_4$, 7% SDS, washing: 1 mM EDTA, 40 mM $NaHPO_4$, 1% SDS).

In some instances, for example when probing gridded libraries, a certain amount of backgroung labelling is required. This is in order for the positive signals to be aligned back onto the master grids and the co-ordinates worked out. In these cases, use Church Gilbert hybridization solution.

SSC washes can be used in conjunction with Church Gilbert hybridization solution if required.

The position of the filter can be marked on the cling film/Saran wrap using a commercial radioactive marker or by mixing a small amount of ^{32}P with permanent writing ink. This radioactivity will develop alongside the signal of the probe on the autoradiograph, showing exactly the position of the filter underneath. This is very useful for determining the size of fragments lit up and the position of well lanes.

Troubleshooting

- If no signal is seen on the developed autoradiograph, this could be due to:
  - Autoradiograph was not left long enough to develop at –80 °C. For single- or low-copy number probes onto genomic Southern blot exposures of 2–3 weeks are not unusual for plants with a large genome size.
  - Probe was not labelled very well and the signal was not strong enough to develop.
  - The probe used was heterologous and the stringency of the hybridization and subsequent washes was too strong and stripped the probe. To remedy, strip filter and rehybridize at a lower temperature with less stringent washes.
  - The photographic developing solution is very old and not working efficiently, so that weak signals are not developed.

- If random spots of intense signal appear on the autoradiograph, these may be due to concentrations of probe which have stuck to the membrane. This often occurs if the hybridization solution was not mixed well enough after the probe was added or the probe was accidently added directly onto the filter. It can also occur if the unincorporated nucleotides were not removed from the labelling reaction before adding to the hybridization solution and long exposures were used.

- A very dark background means that the radioactivity has stuck non-specifically to the filter. This is often caused by insufficient prehybridization. When this occurs, the filter can often be difficult to strip and reprobe. It may also have occurred if the filter dried out at any point during the procedure. Check the levels of blocking agents in the hybridization solution.

1.3.5 Filter Stripping

It is possible to strip the probe and radioactivity from a nylon membrane and re-use it. This can be carried out many times and although the filter may end up looking rather worn, the signal should not be diminished with subsequent probings.

Reagents

- 0.4 M NaOH
- Deprobe buffer: 0.1x SSC/0.1% SDS/0.2 M Tris pH 7.5

Equipment

- Shaking waterbath at 45 °C
- Sandwich boxes
- Plastic bags/Saran wrap
- 3 MM paper

Procedure

1. Incubate the filter for 30 min at 45 °C in 0.4 M NaOH.
2. Incubate the filter for 60 min at 45 °C in deprobe buffer and then check the level of radioactivity with a Geiger counter. For successful stripping, this should be zero.
3. Remove filter and leave to dry on 3-MM paper at room temperature. When dry, store until required in either a plastic bag or Saran wrap at room temperature.

Notes

- An alternative method which does not utilize the rather harsh alkali treatment is to add the filters to boiling 0.1x SSC/1% SDS and leave for 5–10 min. The first method is particularly recommended for filters with high background levels.
- Alkali treatment should be avoided with cosmid filters, as these are very sensitive to harsh treatment and wear easily. In this case, prehybridize the probed cosmid filters for several hours and then replace the hybridization solution just before adding the new probe. This is usually sufficient to strip off old probes.
- Filters can also be stored wet at 4 °C.

References

Cullis CA, Rivin CJ, Walbot V (1984) A rapid procedure for the determination of the copy number of repetitive sequences in eukaryotic genomes. Plant Mol Biol Rep 2:24–31

Feinberg AP, Vogelstein B (1983) A technique for radiolabelling DNA restriction endonuclease fragments to high specific activity. Anal Biochem 132:6–13

Feinberg AP, Vogelstein B (1984) Addendum: a technique for radiolabelling DNA restriction endonuclease fragments to high specific activity. Anal Biochem 137:266–267

Meinkoth J, Wahl G (1984) Hybridization of nucleic acids immobilised on solid supports. Anal Biochem 138:267–284

Rivin C (1986) Analyzing genome variation in plants. Methods Enzymol 118:75–86

Sambrook J, Fritsch EF, Maniatis T (1989) Molecular cloning: a laboratory manual. Cold Spring Harbor Laboratory Press, Cold Spring Harbor

Southern EM (1975) Detection of specific sequences among DNA fragments separated by gel electrophoresis. J Mol Biol 98:503–517

Tautz D, Renz M (1983) An optimized freeze-squeeze method for the recovery of DNA fragments from agarose gels. Anal Biochem 132:14–19

1.4 Non-Radioactive Methods of Detection on Southern Blots

P. LEROY, M. MERLINO, S. NÈGRE, J.C. CAISSARD,
P. SOURDILLE, Y.H. LU, and M. BERNARD

1.4.1 Introduction

Restriction Fragment Length Polymorphism (RFLP) analyses have numerous applications in genetics and plant improvement (Burr et al. 1983; Beckmann and Soller 1988; Tanksley et al. 1989). In the past few years, the genomes of many plant species have been extensively mapped using RFLP markers. These saturated maps can be used to mark useful agricultural genes, to analyze genetic control of qualitative or quantitative traits, and to reveal intervarietal or intergenomic relationships.

While radioactively [^{32}P]-labelled probes represent a general approach in plant RFLP analyses, their limited storage life and specialized handling requirements have limited their routine use. Non-radioactive techniques such as DIG-11-dUTP for plant DNA analysis have now become routine in many laboratories, as they provide a good alternative to the use of radioactive agents such as ^{32}P, for rice (Ishii et al. 1990), potato (Kreike et al. 1990), tomato (Panaud et al. 1993), tobacco (Neuhaus-Url and Neuhaus 1993), maize (Khairallah et al. 1993) and wheat (Khairallah et al. 1993; Lu et al. 1993a,b).

Another important reason for the choice of non-isotopic detection technology is the possibility for transfering molecular techniques to developing countries. In these places, radionuclides are expensive, often not readily available, and proper procedures for the disposal of radioactive waste are frequently absent (Khairallah et al. 1993).

In this chapter we describe the detailed procedures necessary for an efficient, economical, non-radioactive experimental protocol to analyze a complex polyploid genome such as wheat. The protocol uses digoxigenin-11-dUTP (alkali-labile)-labelled probes (DIG-11-dUTP (alkali-labile) provided by Boehringer Mannheim (Lanzillo 1991; Höltke et al. 1992), two-step detection by antibody conjugates (anti-DIG-alkaline phosphatase) and chemiluminescent reaction (Beck and Köster 1990).

This procedure has been tested with many single-copy probes on wheat and involves:

- the extraction of total genomic DNA,
- restriction enzyme digestion of the DNA and Southern blotting,
- probe labelling by PCR-DIG,
- hybridization, washing and detection plus removal of probe from nylon membranes.

1.4.2
Extraction of Total Genomic DNA

DNA extraction protocols have already been extensively described in an earlier section. The problems in isolating DNA, especially from wheat, are the presence of in vivo DNAse activities that degrade the DNA and the presence of other macromolecules that co-purify with or polymerize to the DNA during the isolation procedure. Hence we use a method based on Tai and Tanksley (1991).

1.4.3
Restriction Enzyme Digestion of the DNA and Southern Blotting

These techniques have also been previously covered in detail (Sect. 1.2).

Ten μg of digested total DNA per track are necessary to obtain a good signal with wheat. The digested DNA is purified by phenol: chloroform:isoamyl alcohol (25:24:1) extraction before agarose gel electrophoresis in 1x TAE to obtain a better quality of the signal.

In our laboratory, alkaline transfer is used for Southern blotting. The membrane is stored at 4 °C in 0.1x TE buffer, sealed in a plastic bag.

1.4.4
Probe Labelling by PCR-DIG

Production of labelled vector-free DNA inserts as probes is necessary for Southern analysis. The commonly used method is time-consuming, involving restriction enzyme digestion of plasmids or lambda DNA, isolation of inserts from vector and labelling of the DNA with substituted nucleotides via techniques such as nick-translation and random primer labelling. The PCR (Polymerase Chain Reaction; Saiki et al. 1985) method has several advantages over this: only small amounts of plasmid DNA are required (50 ng) to act as a template for the direct amplification of a large quantity of efficiently labelled vector-free probe (Friedman et al. 1990; McCabe 1990; Simon and Schnorr 1992).

Introduction of non-radioactively substituted nucleotides (biotin-dUTP, DIG-11-dUTP (alkali-labile)) into a PCR reaction permits the synthesis of large quantities (2 to 5 µg in our case) of labelled probe in a few hours, (Dennis Lo et al. 1990; Emanuel 1991; Panaud et al. 1993; Lu et al. 1993a,b).

For non-radioactive RFLP analyses, the procedure of probe labelling, hybridization and the detection of the hybridized sequence is a modification of the supplier's manual (DIG Labelling and Chemiluminescent Detection Kit, Boehringer Mannheim).

For the example shown in Fig. 1.4.1, DIG-11-dUTP was incorporated into DNA by PCR. Two pairs of oligonucleotide sequences, homologous to pUC18 and pBlueScript sequences flanking the inserts psr142 and FBA065 respectively, were used as primers.

Reagents

- LB media: 1% tryptone, 1% NaCl, 0.5% yeast extract, pH 7
- Antibiotics
- 1% Triton X-100
- Sterile ultrapure water
- 50% sterile glycerol
- 10x PCR reaction buffer (Boehringer Mannheim): 0.1 M Tris-HCl pH 8.3; 15 mM $MgCl_2$; 0.5 M KCl; 1 mg/ml gelatin.
- Primers psr142:
- 20 µM solution of reverse primer: 5′ ATTCGAGCTCGGTACC 3′
- 20 µM solution of forward primer: 5′AGGTCGACTCTAGAGG 3′
- Primers FBA065:
- 20 µM solution of reverse primer: 5′ AACAGCTATGACCATG 3′
- 20 µM solution of forward primer: 5′ GTAAAACGACGGCCAGT 3′
- 5 mM solution of dATP, dGTP, and dCTP
- 2.0 mM solution of dTTP
- 0.2 mM digoxigenin-11-dUTP (alkali-labile; Boehringer Mannheim)
- 5 units/µl Taq polymerase (Boehringer Mannheim)

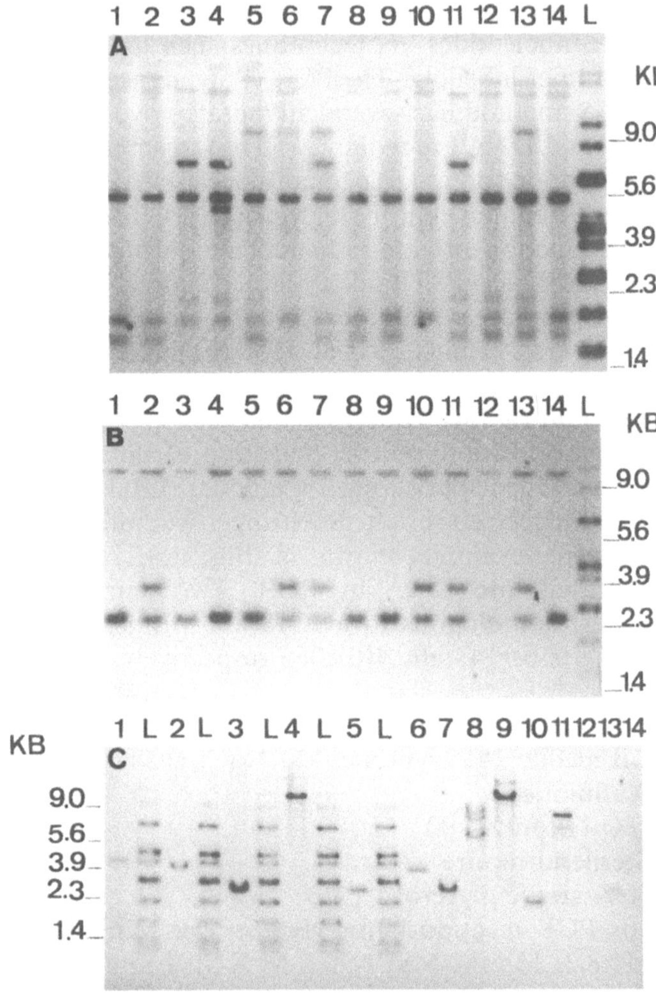

Fig. 1.4.1. A–B Autoradiograph of genomic DNA digested with Hind III and Eco RI repectively of 14 wheat varieties: **1** Courtot; **2** Cappelle; **3** Magnif27; **4** Prinqual; **5** Rex; **6** Apollo; **7** Pernel; **8** Thésée; **9** Eureka; **10** Renan; **11** Bezostaya; **12** RE607; **13** RE714; **14** RE813, hybridized with DIG-11-dUTP labelled probes; **A** FBA065; Pst I genomic clone assigned to the wheat chromosome 2DS, 4AL, 6AS and 7AS (Nelson et al. 1995a,b; Marino et al. 1996). **B** psr142: cDNA which maps to the wheat chromosome 6AL, 6BL, 6DL (Jia et al. 1996). **C** Autoradiograph of genomic DNA digested with Eco RI of several species: **1** barley; **2** rye; **3** *Triticum monococcum* (A genome); **4** *T. tauschii* (D genome); **5** *T. durum* (AB genome); **6** *T. speltoides* (S genome); **7** *Dasypyrum villosum* (V genome); **8** oats; **9** *Lolium*; **10** *Zea mays*; **11** *Orysa sativa*; **12** Brassica; **13** *Arabidopsis thaliana*; **14** *Helianthus annuus*, hybridized with DIG-II-dUTP-labelled probe psr142. The exposure times were 3 to 7 h. **L**: lkb DNA ladder size markers (BRL)

Equipment

- Flow hood
- GeneAmp thermocycler 9600 (Cetus)
- Bulkpack MicroAmp Reaction Tube/Tray/Retainer Assembly (Cetus)
- Gel electrophoresis tank and power pack
- Micropipettes (2–200-μl, 200–1000-μl capacity plus tips)
- Hemolysis tubes 5 ml
- Eppendorf tubes 1.5 ml
- Bench microfuge
- Shaker-incubator 37 °C
- 100 °C waterbath
- Vortex
- Ice
- Freezer –20 °C
- Freezer –80 °C
- Refrigerator 4 °C

Procedure

1. Transfer a single bacterial colony into 1 ml of LB-antibiotic medium in a loosely capped 5-ml hemolysis tube. Incubate the culture overnight at 37 °C with vigorous shaking.

2. Remove 800 μl of the culture into a 1.5-ml Eppendorf tube. Centrifuge at 12 000 rpm (12 500 *g*) for 30 s in a bench microfuge. Store the remaining culture at 4 °C, or mix with 40 μl of sterile glycerol and store at –80 °C.

3. Pour off the supernatant. Resuspend the bacterial pellet in 400 μl of 1% Triton X-100 by vigorous vortexing.

4. Boil the bacterial suspension for 5 min, then incubate the tube on ice for 5 min. Store the bacterial lysate at 4 °C or -20 °C.

5. Add the following reagents for each reaction in a MicroAmp Reaction tube (Cetus) to a final reaction volume of 50 μl:
25 μl sterile distilled water
10 μl of 50% sterile glycerol
5 μl 10x PCR reaction buffer
0.5 μl 5 mM dATP
0.5 μl 5 mM dCTP
0.5 μl 5 mM dGTP
1 μl 2.0 mM dTTP
1 μl 0.2 mM digoxigenin-11-dUTP alkali-labile

0.625 μl 20 μM of primer 1
0.625 μl 20 μM of primer 2
0.25 μl 5 units/μl Taq polymerase
5 μl bacterial lysate

6. Apply the following thermocycle profile:

1 cycle:	94 °C for 1 min
35 cycles:	94 °C for 15 s
	48 °C for 30 s
	72 °C for 3 min
1 cycle:	72 °C for 7 min.

7. Analyze an aliquot of 3 μl of each PCR sample accompanied by a known quantity of a standard reference DNA fragment, by horizontal agarose gel electrophoresis. The concentration of labelled DNA can then be estimated. Store the rest of the PCR product at 4 °C.

Notes and troubleshooting

- For the amplification reaction solution in step **5** we have given the specific volume of each component. Nevertheless, it is strongly suggested to prepare a larger solution containing all the components except the bacterial lysate to facilitate small-volume manipulation. This solution can be stored at –20 °C.
- Addition of 10% glycerol in the reaction mixture improves the efficiency and specificity of PCR amplification (Lu and Nègre 1993c).
- As the digoxigenin-labelled DNA is synthesized approximately half as efficiently as unlabelled DNA, a longer extension time for each thermal cycle is required for the labelling reaction.
- Removal of unincorporated DIG-11-dUTP from the labelled probe is not necessary prior to hybridization (i.e. for prevention of non-specific hybridization). For example, we have used the solution resulting from a failed PCR labelling reaction (no labelled DNA has been amplified as verified by agarose gel electrophoresis) as a probe, in parallel with that of a successful reaction. Neither increased background nor the expected hybridization band were observed by using the failed PCR sample as a probe.
- If the PCR-DIG amplification is not possible directly from bacteria, it is then necessary to use 50 ng of template DNA obtained through a small-scale preparation of plasmid DNA. Nevertheless, for un-

known reasons, it is always possible to find specific RFLP clones which cannot be amplified by this method. In this case, a classical procedure has to be followed: large-scale preparation of plasmid DNA, plasmid DNA digestion with restriction enzymes, isolation of inserts from vector and labelling of the DNA with substituted nucleotides with techniques such as nick-translation and random primer labelling.

1.4.5 Hybridization, Washing and Development Plus Probe Removal

This procedure is given for a Southern blot using a Hybond N+ membrane (Amersham) of 400 cm^2. All steps during signal detection are performed at room temperature and can be performed on the bench. For filter manipulation use two pairs of flat-ended forceps.

Reagents

- Hybridization solution: 5x SSC, 0.5% SDS, 0.1% N-lauroylsarcosine, 1% blocking reagent (Boehringer Mannheim), 200 μg/ml denatured herring sperm DNA.
- Buffer 1: 100 mM Tris-HCl pH 7.5, 150 mM NaCl
- Buffer 2: buffer 1 + 0.5% blocking reagent
- Buffer 2 containing 1:20000 v/v anti-digoxigenin-AP (Boehringer Mannheim)
- Buffer 3: 100 mM Tris-HCl pH 9.5, 100 mM NaCl, 50 mM $MgCl_2$
- Buffer 3 containing 1:200 v/v CSPD^R (Boehringer Mannheim)
- 2x SSC: 30 mM sodium citrate, 0.3 M NaCl pH 7
- Washing solution: 0.1x SSC (1.5 mM sodium citrate, 15 mM NaCl pH 7), 0.1% SDS
- Probe removal solution: 0.2 N NaOH, 0.1% SDS
- 0.1x TE buffer: 1 mM Tris-HCl pH 8, 0.1 mM EDTA

Equipment

- Container suitable for disposal
- Heat-sealable bag
- –20 °C freezer
- 4 °C refrigerator
- Pipettes 20 ml
- Micropipettes (200–1000-μl capacity plus tips)
- Measuring cylinders 100-ml, 500-ml capacity
- Gloves
- Glassware

- Fume hood
- Two pairs of flat-ended forceps (Millipore)
- Scissors
- 65 °C hybridization oven
- Ice
- 100 °C waterbath
- Flat-bottomed plastic container
- Saran wrap
- Amersham MP X-ray film
- Vinyl-covered exposure cassette (Sigma)
- X-ray developer (Amersham)

Procedure

Prehybridization and Hybridization

1. Prepare the amount of hybridization solution appropriate for the task at hand.
2. Rinse the Southern blot at least twice in 2x SSC solution.
3. Transfer the filter to a flat-bottomed plastic box containing at least 40 ml of hybridization solution per filter (up to ten filters can be treated at the same time).
4. Incubate for 5 h at 65 °C with gentle agitation.
5. De-nature the probe in at least 500 µl of hybridization buffer, by heating for 10 min at 100 °C. Quench the probe rapidly on ice.
6. Slip each filter (two filters can be hybridized together if put back to back) into a heat-sealable bag and add 40 ml of new hybridization solution containing the denatured DIG (alkali-labile)-labelled probe (20 ng/ml). Squeeze as much air as possible from the bag. Seal the open end of the bag with the heat sealer so that as few bubbles as possible are trapped in the bag.
7. Incubate the bag overnight at 65 °C with gentle agitation.

Washing

Do not allow the filter to dry out at any stage during the washing procedure. For filter manipulation use two pairs of flat-ended forceps.

8. Prepare enough washing solution for the task at hand.

9. Open the bag by cutting off one corner with scissors. Remove the hybridization solution completely. The hybridization solution is recovered and can be stored at 4 °C or -20 °C over a long period, for subsequent probe re-use. Cut the bag along the length of three sides. Remove the filter and immediately submerge it in a tray containing several millilitres of washing solution. This solution should be kept in a container suitable for disposal.

10. Transfer the filter to a fresh tray containing enough washing solution (120 millilitres for 1 membrane) and incubate for 5 min at room temperature with gentle agitation.

11. Repeat step **10.**

12. Transfer the filter to a flat-bottomed plastic box containing enough washing solution (120 millilitres for 1 membrane) and incubate for 15 min at 65 °C.

13. Repeat step **12** twice.

Chemiluminescent Detection

All the following steps were performed at 37 °C in an oven, with gentle agitation. The following procedure is given for one to ten filters placed in the same flat-bottomed plastic box (steps **14** to **18**). Step **19** is performed using only one or two filters back to back, placed in each heat-sealable bag.

14. Transfer the filter(s) to a flat-bottomed plastic box containing at least 100 ml of buffer 1 per filter and incubate for 5 min.

15. Transfer the filter(s) to a new flat-bottomed plastic box containing at least 40 ml of buffer 2 per filter and incubate for 1 h.

16. Transfer the filter(s) to a new flat-bottomed plastic box containing at least 25 ml of buffer 2 containing anti-DIG per filter and incubate for 30 min.

17. Transfer the filter(s) to a new flat-bottomed plastic box containing at least 100 ml of fresh buffer 1 per filter. Wash the filter(s) three times in these conditions for 30 min.

18. Transfer the filter(s) to a new flat-bottomed plastic box containing at least 100 ml of buffer 3 per filter and incubate for 5 min.

19. Slip each filter (or two if necessary, back to back) into a heat-sealable bag and add 40 ml of buffer 3 containing CSPD^R for 5 min. Squeeze as much air as possible from the bag. Seal the open end of the bag with the heat-sealer so that as few bubbles as possible are trapped in the bag. The buffer 3 + CSPD^R can be kept at –20 °C for reuse.

20. Wrap the damp filter in Saran wrap and leave it overnight at room temperature. Avoid air bubbles.

21. Expose the filter to Amersham MP X-ray film for 4 to 6 h at room temperature in a vinyl-covered exposure cassette (Sigma). If the hybridization signals are too strong with high background, or too weak, a secondary shorter or longer exposure overnight can be carried out.

Removal of Probes from Nylon Membranes

22. Immerse the filter(s) in enough 0.2 N NaOH, 0.1% SDS (100 ml for 1 membrane) solution for 10 min at 37 °C.

23. Repeat step 22. Rinse the filter briefly twice with 2x SSC.

24. If the Southern blot is not re-used immediately, it is essential to keep it in 0.1x TE buffer and store at 4 °C.

Notes and troubleshooting

- The recovered hybridization solution must be boiled for 10 min before reuse.
- Despite the suggestion of Neuhaus-Url and Neuhaus (1993), we do not recommend the use of Tween 20 in buffer 1, because it increases the background under our conditions. If there is significant background, it can be eliminated by increasing the number and length of the washes, especially for the washes following the anti-DIG incubation. It is now also possible to use a new substrate called CDPStar^{TM} (Boehringer Mannheim). This new substrate is an extremely sensitive and fast chlorosubstituted 1,2-dioxetane chemiluminescence for alkaline phosphatase. The signal is approximately ten times faster than CSPD^R on nylon membrane and reaches maximum light emission within minutes. In the case of CDPStar^{TM} use, buffer 2 should contain 1:50 000 v/v anti-digoxigenine-AP and buffer 3 should contain 1:2 500 v/v CDPStar^{TM}.

- The rate of signal production reaches its highest level 4 h after exposure to CSPD^R and remains constant for at least 24 h.
- Special care has to be taken with autoradiographic analysis when using non-alkali-labile DIG-11-dUTP. One problem is the carry-over of the signal from the previous probe used on the same membrane. Because of the nature of the detection reaction, even minimal amounts of the previous probe will be redetected. Thus, the removal of probe from the nylon membranes is crucial and has to be performed carefully in this case. Today, with the DIG-11-dUTP (alkali-labile) (Boehringer Mannheim), this problem is resolved and also allows for a less harsh probe removal procedure to be used than the one previously described in Lu et al. (1993b).
- The various advantages of this non-radioactive method over radioactive techniques are obvious and have been described elsewhere in detail (Khairallah et al. 1993; Neuhaus-Url and Neuhaus 1993). However, in brief, they are:
  - Reduced risk for laboratory workers, who, as a result, can handle the large numbers of samples involved in RFLP analysis with increased confidence and ease.
  - The waste generated does not require special handling. Nevertheless, since the degree and characteristics of toxicity of the DIG-11-dUTP (alkali-labile) are as yet unclear, we suggest keeping, as described, the first washing solution after hybridization for special disposal.
  - We can label as many probes as needed whenever we want and in large amounts. Labelled probes are stored at –20 °C until required (at least 1 year). This ability gives obvious logistical advantages.
  - The possibility of re-using the recovered labelled probe (three to five times), CSPD^R (three to five times), and the Southern blot (five to eight times and sometimes more) increases the cost effectiveness of this chemiluminescent detection system over the traditionally used radioactive method for RFLP analysis.
  - Better planning of experiments, since we do not depend on the availability and timeliness of ^{32}P shipments.
  - Film exposures are carried out at room temperature with no need for intensifying screens or a –80 °C freezer, which reduces initial investment costs.
  - Short exposures (2 h to overnight) mean that every 3rd day the membrane can be re-hybridized.

Acknowledgements. The probe psr 142 has pUC18 as the vector and was kindly provided by Dr. M.D. Gale, Norwich, UK (Sharp et al. 1989). The probe FBA065 has pBlueScript (Stratagene) as the vector and was kindly provided by Dr F. Quétier, Orsay/Paris-Sud University (GENOBLE Club), France.

References

Beck S, Köster H (1990) Applications of dioxetane chemiluminescent probes to molecular biology. Anal Chem 62:2258–2270

Beckmann JS, Soller M (1988) Detection of linkage between marker loci and loci affecting quantitative traits in crosses between segregating populations. Theor Appl Genet 76:228–236

Burr B, Evola SV, Burr FA, Beckmann JS (1983) The application of restriction fragment length polymorphism to plant breeding. In: Setlow JK, Hollaender A (eds) Genetic engineering: principals and methods Plenum, New York, pp 45–59

Dennis Lo YM, Mehal WZ, Flaming KA (1990) Incorporation of biotinylated dUTP. In: PCR protocols: a guide to methods and amplifications. Academic Press Inc, San Diego: pp113–118

Emanuel JR (1991) Simple efficient system for synthesis of nonradioactive nucleic acid hybridization probes using PCR. Nucl Acids Res 19:2790

Friedman KD, Rosen NL, Newman PJ, Montgomery RR (1990) Screening of lambda gt11 libraries. In: PCR protocols: a guide to methods and amplifications. Academic Press, San Diego: pp253–258

Höltke HI, Sagner G, Kessler Ch, Schmitz G (1992) Sensitive chemiluminescent detection of digoxigenin-labelled nucleic acids: a fast and simple protocol and its application. Bio Techniques 12:104–113

Ishii T, Panaud O, Brar DS, Klush GS (1990) Use of non-radioactive digoxigenin-labeled DNA probes for RFLP analysis in rice. Plant Mol Biol Rep 8:167–171

Jia J, Devos KM, Chao S, Miller TE, Reader SM, Gale MD (1996) RFLP-based maps of the homoeologous group-6 chromosomes of wheat and their application in the tagging of *Pm12*, a powdery mildew resistance gene transferred from *Aegilops speltoides* to wheat. Theor Appl Genet 92:559–565

Khairallah M, Acevedo F, Dorregaray F, Feingold S, Lacaze P, Gonzales de Léon D, Hoisington D (1993) RFLP analysis in the Triticeae: scale-up of operations and use of chemiluminescent techniques. In: Hoisington D, McNab A (eds) Progress in genome mapping of wheat and related species. Proc 3rd Public Worksh of the Int Triticeae Mapping Initiative, Mexico, DF, CIMMYT: pp 32–46

Kreike CM, Koning JRA, Krens FA (1990) Non-radioactive detection of single-copy DNA-DNA hybrids. Plant Mol Biol Rep 8:172–179

Lanzillo II (1991) Chemiluminescent nucleic acid detection with digoxigenin-labelled probes: a model system with probes for angiotensin converting enzyme which detect less than one attomole of target DNA. Anal Biochem 194:45–53

Lu YH, Négre S, Leroy P, Bernard M (1993a) PCR-mediated screening and labelling of DNA from clones. Plant Mol Biol Rep 11(4):345–349

Lu YH, Merlino M, Isaac PG, Staccy J, Bernard M, Leroy P (1993b) A comparative analysis between [^{32}P] and digoxigenin-labelled single-copy probes for RFLP detection in wheat. Agronomie 14:33–39
Lu YH, Négre S (1993c) Use of glycerol for enhanced efficiency and specificity of PCR amplification. Trends Genet 8:227
Marino CL, Nelson JC, Lu YH, Sorrells ME, Leroy P, Tuleen NA, Lopes CR, Hart GE (1996) Molecular genetics maps of the group 6 chromosomes of hexaploid wheat (*Triticum aestivum* L. em. Thell.). Genome 39:359–366
McCabe PC (1990) Production of single-strand DNA by asymmetric PCR. In: PCR protocols: a guide to methods and amplifications. Academic Press, San Diego: pp 76–83
Nelson JC, Van Deynze AE, Autrique E, Sorrells ME, Lu YH, Merlino M, Atkinson M, Leroy P (1995a) Molecular mapping of wheat. Homoeologous group 2. Genome 38:516–524
Nelson JC, Sorrells ME, Van Deynze AE, Lu YH, Atkinson M, Bernard M, Leroy P, Faris JD, Anderson JA (1995b) Molecular mapping of wheat: major genes and rearrangements in homoeologous group 4,5 and 7. Genetics 141:721–731
Neuhaus-Url G, Neuhaus G (1993) The use of the nonradioactive digoxigenin chemiluminescent technology for plant genomic Southern blot hybridization: a comparison with radioactivity. Transgenic Res 2:115–120
Panaud O, Magpantay G, McCouch S (1993) A protocol for non-radioactive DNA labelling and detection in the RFLP analysis of rice and tomato using single-copy probes. Plant Mol Biol Rep 11:54–59
Saiki RK, Scharf S, Faloona F, Mullis KB, Horn GT, Erlich HA, Arnheim N (1985) Enzymatic amplification of beta-globin genomic sequences and restriction analysis of sickle-cell anemia. Science 230:1350–1354
Sharp PJ, Chao S, Desai S, Gale MD (1989) The isolation, characterization and application in the Triticeae of a set of wheat RFLP probes identifying each homoeologous chromosome arm. Theor Appl Genet 78:342–348
Simon CJ, Schnorr KM (1992) PCR preparation of DNA inserts from lambda and plasmid vectors for RFLP mapping. Plant Mol Biol Rep 10:367–371
Tai TH, Tanksley SD (1991) A rapid and inexpensive method for isolation of total DNA from dehydrated plant tissue. Plant Mol Biol Rep 8:297–303
Tanksley SD, Young ND, Paterson AH, Bonierbale MW (1989) RFLP mapping in plant breeding: new tools for an old science. BioTechnology 7:257–264

Chapter 2

Cloning from Genomic DNA and Production of Libraries

2.1 Use of PCRs in Plant Molecular Biology

A. Guevara-García, L. Herrera-Estrella, and G. Olmedo-Alvarez

2.1.1 Introduction

Polymerase Chain Reaction (PCR) is an in vitro method of nucleic acid synthesis by which a particular segment of DNA can be specifically replicated. The technique is an ingenious tool for molecular biology that has had more impact on research than the discovery of restriction enzymes and Southern blotting. Since the unveiling of the method by Kary Mullis and colleagues (Saiki et al. 1985, 1988), numerous modifications, improvements and novel applications of PCR have been devised; since then more than 1000 publications involving the use of this method have appeared in the scientific literature.

PCR is typically carried out using two oligonucleotide primers that flank the DNA fragment to be amplified. These primers hybridize to opposite strands of the target sequence and are oriented so that DNA synthesis by the polymerase proceeds across the region between the primers (Fig. 2.1.1). Amplification is achieved through repeated cycles of heat denaturation of the DNA, annealing of the primers to their complementary sequences, and extension of the annealed primers with DNA polymerase. Since the extension products themselves are also complementary to and capable of binding primers, successive cycles of amplification essentially double the amount of target DNA synthesized in the previous cycle. The result is an exponential accumulation of the specific target fragment by the de novo synthesis of the region of

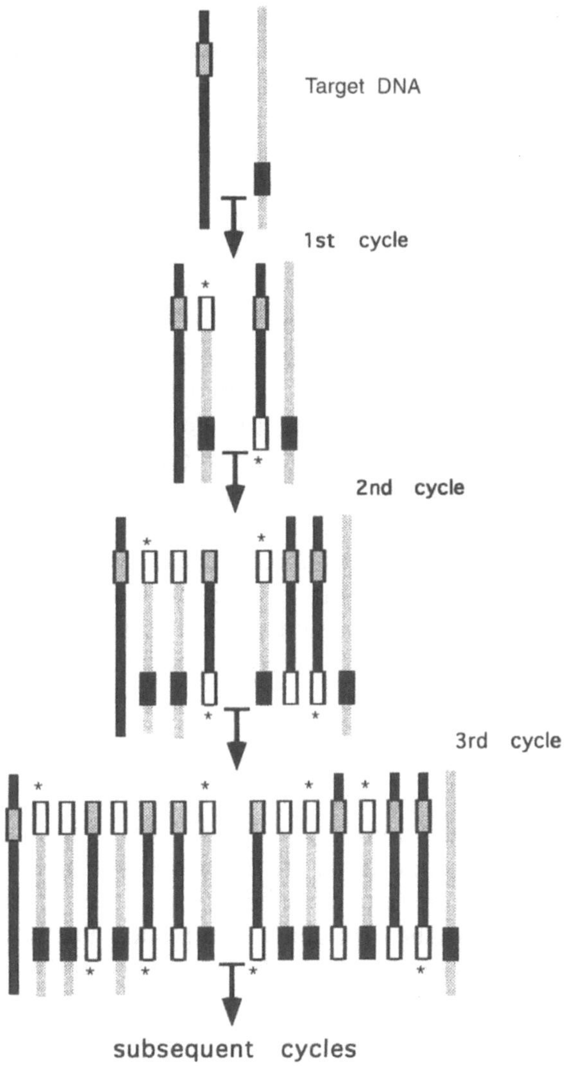

Fig. 2.1.1. Schematic diagram of a PCR process. Using two oligonucleotides as primers (each complementary to opposite strands of DNA on the specified region), new strands (*) are synthesized by primer extension. When the process is repeated (*cycles 1 to 3*), both the original chain of DNA and the newly synthesized strands serve as templates, leading to an exponential increase of products which has its ends limited by the the primers. ▭ and ▬ represent sequences complementary to the primers (▭)

DNA flanked by the two primers (approximately 2^n, where n is the number of cycles of amplification performed).

The basic protocol of PCR is easy and the reaction components are readily available either individually or as a kit from commercial suppliers (i.e. Perkin-Elmer/Cetus: Gene-Amp kit). Additionally, some authors have drawn useful conclusions about the importance or dispensability of certain parameters. Despite this, there is not a single set of instructions that works in all situations, and consequently each new PCR application is likely to require optimization.

Detailed discussion of the full range of experimental approaches utilizing PCR in plant molecular biology is beyond the scope of this chapter. It is rather intended to serve as an introduction to PCR for those new to this area, providing general information about the basic methodology, like obtaining and cloning PCR products, and the specific application of the technique in diagnosis which can be used for the detection of plant pathogens and the screening of transgenic plants. When considered necessary, detailed protocols repeatedly tested in our laboratory are provided, while, in some instances, an overview of a particular field is given with the hope that interested readers will find a reference relevant or useful to their particular research.

The amplification of a specific DNA fragment represents the core, and sometimes the only objective of the PCR methodology. The reaction components can all be assembled to generate DNA fragments of defined length and sequence in a simple automated reaction. Before reaction conditions are described, we briefly describe how components like primers and polymerase and the reaction conditions are selected and how they influence the result. The effect of varying different reaction parameters on the specificity and yield of the amplification has been discussed by many authors. Here, we summarize some of their conclusions, hoping they will be useful when optimizing PCR reactions.

2.1.1.1
Primer Design

Undoubtedly, primer composition, size, and homology with respect to DNA targets determine, more that anything else, the success or failure of a PCR. The approach to the selection of efficient primers remains somewhat empirical and there is not one set of rules that will guarantee the effectiveness of a particular primer pair. Fortunately, guidelines have been worked out by several researchers and, with the aid of software created for this particular application, designing oligonucleotides for PCR is not a difficult task. Some important considerations in primer design are discussed in the compartment following.

- *Primer size and melting temperature* (T_m). A primer size of 20 bases or more is preferred with a T_m of more than 55 °C. Remember that extension is carried out at 72 °C and much shorter oligonucleotides will not remain properly annealed at this temperature. With this primer size, you should not be concerned about the sequence uniqueness. Assuming a random distribution of nucleotides in the genome, the segment length (n) to define a unique sequence can be estimated by:

  $$2\ N(0.25)^n = 1.$$

  If the genome size (N) distribution in plants goes from 5×10^8 to 1×10^{11}, uniqueness would be achieved using oligonucleotides of 15 to 19 bases. Nucleotides are not randomly distributed, but nevertheless you can estimate the minimum size you would want for your oligonucleotides. A longer-sized oligonucleotide, however, will probably have a T_m of 55 °C or higher. Longer oligonucleotides also reduce non-specific background, and therefore are more likely to be successful in the amplification reaction.

- *Primer composition.* The GC content of the primers should be close to 50%. It is better to keep the highest stability distributed towards the 5′ end. Too many Cs and Gs at the 3′ end are more likely to false prime. Primer stability should decrease as the last five bases at the 3′ end are reached. Avoid self-priming, dimer duplexing and hairpins; this is particularly important towards the 3′ end.

- *Use of software.* There are several software programme that will save a lot of time and potential mistakes when designing a pair of primers for PCR. Most programmes perform basic functions and will help identify oligos that have a convenient size and are in the right T_m range, avoid primers that self-anneal, loop or anneal with each other; false priming will also be easily detected. Other programmes go beyond these primary functions. OLIGO (National BioSciences) and Primer Select (DNASTAR, Inc.) are some examples. PrimerSelect, for instance, lets you work with the DNA and amino acid sequence simultaneously, which can be very convenient.

- *Designing restriction sites within the primer.* If you decide to include restriction enzyme recognition sites within your primers to aid in cloning the PCR product, do not forget to allow at least five to ten free bases before the primer's end to allow proper digestion. Engineering restriction sites as part of your oligo is an advantage even if you do not use them in your first cloning, as they may later become useful when sub-cloning into specialized vectors.
- *Degeneracy.* PCR is often applied for isolating similar genes from distantly related organisms. Oligonucleotides homologous to two regions in the gene of interest are used to amplify an intervening gene. A comparison of protein sequences from different taxa usually reveals regions which are conserved and for which primers can be designed. If few other proteins have been sequenced, important regions such as functional sites could be used. Degenerate oligonucleotides are designed considering codon bias and selection of amino acids that require the least amount of degeneracy. Primer degeneracies of 10^5 to 10^6 or greater have been used successfully to isolate the correct gene from diverse species (Gould et al. 1989).

2.1.1.2
Defining the Right Annealing Temperature

To improve yield and specificity, empirical formulas have been developed to calculate optimal annealing temperature (T_a^{OPT}). T_a^{OPT} may be calculated based on the T_m of the less stable primer and the T_m of the PCR product (Rychlik et al. 1990).

$$T_a^{OPT} = 0.3T_m^{primer} + 0.7T_m^{product} - 14.9.$$

Software like OLIGO (National BioSciences) automatically produces a calculated T_a^{OPT} for the primer designed. This gives you a good starting point for homologous templates. In general, you should try to work at a temperature as close to your primer T_m as possible. For heterologous templates you should start with a much lower T_a and then increase it in 2 °C steps to find a band the expected size against a low background of non-specific priming.

2.1.1.3
Fidelity of PCR Products

The use of Taq DNA polymerase for the in vitro synthesis of DNA results in a certain rate of base substitutions and one-base frameshift errors (Tindall and Kunkel 1988).

For most PCR applications, such as direct sequencing or use of a cloned PCR sequence as a probe to screen a library, a low rate of misincorporation will not be a problem. However, when trying to clone a complete gene, maintaining genetic integrity is essential. In this case it is better to use an enzyme having a 3′ to 5′ exonuclease activity (i.e. UltmaTM, Perkin Elmer; Vent, Bio Labs etc.). In general, enzymes that have or lack the 5′ to 3′ and 3′ to 5′ exonuclease activities will perform differently in PCR; therefore, special conditions will have to be set up.

Under certain conditions, the error rate may be reduced when using Taq DNA polymerase; for instance, by increasing the amount of template and reducing the number of cycles.

If you need to amplify long DNA fragments (10 kb or longer), you should mix a commercial enzyme especially designed for this purpose. Expand^{TM} High Fidelity (Boehringer) or Elongase^{TM} Enzime Mix (GIBCO-BRL) are suitable for the amplification of long DNA fragments with high yields and high fidelity (Barnes 1994).

As noted by Eckert and Kunkel (1990), conditions that are optimal for fidelity are not necessarily those that are optimal for synthesis rate and/or product yield. These authors have worked out variations in the reaction components that can reduce mutation rate. It should be noted that if Taq DNA polymerase is used in amplification reactions where integrity is required, it will be necessary to obtain and sequence several independent clones to check for misincorporation events or PCR artefacts.

2.1.2
Procedure for DNA Amplification

The following are standard conditions for setting up a PCR reaction. These conditions should work for most amplifications, though fine tuning will be required to achieve satisfactory specificity and yield. It is used for automated PCR and with thermostable polymerases lacking exonuclease activity, such as Taq polymerase (Boehringer) and AmpliTaq polymerase (Perkin Elmer, Norwalk, Connecticut).

Reagents

- Primers (50 pmol each)
- 0.2 mM dNTPs (it is very important to use good quality dNTPs; freeze in small aliquots, as repeated defrosting results in degradation; care should be taken to have an equal amount of each four in the mix).
- Template DNA: around 0.05 to 1 μg genomic DNA or 0.002 to 0.02 μg plasmid DNA
- Taq DNA Polymerase (Perkin Elmer, Norwalk, Connecticut)
- 10 x PCR Buffer (500 mM KCl, 15 mM $MgCl_2$, 100 mM Tris-HCl, pH 8.3)
- Mineral oil
- Electrophoresis reagents

Equipment

- Thermocycler (MJ research; Perkin Elmer)
- Electrophoresis equipment
- Microcentrifuge
- Micropipettes (1–20 and 20–200 μl)
- 0.5-ml Eppendorf tubes
- UV viewing apparatus and photographic equipment

Procedure

1. Set up the following reaction on ice:

Template	1 μg of genomic DNA
dNTPs	0.2 mM each
10x PCR buffer	10 μl
Primers	50 pmol each
Taq DNA polymerase.	2.5 U

For a 100-μl reaction, mix deionized sterile water and the stated reagents except primers and Taq DNA Polymerase. Overlay 100 μl of mineral oil. Prior to starting the cycles, add the Taq DNA Polymerase and primers on top of the oil

2. Place your reaction tubes in the thermocycler and perform the following program.
   a) A single initial denaturation at 94 °C for 2 min
   b) 30 cycles as follows:
      94 °C for 30 s (denaturation)
      55 °C for 30 s (primer annealing[1])
      72 °C for 30 s (primer extension[2])

[1]Assuming a T_m of 60.

[2]Extension time varies according to product size; 30 s will be enough for up to 500 bp, after that calculate 1 min per 1000 bp.

c) A final extension at 72 °C for 7 min.
After cycles keep reactions at 4 °C.

3. Check for amplification by running up to 5 µl from the reaction on a gel and stain with ethidium bromide. There is no need to eliminate the oil or wax at this step. Simply introduce the pipette tip to the bottom of the tube.

Notes

- Always carry out control reactions to discard obvious problems.
- You should try to obtain as good a resolution as possible to discern the expected product from non-specific products of similar size. Use either a standard agarose gel for large DNA fragments or acrylamide or NuSieve agarose (FMC Bioproducts) for small fragments.
- Once you obtain a PCR product that you are considering cloning, you should always perform a Southern blot on DNA from your source plant to verify that it was actually the source. Cross-contamination from a plasmid or other plant DNA can be a problem, especially if heterologous primers are used. Alternatively, sequence the PCR product to verify identity.
- An evaporation barrier consisting of 100 µl mineral or paraffin oil or wax is placed on top of the mixture. This feature has been used to keep some key reagents separate until the start of the PCR cycles (at the very first step at 94 °C all reagents will mix) and thus reduce some non-specific products. This strategy has been called hot start. We add primers and polymerase last, others add buffer, template and polymerase last.
- If you placed wax on top of your reaction, simply remove it. If you used oil, clean it by adding 1 vol TE, 200 µl chloroform, vortex, spin 30 s and recover upper phase.

2.1.3
PCR Optimization

It may sometimes be tricky to obtain the desired product or it may be necessary to increase yield and specificity. Also, if you plan to carry out routine amplifications in your laboratory, it would be a good idea

Table 2.1.1. Effect of the variation of different reaction components on the specificity and/or yield of PCR products

Reaction component	To increase yield	To increase specificity
Template concentration	Increase	Decrease
Primer concentration	Up to 75 pmol	Down to 15 pmol
Primer size	–	Increase
dNTPs	1 mM each	20 mM each
Tris-HCl (10 mM pH8)	Up to pH 10[a]	Up to pH 10[a]
$MgCl_2$	Up to 4 mM	1.5 mM
DMSO	–	10%[b]
Glycerol	–	2%
Formamide	–	5%
Taq DNA Polymerase[c]	Up to 5 U	0.5 U

[a]The effect can be unpredictable.
[b]When adding DMSO Taq DNA Pol. should be increased (i.e. 5 U) to compensate for 47% inhibition of this enzyme by 10% DMSO.
[c]Perkin Elmer, Norwalk, Connecticut.

to try different conditions that would give you the most reproducible results for you particular application.

To optimize particular PCR reactions, it may be necessary to play with a few variables both in the reaction mix and in the cycling parameters. Table 2.1.1 shows the effect that addition or variation in the amount of a reagent may have in a reaction. It will give you a few pointers on the possible changes you may introduce to increase yield and/or specificity. It is usually enough to modify one or two parameters, for instance, pH and $MgCl_2$. Table 2.1.2 describes the

Table 2.1.2. Cycle parameters that may be varied to increase specificity and/or yield of PCR products

Reaction condition	To increase yield	To increase specificity
Cycles	Maximum of 35	Decrease to 25
Denaturation[a]	Increase to 1 min	–
Primer annealing[b]	Decrease to 45 °C	Increase to 72 °C
Primer extension[c]	Increase in later cycles	Maintain at 30 s

[a]For genomic DNA you may denature at 97 °C for the first few cycles.
[b]Assuming a T_m of 60 °C. See text.
[c]Extension time varies according to product size; 30 s will be enough for up to 1000 bp, after that calculate 0.5 min per 1000 pb. Increasing the extension times in later cycles can improve efficiency.

effects that changes in the cycle parameters will have on your PCR reaction.

2.1.4 Cloning

Gene cloning represents the heart of genetic engineering. Often, the cloning of a particular amplified fragment is the principal experimental objective. Unfortunately, cloning of PCR products has never been an easy task. The DNA fragment to be cloned is frequently part of a mixture of many different fragments with diverse ends, and traditional strategies for cloning are usually very inefficient. In this section we will describe the common practices followed in our laboratory for the purification of PCR products and their subsequent introduction into commercial or prepared vectors. In the last case, we will describe the procedure for the engineering of vectors with good cloning efficiencies.

Over 500 reports and technical tips on the subject provide evidence that cloning a PCR product may not be a straightforward step. PCR products are not blunt-ended and do not behave as such. Clarke 1988 found that an important fraction of all PCR products obtained with Taq DNApolymerase contained one 3′ protruding adenine nucleotide. This certainly explains the difficulty encountered when cloning into blunt-ended vectors. One common way to avoid this problem is to design restriction sites within the oligonucleotides. Another approach is to take advantage of the protruding deoxyadenosine nucleotide to clone into a vector containing a protruding deoxythymidine.

2.1.4.1 *T-A System*

T-A systems exploit the fact that the Taq DNA polymerase acts on dsDNA as a non template-directed transferase leaving a single protruding 3′ dATP; 3′thymidines are therefore placed on the cloning vector, which is called a T-vector. This can be used for the direct and efficient cloning of the PCR product. Vectors already processed to contain a single 3′T are commercially available. The pCR^{TM}II vector from Stratagene, for instance, gives excellent results. However, when a vector with special characteristics is preferred, or when looking for a less expensive alternative, it is possible to prepare your own T-vector.

The following procedure, reported by Marchuk et al. (1990) works well for cloning individual PCR products. It utilizes the terminal transferase activity of Taq DNA polymerase in the presence of high concentrations of dTTP. For a more ambitious task, this method may be improved as described by Khan et al. (1994) by introducing yet another step to isolate the T-tailed vectors from the blunt-ended vector to considerably reduce background. In theory, any vector that has been made blunt-ended should accommodate this modification.

2.1.4.2
Preparation of T-Vectors

Reagents

- 1 µg linearized blunt-ended cloning vector
- Taq DNA polymerase
- 10x PCR buffer
- dTTP
- Tris pH 8-equilibrated phenol.
- chloroform:isoamyl alcohol (24:1)
- 7.5 M ammonium acetate
- Cold 95% ethanol
- Cold 70% ethanol
- TE
- 10x ligation buffer (200 mM Tris-HCl pH 7.6, 50 mM $MgCl_2$, 50 mM DTT, 500 µg nuclease-free BSA)
- 10 mM ATP
- 5 U T4 DNA ligase
- Agarose gel

Equipment

- Waterbath at 37 °C
- Waterbath at 70 °C
- Waterbath at 16 °C
- Micropipettes
- Agarose gel electrophoresis equipment and UV viewing apparatus

Procedure

1. Digest 2–4 µg of the vector of interest with an enzyme that generates blunt ends.
2. Incubate with Taq DNA polymerase (1 unit/µg plasmid/20 µl vol) using PCR buffer conditions (50 mM KCl, 10 mM Tris pH 8.3, and 1.5 mM $MgCl_2$) and 2 mM dTTP at 70 °C for 2 h.

3. Phenol extract and precipitate with 1/2 vol of 7.5 M ammonium acetate and 2 vol ethanol. Centrifuge and wash the pellet with 70% ethanol.

The vector is ready for cloning. However, for increased efficiency you may want to further purify it away from non-T-tailed vectors, in which case proceed as follows:

4. Resuspend the vector in step 3 in 10 µl TE and add ligation buffer, 0.1 mM ATP and 5 U of T4 DNA ligase in a 20 µl reaction mix. Incubate overnight at 16 °C.

5. Run an agarose gel electrophoresis. Non-T-tailed vectors will have formed concatamers and will be observed as slower-migrating species above your linearized vector. Elute the linearized vector. We generally use Gene Clean (Bio RAD) for elution, but other methods will work as well. Just make sure the method is suited for elution of DNA fragments as large as your vector.

Note. Beware of repeated defrosting, which will decrease the efficiency of your vector; you are advised to keep your T-vector in small aliquots at −70 °C.

2.1.4.3
Cleaning of PCR Products

Whether or not to clean and how to clean a PCR product will depend on the cloning strategy followed. For instance, if you are planning to digest with restriction enzymes that have recognition sites (either internal to the product or engineered into your primers) which generate 5′ overhangs or blunt ends, it will be very important to completely remove Taq DNA polymerase and dNTPs. Failing to do this may result in filling-in of the overhanging ends and maybe even the addition of a 3′ overhang base. Bennett and Molenar (1994) have shown that a double phenol extraction will completely remove the enzyme.

On the other hand, if you are using a T-A system for cloning, an old PCR product or one that has been defrosted several times may actually lose its "cloning potency" (presumably through changes to the A-overhang). A "fresh" PCR product used directly for cloning works well in this case.

Frequently, PCR results in various products of different sizes and it is necessary to separate the required fragment(s) from the pool of products. When we confront this problem, we resolve the PCR products on 0.8% agarose gels and then elute "our" fragment with the filtration method described by Zhu et al. (1985). Alternative procedures, including commercial kits, are listed in Chapter 1.3.

2.1.4.4
Removal of Taq DNA Polymerase by Double Phenol Extraction

Reagents

- Tris pH 8-equilibrated phenol
- Chloroform:isoamyl alcohol (24:1)
- 7.5 M ammonium acetate
- Cold 95% ethanol
- Cold 70% ethanol

Equipment

- Microfuge
- Micropipettes
- Eppendorf tubes

Procedure

1. Extract the PCR product twice with an equal volume of Tris-buffered phenol/chloroform (1:1) and once with chloroform:isoamyl alcohol (24:1).
2. Precipitate with a 1/2 vol ammonium acetate and 2 vol 95% ethanol; incubate for 10 min, centrifuge 10 min and wash the pellet with 70% ethanol.
3. Resuspend in the desired volume of TE.

2.1.4.5
Ligation and Transformation

Reagents

- 50 ng T-vector
- PCR product
- Sterile water
- T4 DNA ligase
- 10x ligation buffer (200 mM Tris-HCl, 50 mM $MgCl_2$, 50 mMDTT, 500 μg nuclease-free BSA)
- 10 mM ATP
- Competent cells
- Agar plates plus appropriate antibiotic

Equipment

- 16 °C waterbath
- Micropipettes
- Eppendorf tubes
- 37 °C incubator

Procedure

1. Set up preferably two ligation reactions of 20 μ l consisting of a 1:1 and a 1:3 molar ratio of vector to PCR product. The amount of PCR product (nanograms) to use in a 1:1 molar ratio ligation using 50 ng T-vector is: (PCR product size) (50)/(T-vector size). Use three times as much for the 1:3 reaction.
   Ligation reaction:

10x ligation buffer	2 μl
10 mM ATP	1 μl
Sterile water	14 μl
T-vector (50 ng/μl)	1 μl
PCR product	1 μl
T4 DNA ligase 5 U/μl	1 μl

   Incubate at 16 °C overnight.

2. Five μl of these ligation reactions may be used to transform by heat shock or electroporation. We use cells with a transformation efficiency of about 5×10^7 transformants/μg plasmid. The following section (2.2) on plasmid libraries provides detailed methods on ligation and transformation.

2.1.5 Specific Applications for PCR Other Than Cloning

Since PCR can be used for amplifying and therefore more easily detecting specific DNA sequences, this method permits the screening of plant populations for the identification of members carrying particular genetic information. Therefore, PCR is a good option for the inspection of plant's transgenes (screening of transgenic plants) or for the detection of plants infected with particular plant pathogens (diagnosis of plant diseases).

Application of PCR technology may, in some instances, require high reproducibility in the routine inspection of large numbers of samples. When there is sufficient information about the target DNA sequence and highly specific primers are therefore available, the preparation of DNA samples is often the limiting step in the use of PCR for the scrutiny of large populations. Many protocols are available for this purpose, some use intact tissue and others involve precipitation of DNA from an extraction buffer. However, when reliable PCR results are not obtained, it may be necessary to obtain high-quality plant DNA. Extraction of plant DNA has been thoroughly discussed in the first section (1.1). In our laboratory we routinely use the method of Dellaporta (1983) for large-scale preparations and those of Kliyuk et al. (1993) and Wang et al. (1993) for rapid small-scale preparations.

We will discuss the application of PCR for the detection of transcribed sequences, plant pathogens and the screening of transgenic plants. The influence of factors that affect the specificity, fidelity and yield of the PCR products has been already discussed.

2.1.5.1
RNA Amplification

Detection and analysis of RNA molecules in order to determine the expression of a specific gene are important aspects in many molecular biology studies. Although there are several methods in use for studying specific RNA molecules, such as northern blots, S1 nuclease assays or RNA protection studies, RT-PCR technique is a powerful new method for the analysis of RNA transcripts (mRNA). The RT-PCR technique is very sensitive and allows the analysis and quantitation of mRNA starting with extremely small amounts of template (theoretically from single cells). In studies of tissue- or organ-specific expression in transgenic plants, this detection system is particularly useful. Its high sensitivity allows the testing of different sites of transgene expression without having to sacrifice the transgenic organism. Details of RT-PCR protocols will not be discussed here as they are fully described in the following section (3.1), along with protocols for RNA extraction etc.

2.1.5.2
Screening of Transgenic Plants

The transformation of plant tissues for the generation of transgenic plants has been and will continue to be a routine procedure in all laboratories of plant molecular biology. Although transgenic plants were produced originally only for research purposes, it appears that now many improved crop plants will soon be available for large-scale production. Characterizing plants following transformation is a very important step in obtaining transgenic plants. Therefore, the implementation of methods that permit the screening of large putative transgenic populations for the accurate identification of plants carrying introduced genes is essential. Commonly, Southern blot analysis, assays of enzyme activity and now PCR provide rapid methods to monitor gene integration (Blake et al. 1991; Goyenechea et al. 1991; Chee and Slighton 1992).

Southern blot hybridization has been used for the identification of transgenic plants for over a decade. However, the technique is laborious and incompatible with the rapid analysis of a large number of samples.

Assays of the enzyme activity of reporter genes introduced into plants are effective and facilitate the recognition of transgenic plants. Some enzyme assays are, however, difficult, tedious and expensive. The GUS system (Jefferson 1988), undoubtedly the reporter gene most used, has many properties that make it almost ideal for the analysis of transgenic plants. Perhaps the only disadvantage of the GUS system is that it cannot be used to identify plants carrying inactive transgenes.

PCR provides a rapid and efficient alternative to determine which plants contain accurately integrated target genes. This technique, like Southern blot analysis, makes it possible to estimate the copy number of transgenes integrated in a plant genome. Moreover, PCR permits distinguishing between plants containing one of two different transgenes or both in a single assay.

The limiting step in the application of PCR for transgene detection could be the selection of adequate primers for the amplification of the target gene. However, all plant-transforming vectors contain a selectable marker gene, that confers antibiotic resistance and whose DNA sequence is known. Therefore, oligonucleotides can be designed to those vector markers to check by PCR for the insertion of foreign DNA in the genome of transformed plants. Table 2.1.3 gives a list of primers drawn from the DNA sequences of different selectable marker or reporter genes. Those against NPTII are commonly used in our laboratory for the screening of transgenic plants employing the reaction conditions outlined.

Table 2.1.3. Primers for PCR screening of transgenic plants

Gene target[a]	UP (5′-3′)	LP (5′-3′)	OAT (°C)	Product Length (pb)
NPTII	TCCGGCCGCT TGGGTGGAGAG	CTGGCGCGAG CCCCTGATGCT	63.6	470
CAT	TCTTGCCCGCC TGATGAATGCTC	GCCCCGCCC TGCCACTC	58.0	471
HPT	AGCTGCGCCGA TGGTTTCTACAA	ATCGCCTCGCT CCAGTCAATG	60.5	509
GUS	CTGCGACGCTC ACACCGATACC	TCACCGAAGTTC ATGCCAGTCCAG	59.5	441

[a]From EMBL + GenBank.
UP Upper primer.
LP Lower primer.
OAT Optimal annealing temperature.
NPTII Neomicine phosphotransferase.
CAT Chloramphenicol acetyl transferase.
HPT Hygromycin-B phosphotransferase.
GUS β-Glucuronidase.

2.1.5.3
Detection of Plant Pathogens

Disease diagnosis requires of simple and rapid tests. The detection of infectious agents and the identification of genetic variation associated with disease has been revolutionized by the use of specific nucleic acid hybridization probes. This method is limited, however, by the large amount of DNA or RNA required to show a detectable signal. In this regard, PCR has played a critical role in improving the DNA-based diagnosis by producing a simple, rapid and robust method for the detection and identification of pathogens on samples from a large number of infected plants. Outstanding advances have been made in the application of the PCR methodology in the area of diagnosis of plant pathogens.

Some of the disadvantages of conventional approaches of detection and the advantages of the use of PCR-based plant-pathogen detection are the following.

- Microscopy techniques currently used for detection are not sensitive enough to be used in the detection of pathogens (for plant viruses an electron microscope is required).

- Serological typing and tissue culture techniques currently used for detection lack the specificity and sensitivity required for routine diagnosis. Cross-reactivity between related microbes is frequently a problem.

- Culture-based methods for identification are difficult, lengthy, or unavailable; identification may require several days or weeks depending of the nature of the infectious agent.

- PCR techniques:
  - PCR detection is significantly more sensitive than immune techniques and does not require fresh or properly frozen specimens.
  - Since PCR can be applied to fixed tissues (frozen or formalin-fixed), it is very advantageous in field studies by reducing the potential dangers involved in the transportation of infected vectors.
  - PCR extends the range of specimens that can be analyzed for the presence of the pathogenic organism.
  - Following sample preparation, the same PCR protocols can be used for the detection of plant pathogens of different types in a fast and automated method.

- Specific primers designed based on the sequence of specific genes can discriminate between the pathogen under investigation and related species or even between race or pathovars of the same pathogen.

As can be appreciated, PCR qualities make this technique an excellent option for the diagnosis of microbial plant pathogens. In fact, PCR has already been used for epidemiological studies, research, diagnosis and for the identification of fungus, bacteria, mycoplasma and viruses from infected plant tissues (Langeveld 1991; Navrot et al. 1992; Barker et al. 1993; Hadidi et al. 1993; Johanson and Jeger 1993; Namba et al. 1993; Rojaset et al. 1993; Prosen et al. 1993; Hataya et al. 1994). Plant mycoplasm and virus detection has even been achieved in insect vectors (López-Moya et al. 1992; Vega et al. 1993). The method could also be applied, in principle, to study the complex pathogen-vector relationships. The knowledge generated from these studies will aid to the development of new approaches for the control of plant diseases.

The basic criteria involved in the design of primers for the rapid detection of plant pathogens can be demonstrated with the PCR-based detection of bacterial species. This has been developed based on the conserved regions of the 16S rRNA/rDNA sequences (Bei 1991). The small subunit of rRNA contains segments that are conserved at the species, genus or kingdom level. PCR amplification using such rRNA/rDNA-based universal bacterial primers can potentially be applied for monitoring bacterial DNA or bacterial cells in any type of samples. Additionally, toxin or coat protein genes constitute good candidates as a potential PCR targets for the detection of bacterial or virus plant pathogens (Garzón-Tiznado et al. 1993; Prosen et al. 1993).

To list detailed protocols for the identification of specific microorganisms is beyond the scope of this chapter. However, the general procedures described can be readily applied to the successful detection of microbial plant pathogens. In our laboratory of plant viruses these are being used, among other things, for the diagnosis of plant material infected with HPV, using primers designed against the coat protein gene. Figure 2.1.2 shows an example of this use of PCR. Finally, we hope that the references cited in this section can be used by the interested readers to find relevant information about their particular area of investigation.

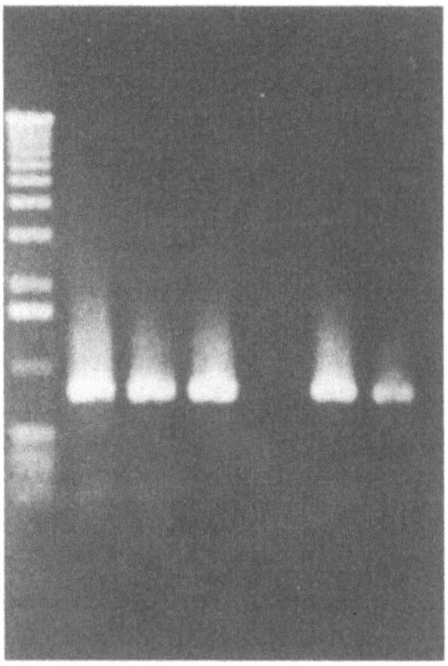

Fig. 2.1.2. PCR screening of tobacco plants for PHV (Pepper Huasteco Virus). Total DNA prepared from tobacco leaves from seven putatively PHV-infected plants was subjected to PCR. The oligonucleotides used are targeted to the capside gene. *Lane 1*, molecular weight marker; *lanes 2 to 8*, PCR products from independent DNA samples. *Lanes 5 and 8* represent uninfected plant samples

References

Barker H, Webster KD, Reavy B (1993) Detection of a potato virus Y in potato tubers: a comparison of polymerase chain reaction and enzyme-linked immunosorbent assay. Potato Res 36:13–20

Barnes WM (1994) PCR amplification of up to 35-Kb DNA with high fidelity and high yield from λ bacteriophage templates. Proc Natl Acad Sci USA 91:2216–2220

Bej AK (1991) Amplification of nucleic acids by polymerase chain reaction (PCR) and other methods and their applications. Crit Rev Biochem Mol Biol 26:301–334

Bennett BL, MolenaarAJ (1994) Cloning of PCR products can be inhibited by Taq polymerase carryover. BioTechniques 16:32–33

Blake NK, Ditterline RL, Stout RG (1991) Polymerase chain reaction used for monitoring multiple gene integration in *Agrobacterium*-mediated transformation. Crop Sci 31:1686–1688

Chee PP, Slighton JL (1992) Transformation of cucumber tissues by microprojectile bombardment: identification of plants containing functional and non-functional transferred genes. Gene 118:255–260

Clarke JM (1988) Novel non-templated nucleotide addition reactions catalyzed by procaryotic and eucaryotic DNA polymerases. Nucleic Acids Res 16:9677–9686

Dellaporta SL, Wood J, Hicks JB (1983) A plant DNA minipreparation: version II. Plant Mol Biol Rep 1:19

Eckert KA, Kunkel TA (1990) High fidelity DNA synthesis by the *Thermus aquaticus* DNA polymerase. Nucleic Acids Res 18:3739–3744

Garzón-Tiznado JA, Torres-Pacheco I, Asencio-Ibauez JT, Herrera-Estrella L, Rivera-Bustamante RF (1993) Inoculation of peppers with infectious clones of a geminivirus by a biolistic procedure. Phytopathology 83:514–521

Gould SJ, Subramani S, Scheffler IE (1989) Use of the DNA polymerase chain reaction for homology probing: isolation of partial cDNA or genomic clones encoding the iron-sulfur protein of succinate dehydrogenase from several species. Proc Natl Acad Sci USA 86:1934–1938

Goyenechea B, Moran R, Gutierrez C, De La Riva G, Perez S (1991) Screening of transgenic plants by PCR. Biotecnol Apl 8:237–241

Hadidi A, Montasser MS, Levy L, Converse RH, Madkour MA, Skrzeckowski LJ (1993) Detection of potato leafroll and strawberry mild yellow-edge luteoviruses by reverse transcription-polymerase chain reaction amplification. Plant Dis 77:595–601

Hataya T, Inoue KA, Shikata E (1994) A PCR-microplate hybridization method for plant virus detection. J Virol Methods 46:223–236

Jefferson RA (1988) Plant reporter genes: the GUS gene fusion system. Genet Eng 10:247–263

Johanson A, Jeger MJ (1993) Use of PCR for detection of *Mycosphaerella fijiensis* and *Micosphaerella musicola*, the causal agents of Sigatoka leaf spots in banana and plantain. Mycol Res 97:670–674

Khan IM, Zammit PS, Green M, Buluwela L (1994) A simple ligation step improves the efficiency of T-overhang vectors. Trends Genet 10:225–226

Klimyuk VI, Carroll BJ, Thomas CM, Jones JDG (1993) Alkali treatment for rapid preparation of plant material for releable PCR analysis. Plant J 3:493–494

Langeveld SA (1991) Identification of a potyviruses using the PCR with degenerate primers. J Gen Virol 72:1531–1541

López-Moya JJ, Cubero J, López-Abella D, D'íaz-Ru'íz J-R (1992) Detection of cauliflower mosaic virus (CaMV) in a single aphids by the polymerase chain reaction (PCR). J Virol Methods 37:129–137

Marchuk D, Drumm M, Saulino A, Collins FS (1990) Construction of T-vectors, a rapid and general system for direct cloning of unmodified PCR products. Nucleic Acids Res 19:1154

Namba S, Kato S, Iwanami S, Oyaizu H, Shiozawa H, Tsuchizaky T (1993) Detection and differentiation of plant-pathogenic mycoplasmalike organisms using polymerase chain reaction. Phytopathology 83:786–791

Navot N, Zeidan M, Pichersky E, Zamir D, Czosnek H (1992) Use of polymerase chain reaction to amplify Tomato Yellow Leaf Curl Virus DNA from infected plants and viruliferous whiteflies. Phytopathology 82:1199–1202

Prosen D, Hatziloukas E, Schaad NW, Panopoulos NJ (1993) Specific detection of *Pseudomonas syringae* pv. *phaseolicola* DNA in bean seed by polymerase chain reaction-based amplification of a phaseolotoxin gene region. Mol Plant Pathol 83:965–970

Rojas MR, Gilberston RL, Russell DR, Maxwell DP (1993) Use of degenerate primers in the polymerase chain reaction to detect whitefly-transmitted geminiviruses. Plant Dis 77:340–347

Rychlik W, Spencer WJ, Rhoads RE (1990) Optimization of the annealing temperature for DNA amplification in vitro. Nucleic Acids Res 18:6409–6412

Saiki RK, Gelfand DH, Stoffel S, Scharf SJ, Higuchi R, Horn GT, Mullis KB, Erlich HA (1988) Primer-directed enzymatic amplification of DNA with a thermostable DNA polymerase. Science 239:487–491

Saiki R, Scharf S, Faloona F, Mullis KB, Horn GT, Erlich HA, Arnheim N (1985) Enzymatic amplification of β-globin genomic sequences and restriction site analysis for diagnosis of sickle cell anemia. Science 230:1350

Tindall KR, Kunkel TTA (1988) Fidelity of DNA synthesis by the *Thermus aquaticus* DNA polymerase. Biochemistry 27:6008–6013

Vega FE, Davis RE, Barbosa P, Dally EL, Purcell AH, Lee IM (1993) Detection of a plant pathogen in a nonvector insect species by the polymerase chain reaction. Phytopathology 83:621–624

Wang H, Qi M, Cutler AJ (1993) A simple method of preparing plant samples for PCR. Nucleic Acids Res 21:4153–4154

Zhu J, Kempenaers W, Van der Straten D, Contreras R, Fiers W (1985) A method for fast and pure DNA elution from agarose gels by centrifugal filtration. Bio/Technology 3:1014–1016

2.2 Plasmid Libraries

M.S. Clark

2.2.1 Introduction

Most plasmid vectors in use today are derived from pBR322, developed by Bolivar et al. (1977). This original vector has since been dramatically transformed such that today many different types of plasmid vector have been commercially produced, tailor-made for a specific purposes: sequencing, transcription and expression systems, apart from their original development as straight cloning vehicles. This chapter concentrates on the latter use.

Plasmid vectors are comparatively small and consequently usually accommodate fragments of DNA smaller than themselves, the efficiency of transformation being inversely related to the size of the insert and, in fact, constrained if the size of vector and insert together exceed 15 kb. Therefore they operate under a certain amount of indirect size selection. Because of their capacity to accommodate only relatively small pieces of DNA, they are impractical vectors for producing representative genomic libraries with full coverage. Hence the term plasmid library is rather misleading, but nonetheless, widely used. Perhaps a more accurate description would be a plasmid clone bank.

Plasmids are therefore used to produce clone banks of short sequences (usually 1000 clones), in particular, repetitive sequences which are usually less than 1Kb in size (Flavell 1980).

The main uses for plasmid library-derived sequences include:

- RFLP probes, in which the "library" is usually produced from PstI digested genomic DNA (Tanksley et al. 1988; McCouch et al. 1988; Kochert et al. 1991; Nodari et al. 1992; Keim et al. 1992; see Chap. 7).

- Species-specific sequences for the analysis of: somatic fusion products (Saul and Potrykus 1984; Imamura et al. 1987; Schweizer et al. 1988; Pehu et al. 1990; Itoh et al. 1991, also see Chap. 10), alien addition lines (Schmidt et al. 1990) and interspecific hybrids, amphidiploids and their derivatives (Hugros et al. 1990).

- Probes to study genetic relationships and evolutionary diversity (Song et al. 1988 and Goffreda et al. 1992 (RFLPs); Iwabuchi et al. 1991 and Solano et al. 1992 (repetitive sequences); Zhao et al. 1989 (genome-specific sequences).

- Repetitive sequences for cultivar identification (Schweizer et al. 1993; Thomas et al. 1993).

- Chromosome-specific probes (Devos et al. 1992).

- Subcloning from bacteriophage, and cosmids for detailed screening and sequencing.

- Cloning of PCR products (discussed in the previous chapter).

General protocols for the production and screening of a plasmid library will be described here, with a bias towards the production of RFLP probes; however, other uses (and screening methods) will be referred to in later chapters.

This chapter describes the use of the Bluescript series of plasmids. These are multi-purpose cloning vectors and are in wide use.

The Bluescript plasmids are small in size (2.96 Kb), contain an ampicillin resistance gene (for selection of transformed *E.coli*) and a multiple cloning site with 21 unique restriction enzyme recognition sites. The polylinker is in the N-terminal of a *lacZ* gene fragment, hence blue/white selection can be made on appropriate strains of bacteria containing the lacZ M15 on an F′ episome. They replicate to high copy number within the bacterial cells.

Along with their cloning function, Bluescript plasmids can be rescued as single-stranded DNA (in the presence of M13 helper phage) for sequencing or site-directed mutagenesis. Flanking the polylinker are the T7 and T3 RNA polymerase promoters which can be used to synthesize RNA in vitro. For PCR amplification of inserts and sequencing, three sets of primer sites are available in the polylinker region: M13 Forward and Reverse, T3 and T7 and KS/SK.

The production of a plasmid library consists of several stages:

- DNA isolation,

- restriction enzyme digestion of genomic DNA and vector,

- ligation into plasmid vector,

- preparation of competent cells,

- transformation
- screening and analysis of plasmids:
  colony blot
  plasmid isolation,
- Storage of clones.

2.2.2
DNA Isolation

This has already been described in the first chapter of this book (Sect. 1.1). DNA must be full length. Approximately 2 μg are required for this technique.

2.2.3
Restriction Enzyme Digestion of Genomic DNA and Plasmid

Reagents

- Genomic DNA
- Plasmid DNA, e.g. any of the Bluescript KS/SK series (Stratagene)
- Restriction enzymes plus digestion buffer (available commercially)
- Sterile distilled water (SDW)
- 40 mM spermidine
- DNA molecular weight marker. e.g. Lambda HindIII
- Glycerol loading buffer (GLB)
- Ethidium bromide (10 mg/ml stock)

Equipment

- 1.5-ml Eppendorf tubes
- Micropipettes and tips (1–20 and 20–200 μl)
- 37 °C water bath
- 65 °C water bath
- Mini-gel tank and power supply
- Apparatus for viewing and photographing gel

Genomic DNA

Procedure

This has already been covered under Southern blotting (Sect. 1.2). Use the same protocol, only scale down the digest reaction for 2 μg of DNA. Check 1 μg of the digested product on a mini-gel. The DNA should be fully digested and appear on the gel as a smear.

Ethanol-precipitate the remaining DNA and resuspend in SDW at a concentration of 100 ng/μl for the ligation protocol.

Plasmid DNA

Procedure

1. Set up the following reaction:
   500 ng plasmid
   1x digestion buffer
   1 U enzyme
   The final volume should be round about 10 μl, but depends on the original dilution of the DNA.
2. Incubate at 37 °C for 1 h. Stop the reaction by incubating at 65 °C for 10 min.
3. Take out the equivalent of 100 ng and run on a mini-gel alongside a DNA molecular weight marker. There should be a single band, which is the cut plasmid. Any extra bands are due to incomplete digestion.
4. Ethanol-precipitate the digest and resuspend plasmid at a concentration of 50 ng/μl in SDW.

Notes and troubleshooting

- Many papers using plasmid libraries (McCouch et al. 1988; Song et al. 1988; Harvey and Muehlbauer 1989; Kochert et al. 1991; Nodari et al. 1992) size-select the fragment to clone. This is carried out either by running the digest out on a gel and excising the fragments with a size of 0.6–2 Kb or selection on a 10–40% sucrose gradient. To a certain extent, this is probably unnecessary, as the cloning procedure itself exerts a certain amount of size selection. However, if required, the gel excision option is much more convenient.
- Restriction enzyme digests have been covered under Southern blotting (see Sect. 1.2), so for any problems check the troubleshooting guide in that section.
- When the ligation mixture is transformed, both blue and white colonies will be produced. Usually, the blue (self-ligated vector) will be in the majority and it can be difficult to successfully pick the whites from the blues without contamination. This can be prevented by de-phosphorylating the digested ends of the vector. This is described in Section 2.3.2.4.

2.2.4 Ligation into Plasmid Vector

Reagents

- T_4 DNA ligase enzyme plus ligation buffer (available commercially, the buffer stock may be 5x or 10x concentration. NB: some buffers require ATP to be added separately)
- Sterile distilled water (SDW)
- DNA molecular weight marker, eg. Lambda HindIII
- Glycerol loading buffer (GLB)
- Ethidium bromide (10 mg/ml stock)

Equipment

- 1.5 ml Eppendorf tubes
- Micropipettes and tips (1–20 and 20–200 µl)
- 15 °C cooled waterbath
- 65 °C waterbath
- Mini-gel tank and power supply.
- Apparatus for viewing and photographing gel

Procedure

1. Set up the following reaction:

4µl	Plasmid (50 ng/µl)
6µl	Genomic DNA (100 ng/µl)
4µl	5x ligation buffer
4µl	SDW
2µl	DNA ligase
20µl	Total volume

2. Incubate overnight at 15 °C. Stop the reaction by incubation at 65 °C for 10 min.

3. Check on mini gel using 10 µl of the reaction mix. Run alongside a DNA molecular weight marker. Store the rest of the reaction mix at –20 °C until required.

Notes and Troubleshooting

- The important point about this reaction is the 3:1 insert to vector ratio as an excess of genomic DNA drives the ligation reaction towards ligating insert to vector. Low ratios favour vector: vector ligation.

- Very small amounts of DNA can be detected on gels. Therefore the check mini-gel after ligation is to ensure that there is no band left which corresponds to the size of the digested plasmid. If this band is present, ligation has not taken place or has been inefficient. This could be due to several reasons:
  - Ligase enzyme inactive or partially active. This can often happen at the end of a batch. Set up reaction with fresh enzyme.
  - Restriction enzyme was not successfully inactivated. Ethanol precipitation should negate this problem. However, if this step was left out and either the genomic DNA or vector was used directly from the restriction digest reaction, it should be noted that not all enzymes can be conveniently heat -killed at 65 °C for 10 min. Consult product profiles supplied with the enzymes for inactivation data. If in doubt, ethanol-precipitate.
  - Contaminants in the original DNA samples which have affected the ligase enzyme. This can often happen if the DNAs to be used in the reaction have not been ethanol-precipitated and re-suspended in SDW, but used straight from the digests.
  - Reaction has not proceeded for long enough. Many protocols state that this reaction can be carried out in a few hours. This may be true, if all that is required is to ligate one insert into the cut vector, but plasmid clone banks with the complexity of genomic DNA available to be ligated into the vector are best left overnight.
  - Non-compatible ends of vector and genomic DNA. Double-check the cutting sites if digesting with two different enzymes.

- The DNA ligase enzyme is temperature-sensitive and the requirement for a cooled 15 °C waterbath may cause problems. Alternatives are PCR machines, "heating" blocks or the fridge/cold room (although these are not at 15 °C, the lower incubation temperature of around 4 °C only means that the reaction proceeds more slowly. Temperatures higher than 15 °C may destroy the enzyme fairly rapidly, before ligation has been efficiently completed.

- Short cuts such as reducing the length of incubation time and not purifying the fragments after enzyme digestion can be taken when a single specific insert is to be cloned into a plasmid; however, to optimize the efficiency of ligation for plasmid libraries, these precautions must not be omitted.

2.2.5
Preparation of Competent Cells

Two protocols are described for the preparation of competent cells. The first, using calcium chloride (based on Cohen et al. 1972), is quick, simple and works well with most strains of *E.coli.* The alternative, based on Hanahan (1983), is more complex and time-consuming, but maintains the competence capacity of cells better in long-term storage, than those produced by calcium chloride.

2.2.5.1
$CaCl_2$ Method

Reagents

- L-Broth media (200 ml): 2 g tryptone/1 g yeast extract/2 g NaCl. Check pH is 7–8. Aliquot 100 ml into a 500 ml conical flask and save the remainder in bottle. Autoclave and store at 4°C.
- 0.1 M $CaCl_2$, store at 4°C
- *E. coli* bacteria, strain e.g. SURE^{TM} (Stratagene)

Equipment

- 37 °C shaker incubator
- Refrigerated centrifuge capable of least 4000 rpm
- 50-ml Oakridge tubes
- Ice
- Micro-pipette (5-ml capacity) plus tips
- Disposable cuvettes for spectrophotometer
- Spectrophotometer

Procedure

1. Set up 5 ml culture of *E. coli* in L-broth. Incubate overnight at 37 °C on a rotary shaker.
2. Add 1 ml of the overnight culture to 100 ml L-broth. Incubate at 37 °C on a shaker until the OD_{650} = 0.3 (about 3 h).
3. Aliquot culture into 2x 50 ml Oakridge tubes. Chill on ice for 10 min. Centrifuge at 4000 rpm for 10 min at 4 °C. Discard supernatant and resuspend the pellets each in 20 ml of cold 0.1 M calcium chloride.
4. Leave on ice for 25 min. Centrifuge at 4 K for 10 min at 4°C. Resuspend each pellet in 2 ml of cold 0.1 M calcium chloride.

5. The cells may then either be used immediately, snap-frozen in liquid nitrogen and stored at –70 °C, or left on ice overnight to increase their competence, and used the following day.

2.2.5.2
Hanahan Method

Reagents

- 1 ml SOB media (20 g bacto-tryptone, 5 g yeast extract, 0.5 g NaCl per 1-l 10 mM $MgCl_2$, 10 mM $MgSO_4$) for overnight culture
- 100 ml SOB medium in 1-l flask
- 50 ml FSB (Frozen Storage Buffer: 10 mM potassium acetate, pH 7.5, 45 mM $MnCl_2$, 10 mM $CaCl_2$, 100 mM KCl, 3 mM hexamminecobalt chloride, 10% glycerol)
- DMSO
- SURE^{TM} cells

Equipment

- 37 °C shaker incubator
- Microcentrifuge tubes
- 50-ml polypropylene centrifuge tubes
- Refrigerated centrifuge capable of 3000 rpm.
- Liquid nitrogen/–70 °C freezer
- Micropipettes

Procedure

1. Set up an overnight culture of appropriate bacterial strain in 1 ml SOB medium. Shake at 37 °C.
2. Transfer 1 ml culture to 1-l flask containing 100 ml SOB medium. Shake at 37 °C for 2–3 h until OD_{600} is between 0.5 and 0.6.
3. Split the 100-ml culture between two 50-ml centrifuge tubes. Pellet the cells at 3000 rpm for 10 min.
4. Remove the supernatant and resuspend the cells in 33 ml of chilled FSB. Leave on ice for 15 min.
5. Pellet the cells, as before, remove the supernatant and resuspend in 8 ml chilled FSB. Add 280 µl DMSO, mix and leave on ice for 5 min.
6. Add a further 280 µl DMSO, mix and leave on ice for 15 min.
7. Aliquot into microcentrifuge tubes and flash freeze by placing tubes immediately into liquid nitrogen or dry ice/alcohol bath. Store at –70 °C.
8. When required, thaw cells on ice and use immediately.

Notes

- $CaCl_2$ cells are more competant if left overnight at 4 °C.
- Competent cells are very suseptable to freezing and thawing. Do not re-freeze. If freezer breaks down, re-make the whole stock.

2.2.6 Transformation Protocol

Selection procedures in this protocol are for use with Bluescript plasmids and SURE™ cells, if other types of vector are used, consult product profile to determine antibiotic resistance, host cell genotype and selection conditions.

Reagents

- Ligation mix
- Competent cells
- Control DNA as appropriate
- L-broth media (10 g tryptone/5 g yeast extract/10 g NaCl, pH 7–8, per 1 l)
- L-broth ampicillin plates: 15 g agar per 1 l L-broth solution). Add ampicillin to a final concentration of 100 µg/ml.
- Ampicillin: stock of 50 mg/ml in SDW
- IPTG: add 2 g to 8 ml DDW. Adjust volume to 10 ml. Filter sterilize. Store at –20°C
- X-Gal: stock = 20 mg/ml in dimethylformamide. Store at –20°C in the dark

Equipment

- Glass spreader
- 1.5-ml Eppendorf tubes
- Micropipettes (1–20 and 20–200 µl capacity) plus tips
- 42 °C water bath
- 37 °C shaking incubator
- Ice
- 37 °C incubator

Procedure

1. Dilute the ligation reaction to 20 ng/µl.
2. Add 1 µl (20 ng) ligated DNA to 50 µl competent cells. Use all the ligation reaction. Leave on ice for 10 min.
3. Heat-shock the cells at 42 °C for 50 s.

4. Add 100 µl of L-broth (pre-warrned to 37 °C) to each tube and incubate with shaking at 37 °C for 1 h.

5. Whilst the cells are incubating, dry the L-broth plates by briefly incubating at 37 °C. Add 44 µl of IPTG/X-Gal (4 µl IPTG and 40 µl of X-Gal) and spread over the surface of the plate. Leave to soak in (about 10 min).

6. After 1 h, add the cells/L-broth solution to the plates . Spread over the surface and leave to soak in.

7. Incubate the plates overnight at 37 °C. Transformed colonies containing vector and insert) are white, re-ligated vector colonies are blue.

Notes

- The multiple cloning site for Bluescript plasmids has been inserted into the gene coding for the amino-terminal fragment of β-galactosidase. This fragment, in the presence of an inducer, IPTG, will complement a defective form of β-galactosidase in the host. If induced, the bacteria produce both enzyme fragments and when plated on media containing X-Gal produce blue-coloured colonies. If a piece of DNA is inserted in the multiple-cloning site, the complementing fragment is inactivated and no colour is produced. Therefore colonies containing a fragment of DNA are white.
- Often the blue colour is not well developed. Leave the plates in the fridge overnight and this will darken the colour. This can be important, as false positives can be produced, and picking colonies whilst the colour is faint will increase the chance of this happening.
- The average number of colonies required for a plasmid library is generally 1000. This figure is deemed manageable.

Troubleshooting

- Controls: The usual control is to reserve some competent cells and transform with uncut vector. This should produce a plate covered in blue colonies and proves that the cells were competent; any problems with the other plates is then due to the ligation reaction.
- *There is nothing on the other plates.* This can be due to inactive ligase enzyme. This can be tested by cutting pUC and religating at the same time as the other ligation reaction. If the cut pUC does not produce any colonies, the ligase is at fault and should be replaced.

- *A lawn of cells is produced.* This could be due to two reasons:
  - There has been no selection pressure on the untransformed *E. coli* cells and these have grown along with the transformants. This is due to lack of antibiotic in the plates or use of low activity antibiotic (this may happen if the antibiotic was added to L-broth which had not been sufficiently cooled).
  - The *E.coli* stock could have picked up an antibiotic resistance gene in a plasmid.

  If this happens, make fresh L-broth plates with fresh ampicillin and without any antibiotic. Streak out the stock bacteria on plates with and without antibiotic. If the stock grows on ampicillin plates, the *E.coli* has picked up a plasmid and must be thrown away and fresh obtained. If the stock grows on the L-broth plates, but not the ones with ampicillin, the ampicillin was missing/below strength in the plates used for the transformation.

2.2.7 Analysis of Colonies

Once the cultures have been transformed and white colonies obtained, they have a limited storage life on the plates (a month maximum). It is best to pick them immediately using sterile toothpicks into microtitre plates, with each well containing 70 µl of TB or similar medium plus antibiotics. The plates are then grown overnight at 37 °C. The following morning, they can either have glycerol added to them for storage (See Sect. 2.2.8) or be used directly in a colony blot for screening purposes.

2.2.7.1 *Colony Blotting*

This is based on the method of Grunstein and Hogness (1975).

Reagents

- LB agar plus antibiotics in large 132-mm petri dishes
- Nylon membrane filters
- Denaturation solution: 1.5 M NaCl/0.5 M NaOH
- Neutralization solution: 1.5 M NaCl/0.5 M Tris pH 7.2/1 mM EDTA
- 2x SSC: 0.3 M NaCl, 0.03 M sodium citrate

Equipment

- 96-well replica plating tool
- Plastic trays
- 3-MM paper
- Forceps
- 80 °C oven or UV transilluminator for fixing blots

Procedure

1. The 96-well replica plating tool is sterilized and dipped into the microtitre cultures. This is then stabbed onto a similarly sized rectangle of nylon membrane. Make duplicates.
2. The membrane is placed culture side up on a large petri dish containing L-B agar plus antibiotics. The filter is incubated at 37 °C overnight.
3. Remove filter from plates and place colony side up on a pad of absorbant filter paper (3 MM) soaked in denaturation solution. Leave for 7 min.
4. Repeat process using neutralizing solution for 3 min. Repeat once.
5. Wash briefly in 2x SSC.
6. Air dry and fix DNA to filters. This can be done either by baking for 2 h at 80 °C or exposing to UV (the length of time varies according to the type of transilluminator, but is generally round about 4 min). The filters can now be stored either sealed in plastic bags or Saran wrap at room temperature until required.

Notes

- Not all colonies transfer equally well, this must be borne in mind when screening the colony blots, i.e. a low signal may be a result of poor transfer and not necessarily due to a low copy number insert. Hence it is always a good idea to use duplicate filters, as this reduces the possibility of false positive results arising due to poor transfer of colonies.
- In our laboratory, we have a home-made device which allows eight microtitre plates to be spotted onto the same filter. It may be worth investigating the possibility home-made or commercial versions of a similar system if intending to carry out many such assays. It does save time and reagents.
- It is possible to screen the microtitre libraries directly using PCR. With this, a 96-well PCR is carried out using a microtitre PCR

machine. This is quite a tedious procedure (especially loading 96 samples onto a gel) but will give instant information on insert status and size, and the gel can be blotted directly for screening purposes.

Generalized screening notes

Having obtained the blots, the library can then be screened with either radioactive or non-isotopic probes. Hybridization of colony blots can be carried out in a hybridization oven, but sandwich boxes may be preferable if a large numbers of blots have to be screened. Prehybridization times should be extended slightly, with colony blots used for the first time and the solution changed once or twice before adding the probe. This is because the colony blots still have lots of cell debris attached to them, which will dissociate easily when are added to the heated hybridization solution and agitated, producing a brown, slightly viscous hybridization solution.

The blots can be screened for:

- specific sequences or overlapping clones (e.g. for sequencing and building up contigs from sub-cloned phage and cosmids),
- repetitive DNAs, general or species-specific,
- single-copy DNAs.

For the first objective, hybridization and screening is obviously carried out with either a homologous or heterologous probe. Plasmid libraries are not very effective at isolating specific genes from genomic DNA and are more commonly used for subcloning purposes.

For the latter two, total genomic DNA is used for the screen, and selection is carried out on the strength of signal. The only type of DNA to give a signal with a total genomic probe will be either repetitive or extrachromosomal sequences and therefore any colonies lighting up will be repetitive to a certain extent (Saul and Potrykus 1984).

The primary requirement for RFLP probes is that they are either single or low-copy number sequences. This is because probes of a more repetitive nature will produce multiple bands on Southern blots, making the subsequent statistical analysis of inheritance unmanageable. But how effective is a plasmid library at producing single copy sequences and what type of sequences can one expect to obtain in a plasmid library?

This is dependent upon:

- genome size,

- restriction enzyme used to produce the library,
- methylation status of the genomic DNA.

Genome size. As the DNA amount increases, so does the proportion of repetitive sequences, such that many plants have between 65 and 95% repetitive sequences (Flavell et al. 1974). As a general rule, plants with haploid genomes of more than 2 pg, 20–35% of the sequences are single-copy. When this is less than 2 pg, the proportion of single copy sequences is inversely related to the genome size (Flavell et al. 1980). For example

	1C	% Repetitive	Ploidy
Secale cereale	9.4	92	2x
Triticum aestivum	15.7	83	6x
Daucus carota	1.05	62	2x
Beta vulgaris	1.35	63	2x
and at the lowest level			
Arabidopsis	0.07	14	2x

(Leutwiler et al. 1984)

A plasmid library will thus be more effective at isolating single-copy sequences from plants with small genomes.

Choice of restriction enzyme used to cut the genomic DNA for the library. In general, most plasmid libraries are made from PstI cut genomic DNA (McCouch et al. 1988; Song et al. 1988; Tanksley et al. 1988; Kochert et al. 1991; Goffreda et al. 1992; Keim et al. 1992; Kleinhofs et al. 1993). PstI recognizes the sequence 5′ CTGCAG 3′ with sensitivity to methylation of the 5′C.

The reasoning behind this is that much of the nuclear DNA in eukaryotes contains methylated cytosines, with repetitive sequences being affected by this phenomenon to a greater extent than single-copy sequences. Thus, by cutting genomic DNA with a methylation-sensitive enzyme, e.g. PstI, the libraries produced are enriched for single-copy sequences. Experiments by Tanksley et al. (1988) on tomato showed that plasmid libraries produced by PstI digestion produced 92% single-copy sequences, whereas with EcoRI, this figure was reduced to 35%.

If plasmid libraries are to be the main source of clones, a range of enzymes can be used to cut the genomic DNA. This will result in different sets of clones, as restriction enzymes always exhibit a certain amount of bias in the sequences they cut (Zamir and Tanksley 1988). For non-sequence-biased libraries, sheared genomic DNA should be used.

2.2.7.2
Mini-Prep Isolation Method

Once the primary screen has been carried out on the colony blots and the number of colonies has been narrowed down, further characterization of clones can be carried out. Individual plasmids can be isolated and further selective screening carried out either on blots of plasmid digests, PCR screens or slot/dot blots. This further step is often necessary to check the validity of the signal obtained with the colony blots (due to the possible problems of bad transfer etc.) or to check for duplication of inserts within the colonies selected.

Colonies, hybridization-positive on blots, can be PCRed directly from a culture or glycerol stock and then used for sequencing (Chap. 4) and making probes. This is fine for a one-off approach, but it is better to bulk it up via a mini-prep than constantly use the glycerol stock, as these do deteriorate with time, especially when they have been frozen and thawed many times.

There are currently many different kits on the market for rapid isolation of plasmids. These do tend to be expensive, and presented here is a quick and reliable method based on Birnboim and Doly (1979).

Reagents

- Culture medium, e.g. TB
- Solution I: 50 mM glucose/25 mM Tris/HCl pH 7.5/10 mM EDTA pH 8.0
- RNase A: stock solution 10 mg/ml
- Solution II: 0.2 M NaOH/1% SDS
- Absolute ethanol
- 70% ethanol
- 3 M NaAc, pH 5.2
- TE buffer (10 mM Tris/HCl/1 mM EDTA, pH)

Apparatus

- 1.5-ml microcentrifuge tubes
- Micropipettes (1–20, 20–200 and 200-μl–1 ml capacity) plus tips
- Microcentrifuge
- Ice

Procedure

1. Grow up a 2–3 ml culture overnight at 37 °C. Transfer 1.5 ml to an Eppendorf tube. Centrifuge for 3 min. Discard supernatant and resuspend pellet in 200 μl Solution I, add 5 μl RNase A stock solution and incubate at room temperature for 10 min.
2. Add 400 μl Solution II. Mix by inversion. Place on ice for 5 min.
3. Add 300 μl 3 M NaAc pH 5.2. Invert to mix. Place on ice for 5 min.
4. Microfuge for 5 min. Decant supernatant into fresh Eppendorf containing 800 μl cold ethanol, vortex briefly and allow to stand for 1 min. Centrifuge for 10 min, remove ethanol, wash pellet in 70% ethanol and dry pellet.
5. Resuspend the pellet in 40 μl TE buffer.

Notes

- This method results in a fairly impure preparation but which will still restrict and enable excision of the insert or amplification by PCR. If a purer preparation is required, a phenol chloroform step can be added.
- The insert can be isolated either by restricting the plasmid or by PCR amplification using primers for the multiple cloning site of Bluescript.
- It is possible to screen the plasmids on a Southern blot. This is carried out either by digesting the plasmids with the appropriate restriction enzyme, which will excise the insert, or by PCR amplification. The products of both can be run on a gel and then blotted. It is possible if the blot only contains small fragments, i.e. a few Kb to blot the gel straight away without any HCl pretreatments. The blot also does not need to be carried out overnight.

Troubleshooting

- The main problem which arises with this technique is lack of a plasmid at the end of the preparation. This could be due to the plasmid being lost from the bacteria. This can happen at random, particularly if the transformed bacteria are stored for a long time. Replica plates should always be made as a back-up.

- *An insert is not present in the plasmid*: the colony picked was a false positive and did not, in fact, contain an insert. This often happens if colonies are picked when the blue colour has not developed sufficiently.
- *If the plasmid will not cut with a restriction enzyme*, try to purify further with a phenol chloroform step.
- When the plasmid is cut, it is possible that several bands are obtained. One will be the cut vector and the others will be inserts of genomic DNA, occasionally more than one fragment will ligate into the same vector. Production of several bands could also be due to star activity, if too much enzyme was added.

2.2.8 Storage of Library/Plasmids

The library can easily be stored for future use or further screening and, in fact, it is often useful to keep a backup in case anything goes wrong.

Library. Add 30 µl of 50% glycerol to each well of the microtiter plate. Store at –70 °C.

Individual plasmids. If specific plasmids require saving, innoculate a 2-ml culture and incubate overnight with shaking at 37 °C. Remove 850 µl of the culture to a screw-capped microcentrifuge tube, add 150 µl of 50% glycerol. Vortex to mix and snap freeze in liquid nitrogen. Store at –70 °C.

To re-grow these stored cultures, scrape the frozen cultures with either a wire innoculating loop or sterile toothpick, and innoculate either an agar plate or liquid culture containing the appropriate antibiotics. Grow overnight at 37 °C. Do not thaw the frozen stocks, as this dramatically decreases their lifespan.

References

Birmboim HC, Doly J (1979) A rapid alkaline extraction procedure for screening recombinant DNA. Nucleic Acids Res 7:1513–1523

Bolivar F, Rodriguez RL, Greene PJ, Betlach MC, Heyneker HL, Boyer HW (1977) Construction and characterisation of new cloning vehicles II a multipurpose cloning system. Gene 2:95–113

Cohen SN, Chang ACY, Hsu L (1972) Non-chromosomal antibiotic resistance in bacteria: genetic transformation of *Escherichia coli* by R-factor DNA. Proc Natl Acad Sci USA 69:2110–2114
Devos KM, Atkinson MD, Chinoy CN, Lui CJ, Gale MD (1992) RFLP-based genetic map of the homologous group 3 chromosomes of wheat and rye. Theoret Appl Genet 83:931–939
Flavell RB (1980) The molecular characterisation and organisation of repetitive plant chromosomal DNA sequences. Annu Rev Plant Physiol 31:569–596
Flavell RB, Bennett MD, Smith JB, Smith DB (1974) Genome size and the proportion of repeated nucleotide sequence DNA in plants. Biochem Genet 12:257–269
Goffreda JC, Burnquist WB, Beer SC, Tanksley SD, Sorrells ME (1992) Application of molecular markers to assess genetic relationships among accessions of wild oat *Avena sterilis*. Theoret Appl Genet 85:146–151
Grunstein M, Hogness DS (1975) Colony hybridization: a method for the isolation of cloned DNAs that contain a specific gene. Proc Natl Acad Sci USA 72:3961–3965
Hanahan D (1983) Studies on transformation of *Escherichia coli* with plasmids. J Mol Biol 166:557–580
Havey MJ, Muehlbauer FJ (1989) Linkages between restriction fragment length, isozyme, and morphological markers in lentil. Theor Appl Genet 77:395–401
Hugros G, Monte JV, Ferrer E (1990) *Hordeum chilense* repetitive sequences. Genome characterisation using biotinylated probes. Theor Appl Genet 80:24–32
Imamura J, Saul MW, Potrykus I (1987) X-ray irradiation-promoted asymmetric somatic hybridization and molecular analysis of the products. Theor Appl Genet 74:445–450
Itoh K, Iwabuchi M, Shimamoto K (1991) In situ hybridization with species-specfic DNA probe gives evidence for asymmetric nature of *Brassica* hybrids obtained by X-ray fusion. Theor Appl Genet 81:356–362
Iwabuchi M, Itoh K, Shimamoto K (1991) Molecular and cytological characterisation of repetitive DNA sequences in *Brassica*. Theor Appl Genet 81:349–355
Keim P, Beavis W, Schupp J, Freestone R (1992) Evaluation of soybean RFLP marker diversity in adapted germplasm. Theor Appl Genet 85:205–212
Kleinhofs A, Kilian A, Saghai Maroof MA, Biyashev RM, Hayes P, Chen FQ, Lapitan N, Fenwick A, Blake TK, Kanazin V, Ananiev E, Dahleen L, Kundrna D, Bollinger J, Knapp SJ, Lui B, Sorrells M, Heun M, Franckowiak JD, Hoffman D, Skadsen R, Steffenson BJ (1993) A molecular, isozyme and morphological map of the barley (*Hordeum vulgare*) genome. Theor Appl Genet 86:705–712
Kochert G, Halward T, Branch WD, Simpson CE (1991) RFLP variability in peanut (*Arachis hypogaea* L.) cultivars and wild species. Theor Appl Genet 81:565–570
Leutwiler LS, Hough-Evans BR, Meyerowitz EM (1994) The DNA of *Arabidopsis thaliana*. Mol Gen Genet 194:15–23
McCouch SR, Kochert G, Yu ZH, Wang ZY, Khush GS, Coffman WR, Tanksley SD (1988) Molecular mapping of rice chromosomes. Theor Appl Genet 76:815–829

Nodari RO, Koinange EMK, Kelly JD, Gepts P (1992) Towards an integrated linkage map of common bean. I. Development of genomic DNA probes and levels of restriction fragment length polymorphism. Theo Appl Genet 84:186–192

Pehu E, Thomas M, Poutala T, Karp A, Jones MGK (1990) Species-specific sequences in the genus *Solanum*: identification, characterisation, and application to study somatic hybrids of *S. brevidens* and *S. tuberosum*. Theor Appl Genet 80:693–698

Saul M, Potrykus I (1984) Species-specific repetitive DNA used to identify interspecific somatic hybrids. Plant Cell Rep 3:65–67

Schmidt T, Junghans H, Metzlaff M (1990) Construction of *Beta procumbens*-specific DNA probes and their application for the screening of *B. vulgaris* x *B. procumbens* (2n = 19) addition lines. Theor Appl Genet 79:177–181

Schweizer G, Borisjuk N, Borisjuk L, Stadler M, Stelzer T, Schilde L, Hemleben V (1993) Molecular analysis of highly repetitive genome fractions in *Solanum* and their use as markers for the characterisation of species and cultivars. Theor Appl Genet 85:801–808

Schweizer G, Ganal M, Ninnemann H, Hemleben V (1988) Species-specific DNA sequences for identification of somatic hybrids between *Lycopersicon esculentum* and *Solanum acaule*. Theor Appl Genet 75:679–684

Solano R, Heuros G, Fominaya A, Ferrer E (1992) Organisation of repeated sequences in species of the genus *Avena*. Theor Appl Genet 83:602–607

Song KM, Osborn TC, Williams PH (1988) *Brassica* taxonomy based on nuclear restriction fragment length polymorphisms (RFLPs). 1 Genome evolution of diploid and amphidiploid species. Theor Appl Genet 75:784–794

Tanksley SD, Bernatzky R, Lapitan NL, Prince JP (1988) Conservation of gene repertoire but not gene order in pepper and tomato. Proc Natl Acad Sci USA 85:6419–6423

Tanksley SD, Miller J, Paterson A, Bernatzky R (1988) Molecular mapping of plant chromosomes. In: Gustafson JP, Appels R (eds) Chromosome Structure and Function: Impact of New Techniques. Plenum D, NY London, 157–173

Thomas MR, Matsumoto S, Cain P, Scott NS (1993) Repetitive DNA of grapevine: classes present and sequences suitable for cultivar identification. Theor Appl Genet 86:173–180

Zamir D, Tanksley SD (1988) Tomato genome is comprised largely of fast-evolving low copy number sequences. Mol Gen Genet 213:254–261

Zhao X, Wu T, Xie Y, Wu R (1989) Genome-specific repetitive sequences in the genus *Oryza*. Theor Appl Genet 78:210–209

2.3
Lambda Genomic Cloning

G. ELGAR

2.3.1
Introduction

The most important decision to make before construction of a genomic library from any organism is the choicc of cloning vector. With the ever-increasing array of plasmids, phages (lambda and P1), cosmids and artificial chromosomes [BACs , PACs and YACs (Bacterial, P1 and Yeast Artificial Chromosomes)], it is important to tailor the choice of vector to the objectives behind making the library.

As insert size increases, so the resolution and ease of manipulation of individual clones decreases and the labour intensity shifts from the screening of the library to the mapping and characterization of individual clones. Gross mapping of large areas of a genome therefore favours a large insert library, whereas higher resolution sequencing and mapping of individual genes, for instance, favour smaller inserts. It is wise, therefore, to review the advantages and disadvantages of various vector systems whilst remembering the old adage that size isn't everything.

The smaller the average insert, the greater the number of clones needed to cover a whole genome. However, it is much easier to screen 1×10^6 phages than colonies. It is also easier to store phages on plates over long periods. In addition, because of the ability of cosmids and phages to package specific sizes of inserts, it is not always necessary to size-select DNA for these vectors.

An important consideration is the level of redundancy that each type of library affords, i.e. non-recombinant, deleted and chimaeric clones. The frequency of non-recombinants can be most effectively reduced by genetically selecting against them, and cloning in lambda vectors provides a powerful means to do this (discussed later). Size selection from gels or sucrose gradients can reduce the number of co-ligations in packaged libraries but a useful alternative is to extensively de-phosphorylate the ends of the insert DNA, thus excluding the necessity of a size-selection step altogether, and this is outlined later in the chapter. The problems encountered in either size-selecting or de-phosphorylating very large (>50 kb) fragments of DNA are considerable and may reduce the overall efficiency of large fragment cloning. A further consideration with very large fragment libraries is the instability of some sequences, resulting in recombination of heterologous DNA fragments during construction of a library and the deletion of certain sequences during growth of clones. Ultimately, all libraries will have a degree of redundancy but, in general, the larger the insert size, the greater this redundancy becomes. The only way to overcome redundancy is to have a good coverage library, with each stretch of DNA represented by a number of independent clones.

Finally, the choice of vector may be limited to the size and quantity of the source genomic DNA. This should be at least three to five times the intended average size of insert. This means that inserts greater than 50 kb should be derived from agarose-embedded DNA, once again complicating construction of a library. By de-phosphorylating DNA rather than using a size-selection step, small quantities (less than 1 μg) of genomic DNA may be used.

2.3.1.1
General Characteristics of Cloning in Lambda Replacement Vectors

Lambda vectors must now be considered at the lower end of the size range for genomic library construction but they present a simple, efficient way of producing good coverage, low redundancy genomic libraries from small amounts of genomic DNA which does not have to be of particularly high quality.

Lambda vectors used in genomic library construction are termed replacement vectors in that during the cloning procedure an inessential "stuffer fragment" is replaced with genomic insert DNA. The vector consists of two "arms" flanking the stuffer fragment and these arms contain all the DNA necessary for propagation of the phage (see Fig. 2.3.1). Generally, fragments of 10–23 kb can be accommodated by these vectors because the phage heads will only package phage genomes of between 39 and 52 kb (Hendrix et al. 1983) including about 29 kb of arms. Multiple-cloning sites, usually head to head, flank each arm and these vary in the different vectors available.

The fact that DNA fragments of less than 10 kb are not packaged is particularly useful, as this provides a means to exclude small fragments from preferentially cloning. By far the most efficient way of

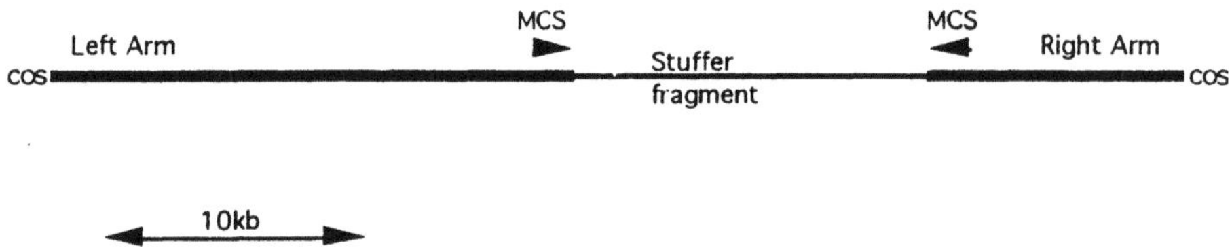

Fig. 2.3.1. Schematic diagram of a typical lambda replacement vector, the *thick line* showing the retained portaion of the vector (the arms) and the *thinner line* showing the stuffer fragment which is lost during cloning. *MCS* Multiple cloning sites which face each other at the ends of the arms; *cos* cohesive ends

introducing the phage DNA into bacterial cells is to package the DNA in vitro into phage heads. These packaging mixtures are commercially available as the protein products of two deficient phages that complement each other on mixing and allow accurate packaging of DNA.

The optimum packaging efficiencies are achieved when DNA is at high concentration which promotes the formation of concatemers. Concatemers are strings of DNA molecules, formed because lambda DNA possesses cohesive (cos) ends at the end of each arm (see Fig. 2.3.1) which will ligate together. Phage heads will only package DNA flanked by two cos sites (see Fig. 2.3.2). If the order of fragments (L.arm-insert-R.arm) and the distance between two cos ends (39–52 kb) is correct, then the DNA will be processed into the phage head.

2.3.1.2
Generation of Insert DNA Representative of the Whole Genome

In order to generate a truly random library, each part of the genome must be represented in equimolar amounts. The most effective way to do this is by shearing the DNA; by sonication, for instance. Unfortunately, this makes ligation of these fragments somewhat troublesome without the addition of adaptors. An easier way of making a representative library is to choose a frequently cutting restriction enzyme which generates ends compatible with those of one of the polylinker enzymes. The enzyme *Sau3AI* is commonly used for this purpose as its ends are compatible with *BamHI*, an enzyme often found in the multiple cloning site of lambda vectors.

Sau3AI recognizes the 4-base pair sequence /GATC, which occurs on average every 256 (4^4) base pairs in DNA with a base composition of 50% [G+C]. The probability of this site not occurring in large

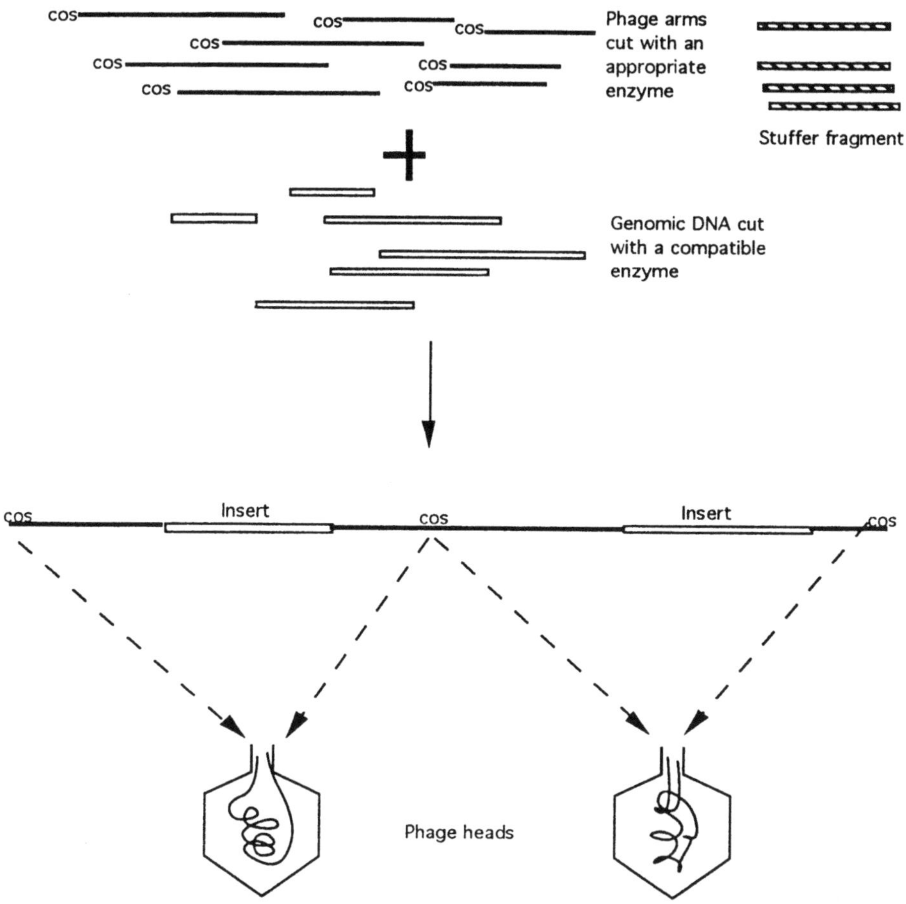

Fig. 2.3.2. Basic approach to cloning genomic fragments into lambda replacement vectors. Genomic DNA and vector are cut with enzyme/s with compatible ends. After ligation at high concentrations, the correct arm-insert-arm combinations will be packaged, providing the distance between cos sites is between 39 and 52 kb

stretches (i.e. greater than 20 kb) is therefore quite low although there may well be regions of highly repetitive DNA that do not contain this specific sequence. By generating a number of different partial digests and pooling these before ligation, areas of the genome both rich and poor in *Sau3AI* sites will be represented. An alternative way of producing partial digests is to use a competitive assay between a methylation-sensitive enzyme and its methylase. The enzyme *MboI*, which, like *Sau3AI*, cuts the recognition sequence GATC, is sensitive to *dam* methylation, which methylates the adenine. By incubating DNA with a ratio of methylase and *MboI*, a partial digest is generated (Nizetic et al. 1991).

The number of clones necessary to provide good coverage can be reduced if an average insert size of about 15–20 kb is achieved. The clone number can be calculated from the equation of Clarke and Carbon (1976) assuming a random insert library where the number of clones, N, gives a probability, p, of any sequence being represented in the library with an average insert size x and genome size y.

$$N = \frac{ln(1 - p)}{ln(1 - x/y)}.$$

With an average insert size of 20 kb, the probability of finding any sequence from a genome of, for instance, 400 Mb, in a library of 1×10^5 clones would be greater than 0.99. This is equivalent to a five times coverage library. This assumes a random distribution of fragments, and it must be remembered that, in reality, using a restriction enzyme will bias the library and it is therefore worth trying to achieve at least ten times coverage.

2.3.1.3
Plating Out Libraries and Selecting for Recombinants

Plating out of phage libraries is very straightforward and represents one of the major advantages of this form of library. Up to 50000 phages may be plated on a single large petri dish, a representation of up to 1000 Mb of insert DNA. Even at this density, hybridization filters may be made and individual plaques identified, although further screening is necessary to isolate individual clones.

As size selection is a key feature of these vectors, the only form of selection necessary is between recombinants an non-recombinants. There are a number of ways in which this can be achieved.

During vector preparation, it is possible to purify the stuffer fragment away from the arms during gel electrophoresis. An easier way, however, is to double-digest the vector with two polylinker enzymes and remove the small linker fragment by selective precipitation. This leaves vector arms and stuffer fragment with incompatible ends, so that only insert DNA is compatible with the arms.

A second and extremely effective form of selection uses a strain of host bacteria that is lysogenic for phage P2. Bacteria containing this

Spi- genotype are unable to grow when infected with phage carrying the *gam* gene, which is present in the stuffer fragment. Thus, any non-recombinant phage containing stuffer fragment will not be propagated. As well as using a *Spi-* bacterial strain, it is essential to use a strain deficient in restriction/modification systems that might cleave methylated DNA from eukaryotic origin. There are now a number commercially available strains that lack these *mcr* and *mrr* systems.

2.3.2 Protocol for Construction of a Lambda Genomic Library in the Replacement Vector λ2001

The following protocol describes the construction of a genomic library in λ2001, but it can be adapted for use with any replacement vector containing a BamHI polylinker cloning site.

Lambda 2001 (Karn et al. 1984) is a selective cloning vector which can accommodate fragments of between 10 and 23 kb. The non-recombinant phage consists of a right (9.129 kb) and left arm (19.985 kb) and a central stuffer fragment (12.785 kb) which is non-essential for propagation of the phage, and is replaced in recombinants. Recombinants are Spi-, and are able to grow on bacterial strains carrying P2 lysogens. At the ends of both the left and right arms there is a polylinker containing sites for

5′ XbaI SacI XhoI BamHI HindIII EcoRI XbaI 3′.

Reagents

- BamHI, HindIII and Sau3AI restriction enzymes
- *dam* methylase (New England Biolabs)
- Calf intestinal alkaline phosphatase (CIAP; Boehringer Manheim)
- T4 DNA ligase
- DNAse (Sigma chemicals)
- RNAse A (Sigma chemicals)
- Lambda packaging extracts
- 10x TAKS buffer (300 mM Tris-HCl pH 7.9, 600 mM KOAc, 90 mM $Mg(OAc)_2$ and 5 mM dithiothreitol)
- Suitable phage replacement vector (at least 50 μg; e.g. lambda 2001, Embl3)
- High molecular weight genomic DNA(> 50 kb average size)

- *E. coli* strains for propogation of recombinant and non-recombinant phages, these must be $RecA^+$ [e.g. Q358, Q359(Spi^-), CQ6(Spi^-, $McrBC^-$)]
- Agarose for gel electrophoresis
- 10 mg/ml ethidium bromide
- 10x gel loading dye (20% Ficoll 400, 0.4% bromophenol blue and 0.1 M EDTA)
- 50x TAE gel running buffer (242 g Tris base, 57 ml glacial acetic acid and 100 ml 0.5 M EDTA made to 1 l with distilled water)
- PEG precipitation mix (6.2% PEG 8000, 6.6 mM $MgCl_2$ and 20 ml 3M potassium acetate solution made up to 100 ml with ddH_2O
- STE (100 mM Tis-HCl pH 7.5, 100 mM NaCl and 25 mM EDTA and 5 μg/ml boiled RNAse)
- Nitriloacetic acid (tri-sodium salt Sigma chemicals)
- S-adenosylmethionine (Sigma chemicals)
- 75% and 100% ethanol (made up with ddH_2O)
- TE (10 mM Tris-HCl pH7.5/1 mM EDTA) (autoclaved)
- Sterile distilled water (ddH_20)
- Chloroform
- CY (10 g casamino acids, 5 g yeast extract, 3 g NaCl and 2g KCl made to 1 l with distilled water)
- CY + supplements (CY plus 10 mM $MgSO_4$, 10 mM $MgCl_2$ and 0.2% maltose)
- λ dil (100 mM NaCl, 10 mM $MgSO_4$, 50 mM Tris-HCl pH 7.5 and 0.01% w/v gelatin)
- λ plates (12 g agar, 2.5 g NaCl and 10 g tryptone made to 1 l with distilled water)
- λ top agar (as for λ plates but with 7 g agar/l)
- λ top agarose (as for λ plates but with 4 g agarose/l. Agarose gives a better quality surface than agar for plating out libraries and making filters)
- 20x SSC (3 M NaCl, 0.3 M Na citrate pH 7)
- 4% sodium dodecyl sulphate (SDS; prepare in ddH_2O in hood)
- 3 M potassium acetate(KAc). Make up 60 ml of 5 M potassium acetate; add 11.5 ml of glacial acetic acid and make up to 100 ml with ddH_2O
- Isopropanol
- Absolute ethanol
- Whatman 3 MM chromatography paper
- Sterile distilled water (ddH_2O)

- Hybond-N (Amersham) nylon membranes (80 and 130 mm diameter)

Equipment

- Horizontal gel electrophoresis kit and UV transilluminator
- Microfuge
- adjustable temperature/water bath/heating block
- 37 °C incubator
- Spectrophotometer

2.3.2.1
Preparation of Vector DNA Arms

Commercial preparations of phage DNA are of sufficient quality to use directly for digestion of arms and, in some cases, suppliers provide arms pre-cut, in which case this section may be skipped. However, if phage DNA is to be made in-house, it is suggested that it is prepared by CsCl density centrifugation (Sambrook et al. 1989).

As mentioned in the introduction, it is advantageous to use high concentrations of DNA in the ligation reactions during construction of a phage library. This means that, especially if trial ligations are to be carried out, a large quantity of vector DNA is required. To be sure that the arms are of a consistent quality, they should all be prepared in one large digest and then used as required. In order to generate non-compatible stuffer fragment end, the following digest is prepared.

Preparation

1. Set up the following mixtures

10x BamHI restriction buffer (as supplied with enzyme)	20 µl
Lambda 2001 DNA	20 µg
BamHI	40 units
HindIII	40 units
Sterile ddH_2O	to 200 µl

2. Incubate at 37 °C for 2 h. This equates to a fourfold overdigestion and should be sufficient to cut all the DNA without damaging the ends.

3. Heat-kill the digest at 68 °C for 15 min and place on ice. Although HindIII is effectively heat-killed, BamHI is not completely inactivated, and so the reaction mix should be kept on ice until PEG is precipitated.

Alternatively, EDTA can be added to the reaction to a final concentration of 20 mM.

4. Run out a small aliquot (0.5–1 μg) on a 0.5% agarose TAE gel after addition of 1/10 vol gel loading dye (agarose gels of this concentration resolve better in TAE). Run the gel slowly overnight.

If the vector has cut to completion there should be three distinct bands. If there are four or more bands, there is still some uncut vector remaining.

Selectively precipitate the large fragments using the PEG mix described in reagents, as follows:

Procedure

1. At room temperature, add an equal volume of PEG mix.

2. Vortex thoroughly and leave at room temperature for 5 min.

3. Spin in a microfuge for *at least* 15 min.

4. Decant supernatent and wash pellet (which may or may not be visible) with 500 μl of 75% ethanol. Vortex this before spinning to ensure thorough washing. Spin in a microfuge for 5 min. Remove ethanol and air-dry pellet.

5. Resuspend pellet in 20 μl of TE (see Reagents). This should give (assuming reasonable recovery) a concentration of between 0.5 and 1 μg/ μl.

To test the quality of the arms, the following experiments should be carried out.

Test

In a total volume of 5 μl, ligate 0.5–1 μg of arms together (Sect. 2.3.2.5).

Run 4 ul of this reaction out on a 0.5% agarose gel (as above) next to 0.5 μg of unligated arms (for comparison) and assess whether ligation has taken place and to what extent. A good ligation should result in the generation of an additional high molecular weight band or smear right at the top of the gel together with a significant reduction in intensity of the bands representing individual arms and stuffer (the stuffer should not ligate to the arms but will, of course, ligate to itself).

The remaining 1 μl of ligated vector should be packaged as described later in the chapter. Aliquots of the packaging reaction should be plated on one *E. coli* strain lysogenic for phage P2 (Spi-) and one strain that does not harbour P2 (as described later). Titres should be very low for both strains, in particular on the Spi- strain. On the permissive Spi+ strain, any vector that was not cut in the original reaction will produce plaques and if the titre is high on this strain (greater than 10^4–10^5 pfu/μg), then it is either as a result of poor digestion or incomplete removal of the linker fragment allowing re-ligation of the stuffer to the arms. The latter is unlikely, however, as the PEG step is extremely efficient at removing fragments of less than 100 bp. As long as the titre on the Spi- strain is very low, however, this is not a problem. If the titre is high on the Spi- strain, then the bacterial strain has become leaky and may not harbour the P2 phage. A new bacterial stock should be taken and the titering repeated.

If there is still a doubt as to how well the ligation has worked, then it is advisable to take a small aliquot of lambda 2001 and in separate reactions, digest with HindIII and BamHI and re-ligate. These ligation reactions should give titers of around 10^7–10^8 pfu/μl on good commercial packaging extracts.

2.3.2.2
Preparation of Insert DNA by Partial Sau3AI Digestion

This protocol is the simplest to perform, but requires a little more DNA than when using MboI and *dam* methylase. As mentioned earlier in the introduction, it is necessary to carry out a number of different partial digestions of genomic DNA in order to accommodate regions of the genome with many sites (represented in very partial digests) and few sites (represented in almost complete digests). For instance, if a digest is performed that cuts one in every 20 Sau3AI sites, then resulting DNA fragments of 20 kb will contain on average one Sau3AI site every 1 kb. If a stretch of DNA has one site every 100 bp, it will be cut into fragments of 2 kb and these will not be cloned. In the same way, regions of genomic DNA with one Sau3AI site every 5 kb will only be cut to 100 kb and will also not be included in the clonable fraction with this particular partial digest.

The best range of partials is produced by using serial dilutions of enzyme and a constant time of incubation. After running out a fraction of the digests on a gel, the appropriate digests are pooled and dephosphorylated with alkaline phosphatase.

The efficiency of cleavage of DNA depends on its purity. The definition of one unit of Sau3AI is that it will cut 1 μg of lambda DNA to completion in 1 h at 37 °C. However, it is recommended that small test digestions are carried out initially to assess the amount of enzyme used. The following example should be used as a guide only.

Preparation

1. For digestion of genomic DNA set up the following mixture:

10x Sau3AI buffer (as supplied by manufacturer)	20 μl
Genomic DNA (> 50 kb average size)	40 μg
Sterile ddH_2O	to 190 μl

2. Pre-incubate this mixture at 55 °C for 15 min to fully disperse the DNA.

3. Aliquot into ten tubes of 19 μl each.

4. Add the following amounts of Sau3AI in approximately 1 μl.

Tube number	*Units of Sau3AI added*
1	4
2	2
3	1
4	0.5
5	0.25
6	0.125
7	0.06
8	0.03
9	0.015
10	0.0075

The dilutions may be made in TE but must be added to the DNA immediately as they are not stable.

5. Incubate at 37 °C for 20 min. Heat-kill at 68 °C for 5 min and place on ice.

6. Run 4 μl out of 20 of each digest on a 0.5% agarose TAE gel slowly overnight to assess the size ranges of the different digests.
Figure 2.3.3 shows a series of digests as detailed in the protocol above.

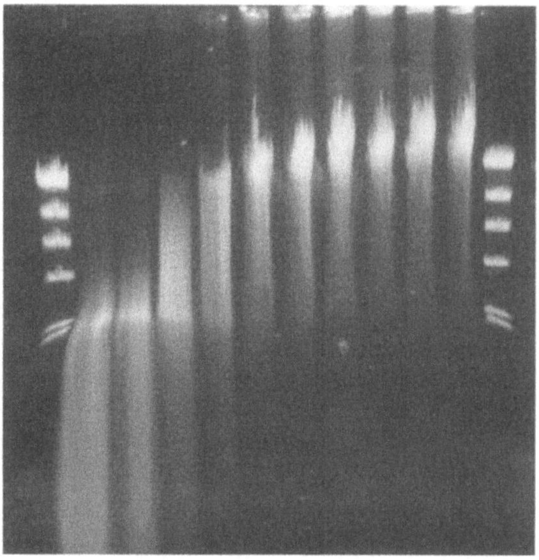

Fig. 2.3.3. Partial Sau3 AI digests of genomic DNA. Concentrations of enzyme and DNA are detailed in the text. Lambda DNA cut with HindIII is used as a size marker on either side of the digestions with bands corresponding to 23, 9.5, 6.5, 4.4, 2.3 and 2 kb

If there are digests present which have little or no DNA in the size range required for cloning, then these may be discarded. However, it is worth remembering that although these digests do not have much DNA in the clonable size range, the few clonable fragments that are present may not be represented in any of the other dilutions. It is a fine balance between including all of the clonable DNA and reducing the amount of DNA which is not of the right size. This DNA will not package but will absorb vector arms in the ligation reaction.

The remaining 16 µl of each of tubes 3–8 shown in Fig. 2.2.3 above, containing a total of approximately 20 µg of DNA, were dephosphorylated as described in Section 2.3.2.4 below.

2.3.2.3
Preparation of Insert DNA by MboI/dam Methylase Titration

This protocol can be performed using very small amounts of DNA, which may be particularly important when making libraries from flow-sorted chromosomes, for instance, where the avalability of starting materials is limited. Because the quality and activity of *dam* methylase can vary considerably from batch to batch, it is important to carry out a number of controls and to use the same batch of enzyme in each digest. *dam* methylase transfers a methyl group from S-adenosylmethionine (SAM) to the adenine in the sequence GATC. This prevents cleavage by MboI, which can only cut unmethylated DNA.

The following reactions are provided as a guide and if possible, small trial digests should be performed first. However, if these turn out to provide the correct degree of digestion, they can be used in ligation reactions.

Preparation

1. Set up a series of four dilutions containing 0.01 units of MboI and 15, 30, 45 and 60 units of dam methylase (New England Biolabs). This is equivalent to enzyme ratios of 1:1500, 1:3000, 1:4500 and 1:6000.

2. Set up reactions as follows and incubate at 37 °C for 20 min

10x DAM buffer (see reagents)	2 µl
3.2 mM SAM	0.5 µl (80 µM final concentration)
dam methylase	15, 30, 45 or 60 units
MboI	0.01 units
Genomic DNA	0.1–0.5 µg
Sterile ddH_2O	to 20 µl

3. In addition, set up one digest containing no MboI but 60 units of *dam* methylase, one digest containing 5 units of MboI and no methylase and one digest with no enzymes present at all. The MboI digest should confirm activity of the MboI enzyme by cutting the majority of the DNA into very small fragments. The methylase digest should look identical to the DNA with no enzymes added, otherwise there is some degrading activity in the *dam* methylase and a new batch should be tested.

2.3.2.4
De-Phosphorylation of Insert DNA

This part of the procedure is the most critical in obtaining good-quality, high-titer genomic libraries. Calf intestinal alkaline phosphatase (CIAP) is an extremely active enzyme which removes the 5 prime phosphate group from the ends of DNA fragments. Because T4

DNA ligase requires at least one phosphate group in order to ligate two DNA fragments together, selective removal from one species of DNA prevents its self-ligation. Thus genomic insert DNA may be de-phosphorylated with CIAP so as to prevent self-ligation. However, as the vector arms retain their 5 prime phosphates, T4 DNA ligase will vector to insert. One of the inherent problems with CIAP is that it is difficult to inactivate completely. This means that any residual phosphatase activity will remove phosphates from vector as well as insert DNA, thus significantly reducing the efficiency of ligation.

Whilst it is important to remove all CIAP from a reaction, it is also critical that the insert DNA is completely de-phosphorylated, otherwise small fragments from the Sau3AI digestions will ligate together and attain a packageable size. A library containing even a small proportion of these co-ligated clones can become a nightmare when trying to isolate and map genomic regions.

Nitrilo acetic acid (tri-sodium salt; TNA) is an extremely powerful chelating agent, that, in combination with high temperature, completely and irreversibly inactivates CIAP. It is many times more efficient than phenol/chloroform extraction and a great deal simpler. After treatment with TNA, DNA may simply be precipitated with salt and ethanol.

The following protocol has been carefully titrated using Boehringer's CIAP, but as the unit definition for this enzyme is universal, it should apply to any manufacturer.

Procedure

1. To the digested, pooled DNA samples from either the Sau3AI or MboI/*dam* methylase partials in the sections above add 1/9 volume of the manufacturers 10x CIAP reaction buffer.

2. Add 0.1 units of CIAP per 1 µg of DNA, vortex and incubate at 37 °C for 20 min.

3. Prepare 10 ml of 150 mM TNA in sterile ddH_2O.

4. Add 1/9 volume of 150 mM TNA, vortex and heat-kill at 68 °C for 30 min.

5. Add either 1/10 volume of 3 M potassium acetate (see Reagents) or 1/50 volume of 5 M NaCl.

6. Add 2.5 volumes of 100% ethanol and vortex. Incubate on dry ice

or at −20 °C for 20 min and spin at 13 000 rpm in a microfuge for at least 10 min to precipitate DNA.

7. Remove the supernatent carefully and add 500 μl of 75% ethanol. Vortex and spin for 5 min at 13 000 rpm. Remove supernatent and air-dry pellet.

2.3.2.5
Ligation and Packaging of Phage Libraries

Ligation conditions are critical for the production of good titer phage libraries. In particular, high concentrations of DNA favour the formation of concatemers where *cos* ends have come together to form long strings of arm-insert-arm-arm-insert-arm. Where these combinations of insert and arms are in the correct order and the distance between two *cos* sites is between 38 and 52 kb, the phage head will package the DNA. At low concentrations, circular DNA molecules are generated with only one *cos* site and these are not packaged efficiently. With this consideration, all ligations are carried out at high concentrations of DNA (> =200 ng/μl).

Test ligations and packagings of vector arms are performed as outlined in Section 2.3.2.1. Because the molecular weight of the insert DNA is similar to that of the two arms, ideally a mass ratio of 2:1 arms:insert should be used in the ligation reaction. However, it is advisable, if enough DNA is available, to carry out a series of arms:insert ratios around this figure.

Procedure

1. Ligations are carried out in a volume of 5 ul using approximately 1 μg of insert and 2 μg of arms as follows:

Insert DNA	1 μg
Vector arms DNA	2 μg
T4 DNA Ligase (New England Biolabs)	200 units
10x ligase buffer (including ATP supplied by manufacturer)	0.5 μl
ddH_2O	to 5 μl total volume

2. Incubate at 12 °C overnight.

3. Packaging of libraries should be performed in commercially avail-

able packaging extracts such as Gigapack Gold Packaging Extract (Stratagene) and according to their protocol. Briefly, 2–5 μl of ligation is added to the phage extracts as they thaw and incubated at room temperature for 2 h. After this, the volume is made up to 500 μl with TE and 20 μl of chloroform added. After a brief spin, packaged ligations should be stored at 4 °C where they are stable for long periods of time (unless the titre is very low).

2.3.3 Titering of Lambda 2001 Libraries

Lambda libraries should be test plated on a non-selective (such as Q358) as well as a Spi⁻ strain of *E. coli*. Where the source genomic DNA may be methylated, it is essential to use a strain deficient in restriction/modification systems. It is also essential that strains used are $RecA^+$, as this gene is essential for propagation of lambda. Q359 is Spi^- and will therefore accommodate only recombinant phages. CQ6 is Spi^- and also $McrBC^-$, and as such does not restrict cytosine-methylated DNA and will only accommodate recombinant phages; 0.2- and 2μl-aliquots of libraries are plated as follows.

Procedure

1. Set up 2 ml bacterial cultures from single colonies in CY + supplements and grow to saturation overnight.
2. After spinning at 2000 rpm for 5 min in a benchtop centrifuge, resuspend cells in an equal volume of sterile 10 mM $MgSO_4$ and measure the OD_{600}.
3. Adjust the OD to 0.5 with 10 mM $MgSO_4$, these can be stored at 4 °C for up to 5 days.
4. Add 0.2- and 2 μl aliquots of the library, diluted in λ dil if necessary, to 100 μl of prepared cells and incubate at 37 °C for 10–15 min.
5. After incubation, add the cells to 3 ml of liquid λ top agar at 50 °C, vortex and pour onto dry, pre-warmed λ plates.
6. Incubate at 37 °C for 12–16 h, plaques can then be counted and the titre of the library calculated.

2.3.4
Amplification of Lambda 2001 Genomic Libraries

Aliquots of phages may be used to amplify part or all of the primary library. The advantage of amplifying a library is that it may then be plated many times, and this should be a consideration, particularly if the library is to be screened by many users or many times. The disadvantage of amplifying libraries is that some phages grow more quickly or more slowly than others and so the representation of some clones becomes biased. There are two ways of amplifying a library: by plate lysate or liquid culture.

The degree of amplification obtained through plate lysates is generally lower but more even than in liquid culture. This procedure is described in Section 2.3.5. Generate an amplified library from liquid culture as follows.

Procedure

1. Inoculate 50 ml of CY + supplements (this can be Q358, as the DNA has already been passaged through a selective host) with 100 µl of an overnight culture of host strain.
2. After 60 min, when the culture should be at an OD_{600} of about 0.1 to 0.2, add an appropriate aliquot of primary library (this should represent at least 5x coverage of the genome, if possible) and grow the culture to lysis, when clumps of bacterial debris will be seen floating in the media. This should take about 3 h.
3. Add 500 µl of chloroform, incubate for a further 15 min to complete lysis and spin down the cell debris.
4. The amplified library can be stored at 4 °C indefinitely.

Amplified library aliquots may be re-amplified, but this will further affect the representation of the library.

2.3.5
Plating Out Lambda Libraries for Amplification or Screening by Hybridization

Aliquots of libraries are plated out for screening on 14-cm circular λ plates. Phages are generally plated out for both amplification and screening at a density of 20–50 000 phages per plate, depending on how many phages are needed in total.

Procedure

1. Incubate phages with 600 µl OD_{600} cells per plate (CQ6 for unamplified and Q358 for amplified libraries) in 10 mM $MgSO_4$ for 15 min at 37 °C.
2. Add the cells to 7 ml of molten λ top agarose (melted in a microwave with care and kept at 50 °C in a waterbath), vortex and plate.
3. Limit incubation at 37 °C to 8–12 h to prevent over-growing of plaques.
4. Cool plates to 4 °C before either membranes are lifted from them (Sect. 1.2.9) or harvesting as an amplified library. Harvesting of amplified phages is performed as follows:
5. Add 10 ml of λ dil to each 14 cm plate; the plates are then gently shaken at room temperature for 30 min.
6. Flame a glass microscope slide and scrape off the top agarose together with the λ dil into a 50-ml Falcon tube (or similar).
7. Add 500 µl of chloroform and, after vortexing thoroughly, spin down the debris in a benchtop centrifuge and decant the supernatant containing the phages to a fresh tube, ultimately to be pooled together with the other plate lysates to generate the amplified library.
8. Store at 4 °C.

2.3.6
Preparation of Membranes for Plaque Hybridizations

Hybond-N (Amersham) nylon membranes (130 mm) or similar are used for transfer of DNA from plaques. These days, most membranes are double-sided, but this should be checked if there is any doubt. By transferring some of the phages from each plaque onto membranes, exact copies of the plates are made. After hybridizing the mebranes with a probe, positive phages can be identified, and these can be found on the original plates. It is important in this procedure to take care to accurately mark how the membranes lie on the plates when phages are being transferred. This is done as follows:

Procedure

1. Before starting the transfer of phages, mark all membranes asymmetrically in three or more places around the edges using a small

hole punch, and number them according to the plates; each plate is then transferred onto duplicate membranes (e.g. 1a and 1b for plate 1, 2a and 2b etc.).

2. Place the membranes face down on the surface of the plates and leave for 1 min to allow transfer of phage DNA. During this time, mark the positions of the holes punched in each membrane on the underside of the plate, the duplicate positions marked in different colours.
3. Place the membranes face up on Whatman 3-MM chromatography paper soaked in denaturant (0.5 M NaOH, 1.5 M NaCl) for 5 min.
4. Immediately neutralize the membranes twice for 3 min on 3-MM paper soaked in 0.5 M Tris-HCl pH 7.4, 1.5 M NaCl and 10 mM EDTA.
5. Transfer membranes to a beaker containing 2x SSC to remove any debris that may still be attached (do not worry if this does not all cone off).
6. Dry the membranes on filter paper and then bake for 2 h at 80 °C (or UV cross-link them) to fix the DNA permanently ready for hybridization.

2.3.7
Isolation of Positive Phage and Generarion of High-Titer Stocks

Because of the high number of phages in a plaque (around 1000), it is very easy to pick positive phages from plates. However, due to the density of plating, it is not possible to identify just one phage from primary screens and so a 5-mm area around the positive should be picked (i.e. gently stippled) to ensure the positive plaque is included (this allows for a small error in the lining-up of the membranes on the plates). Alternative protocols suggest removal of an agar plug in the region, but this precludes any possibility of going back to the area and re-picking in the event of a disaster or another positive being in the same region.

Procedure

1. With a toothpick, gently stipple the area where the positive lies and transfer to 20 μl of λ dil in a 0.5-ml Eppendorf tube, mixing thor-

oughly. (If sequence-specific primers are available for the probe, PCR may be performed directly from 1 µl of this).

2. Make 1/10 and 1/100 dilutions of the phage picks in λ dil and incubate 2 µl of these together with 2 µl of undiluted phage at 37 °C with 100 µl of cells each (see Sect. 1.2.8).

3. Plate out on 90-mm λ plates plus 3 ml molten λ top agarose and incubate at 30 °C overnight or 37 °C for 8–12 h. Plates with between 200 and 1000 plaques are used for secondary screening, 80 mm membranes prepared as in Section 1.2.9. On some occasions, it may be necessary to screen a third time in order to isolate single plaques which are well defined and isolated from any others.

Once a single plaque had been identified, it is picked into 20 µl of λ dil; 2 µl of this is plated out as above (except that it may be grown overnight) and this usually contains enough clonal phage to achieve confluent lysis. A control plate should be incubated containing just plating bacteria to check that this has grown and to provide a contrast to plates showing completely confluent lysis as these just appear clear.

Confluent phage plates are harvested to generate a high-titer stock as in Section 2.3.7 but using a 90-mm plate and 3 ml of λ dil. This can be stored at 4 °C in 100 µl of chloroform.

2.3.8
Preparation of DNA from Hybridization-Positive Phage

Phages are grown up in 200–500-ml cultures, lysed, and DNA-purified using caesium chloride centrifugation as described in Sambrook et al. (1989). Alternatively, DNA can be prepared using an adaptation of the Qiagen protocol.

Procedure

1. After lysis, spin 100–500 ml cultures 4000 rpm for 10 min.

2. Transfer phage supernatent to a fresh tube/flask and add DNAse and RNAse to a final concentration of 10 µg/ml each. Incubate at 37 °C for 30 min.

3. After cooling on ice, add 1/5 volume (30% PEG6000, 3 M NaCl) and mix thoroughly.

4. Leave at 4 °C for at least 1 h but preferably overnight.

5. Spin for 10 min at 10000 rpm to pellet the phage/PEG complex and after decanting the supernatant, resuspend the pellet in 2–5 ml STE.

6. After a brief room temperature incubation (5–10 min) add an equal volume of 4% SDS and incubate at 70 °C for 20 min.

7. Cool to 4 °C and add a further volume of 3 M potassium acetate (i.e. another 2–5 ml).

8. Mix gently and leave on ice for 10 min.

9. Spin for 10 min at 10000 rpm and transfer the superanatant to a fresh tube.

10. Precipitate the supernatent with 0.6 vol of isopropanol, 2 vol of ethanol or an equal vol of PEG precipitation mix.

11. Phage DNA is stored in TE at 4 °C.

2.3.8.1 *Crude Phage Mini-Preps for the Digestion and Southern Blot Analysis of Multiple Clones*

In order to reduce the amount of work involved, multiple phage positives can be checked by Southern blot analysis before large-scale purification to partially determine their restriction maps. This is done by digestion and probing of very crude mini-preps from high-titre phage lysates.

Procedure

1. Each phage, after isolation by secondary or tertiary screening, is grown to confluent lysis on one 90 mm λ plate and high-titer stocks prepared as described in Section 1.2.10.

2. 50 µl of high-titer stock is taken and 2.5 µl 10 mg/ml RNAse A, 2.5 µl 4% SDS and 1 µl 0.5 M EDTA added. For multiple phage, the whole procedure can be carried out in microtitre trays.

3. Incubate at room temperature for 20 min and then add 7 µl of 3 M potassium acetate and 143 µl of 100% ethanol.

4. Vortex thoroughly and spin in a microfuge or at 4000 rpm in a microtitre plate centrifuge for 10 min.

5. Remove the supernatent and, after a 75% ethanol wash, resuspend the DNA in 10 μl of TE.

6. Check a 2-μl aliquot on a 0.8% TAE agarose gel to assess yield.

7. There should be enough DNA for one or possibly two digests, and these can be Southern blotted and probed to aid in characterization of multiple phages.

2.3.8.2
Isolation of Total Insert from Phage DNA

Additional Reagents

- Dynabeads M-280 (Dynal)
- Bio-11-dCTP (Sigma chemicals, Enzo diagnostics)
- dATP, dGTP, dTTP
- Klenow fragment
- Restriction enzymes (+buffers; e.g. XhoI, XbaI, SacI etc.)

Additional Apparatus

- Tube magnet (Dynal, Promega)

In order to work directly from phage DNA, it is often advantageous to be able to isolate insert DNA away from the phage arms. This greatly simplifies shotgun projects, for instance, as there is no need to discriminate between vector and insert after sub-cloning. Gel purification of insert away from vector arms is extremely difficult, the two arms being approximately 10 and 20 kb and the insert 15–20 kb. Even if separation is possible, there is some carry-over of arms, and recovery is very low.

An alternative technique isolates pure insert DNA by biotinylation and removal of the vector arms, thus leaving the restriction enzyme-liberated insert in the supernatant (Elgar and Brenner 1992).

Procedure

1. Heat 0.2–1 μg of phage DNA to 70 °C for 5 min and then rapidly cool on ice to release any annealed *cos* ends.

2. Add the DNA to a reaction containing 0.3 μM biotin-11-dCTP, 0.5 μM each dATP, dGTP and dTTP and 1x TAKS buffer in a final volume of 50 μl.

3. Add 5 units of Klenow fragment and reaction incubate at 37 °C for 45 min.

4. Heat-kill at 70 °C for 5 min.

5. Add streptavidin beads (100 µl of Dynabeads M-280, pre-washed three times in 10 mM Tris-HCl pH 7.5, 2 M NaCl and resuspended in TE) and incubate at room temperature for 30 min with gentle mixing.

6. Using a magnet, pellet the beads and wash three times in 100 µl of TE or ddH_2O.

7. Resuspend in 1x TAKS buffer for restriction digestion using an enzyme that liberates insert but does not cut internally in the arms (e.g. Xho I, Sac I or Xba I in lambda 2001). Cleavage within the insert is not a problem, as this simply liberates more than one fragment.

8. After 60 min digestion at 37 °C with 20 units of enzyme, pellet the beads once again with the the magnet and transfer the supernatant containing the insert to a fresh tube.

9. The quality and yield of the liberated insert may be checked on a 0.8% TAE agarose gel.

This method has also been adapted to liberate end clones (Sandford and Elgar 1992). Liberated end fragments can be directly labelled to allow simultaneous bi-directional walking, or cloned to allow separation and uni-directional walking.

References

Clarke L, Carbon J (1976) A colony bank containing synthetic Col E1 hybrid plasmids representative of the entire *E. coli* genome. Cell 9:91–99

Elgar GS, Brenner S (1992) A novel method for the isolation of large insert DNA from recombinant lambda DNA. Nucleic Acid Res 20:4667

Hendrix RW, Roberts JW, Stahl F, Weisberg RA (eds) (1983) Lambda II. Cold Spring Harbor Laboratory Press, Cold Spring Harbor

Karn J, Matthes HWD, Gait MJ, Brenner S (1984) A new selective cloning vector, λ2001, with sites for BamHI, HindIII, EcoRI, SstI and XhoI. Gene 32:217–224

Nizetic D, Zehetner G, Monaco AP, Gellen L, Young BD, Lehrach H (1991) Construction, arraying, and high-density screening of large insert libraries of human chromosomes X and 21: their potential use as reference libraries. Proc Natl Acad Sci USA 88:3233–3237

Sambrook J, Fritsch EF, Maniatis T (1989) Molecular cloning: a laboratory manual, 2nd edn. Cold Spring Harbor Laboratory Press, Cold Spring Harbor

Sandford RN, Elgar GS (1992) A novel method for rapid genomic walking using lambda vectors. Nucleic Acid Res 20:4665–4666

2.4
Cosmid Libraries

C.-N. Liu

2.4.1
Introduction

A cosmid is a cloning vector used for isolating genomic DNA of a relatively large size (35–45 kb). The basic features of all cosmid vectors include *cos* sequence(s), a plasmid origin of replication (usually but not always the ColE1 origin), and an antibiotic resistance marker. *cos* sequences (cohesive end sites) are cleaved by lambda terminase, producing 12-base 5′ overhangs. A *cos* sequence is required for packaging of DNA into bacteriophage λ particles. The origin of replication and the antibiotic resistance gene make the vector capable of being transformed into *E. coli* and being propagated as plasmid.

On the one hand, unlike bacteriophage λ phage and some plasmid vectors, large segments of genomic DNA can be cloned into cosmid vectors. The large DNA capacity of cosmid vectors significantly reduces the labour involved in the construction, propagation and screening of genomic DNA libraries of complex plant genomes such as tomato (1×10^6 kb), maize (2.5×10^6 kb) and tobacco (4.3×10^6 kb). The large capacity of cosmid vectors has proved useful in two particular circumstances. First, an entire plant gene including 3′ and 5′ regulatory regions can often be cloned and propagated within a single recombinant cosmid. Secondly, a region of plant genomic DNA containing a family of genes can be cloned and analyzed within a single or smaller number of recombinant cosmids. This was demonstrated by cloning and analyzing a cluster of five maize α-zein genes (Geraghty 1985; Liu and Rubenstein 1992a,b, 1993).

On the other hand, unlike the libraries constructed in yeast artificial chromosomes (YAC), one can extract, in large quantity, recombinant cosmid DNA basically free of contamination of host chromosomal DNA. A cosmid library alleviates the problem of the high percentage (40–60%) of clones that are chimaeric in YACs. In addition, the time required for construction of a cosmid library is much shorter than that required to construct a YAC library, due to the higher transformation efficiency in generating libraries.

Procedures for cosmid cloning have been greatly improved since the late 1970s to solve some technical problems that originally beset the construction and propagation of genomic DNA libraries in cosmid vectors. Examples of these problems are concatenation of the vector, insertion of more than one segment of foreign DNA into the same recombinant cosmid, and instability of foreign DNA sequences. Despite these improvements, there are still difficulties associated with construction and characterization of cosmid libraries. For example, amplification of a cosmid library is more difficult than a bacteriophage λ library or a simple plasmid library due to disproportionate growth of recombinant bacteria. The process of making replica filters used for cosmid library screening is laborious and time-consuming.

2.4.2 Cloning Strategy

There are different procedures for construction of cosmid libraries depending upon methods of preparing plant genomic DNA, and the type of cosmid vector to be used. The procedure described here was modified from protocols previously described (Wahl et al. 1987; Evans et al. 1989). The basic steps involved in cosmid cloning are diagrammatically shown in Fig. 2.4.1. Size selection of genomic DNA, together with the use of a cosmid vector with two *cos* sites, effectively alleviates the problems of concatenation of the vector and insertion of non-contiguous fragments of plant DNA into the same recombinant cosmid.

The size of a cosmid library of completely random fragments of genomic DNA is determined by the average size of recombinant clones and the size of the plant genome. The likelihood that a sequence of interest is present in a random cosmid library can be estimated by statistics based on the Poisson distribution (Clarke and Carbon 1976). The number of independent clones in a cosmid library, N, that must be generated to isolate a particular single-copy sequence of plant DNA with probability P is given by

$$N = \mathrm{In}(1 - P)/\mathrm{In}[1 - (I/G)] \ ,$$

where I is the size of the average cloned fragment, in base pairs, and G is the size of the target genome, in base pairs.

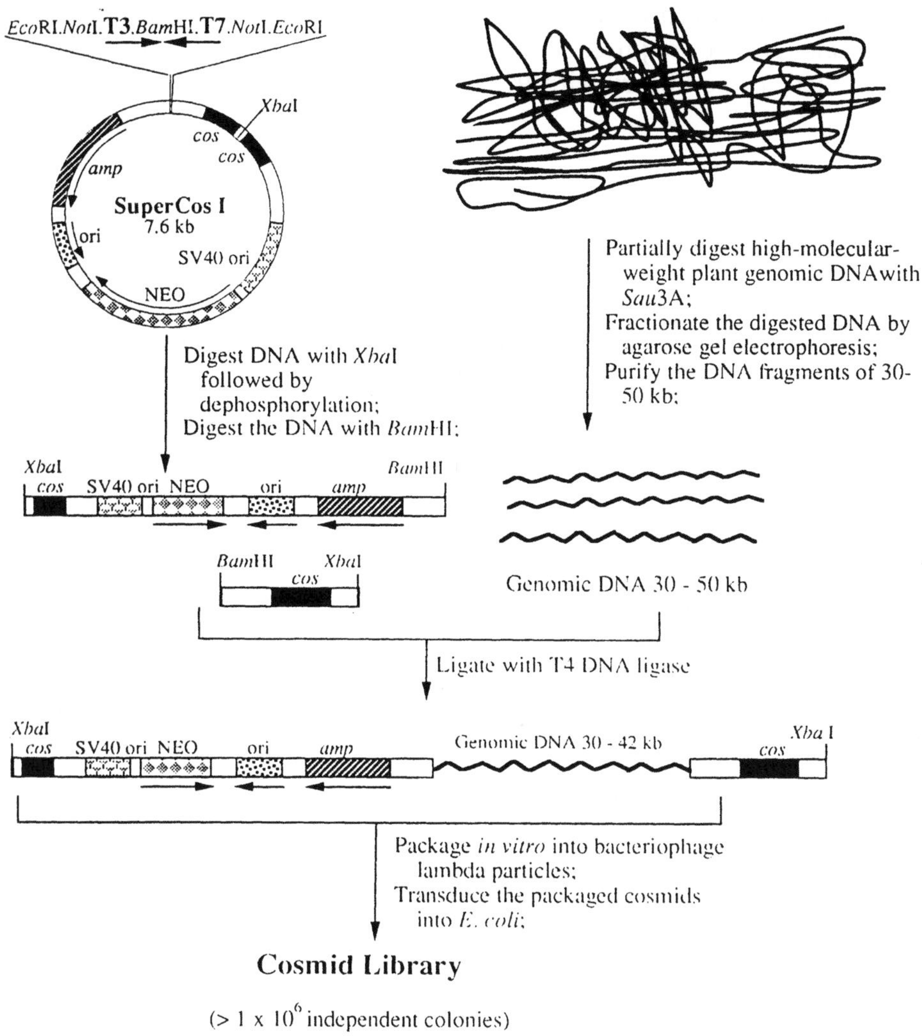

Fig. 2.4.1. Schematic diagram outlining the steps involved in construction of a cosmid library. Each step is discussed in more detail in the text

Using this formula, with an average clone size of 40 kb, and an estimated rice genome size of 4×10^8 bp (Arumuganathan and Earle 1991),

$$N = \ln(1 - 0.99)/\ln[1 - (4\text{x}10^4/4\text{x}10^8)] = 46047 \; .$$

Forty six thousand independent cosmid clones, four rice genome equivalents, are required for having any rice DNA sequence represented in the library with 99% probability.

2.4.3
Library Production

Reagents

- High-molecular-weight plant genomic DNA (≥200 kb)
- Restriction endonucleases for DNA digestion (New England Biolabs), e.g. *Xba*I, *Sau*3AI, *Bam*HI
- CIAP (calf intestinal alkaline phosphatase; New England Biolabs)
- High-molecular-weight DNA size markers (BRL, HMW DNA markers)
- 10x restriction enzyme buffer (New England Biolabs)
- T_4 DNA ligase (New England Biolabs)
- 10x T_4 DNA ligase buffer (New England Biolabs)
- 0.5 M EDTA
- Bacteriophage λ Packaging Extract
- SM media: 100 mM NaCl, 10 mM $MgCl_2$, 50 mM Tris/Cl pH 7.5, 0.01% gelatin
- LB (Luria-Bertani) medium (per l): bacto-tryptone 10 g/bacto-yeast extract 5 g/NaCl 10 g
- LB/ampicillin(100 µg/ml) agar plate
- *E. coli* cells (NM554) for transduction of packaged cosmids
- Cosmid vector DNA SuperCos I (Stratagene)
- DMSO: dimethyl sulfoxide
- 0.2% maltose/10 mM $MgCl_2$
- 10 mM $MgCl_2$
- Ethanol
- 70% ethanol
- Phenol/chloroform (1:1)
- Chloroform
- Chloroform:isoamyl alcohol (24:1)
- Sterile H_2O (ddH_2O)
- 1x agarose gel-loading buffer: 36 mM Tris-HCl pH 7.7/30 mM NaH_2PO_4/50 mM EDTA/50% glycerol/0.1% Bromophenol blue

- 1x agarose gel electrophoresis running buffer: 36 mM Tris-HCl, pH 7.7/30 mM NaH_2PO_4/1 mM EDTA
- Agarose gel: various percentages, e. g. 0.4
- 10x TBE (Tris-borate/EDTA electrophoresis buffer), 0.9 M Tris-borate, 0.02 M EDTA, pH 8
- Ethidium bromide (10 mg/ml)
- Neutralization solution: 1.5 M NaCl, 0.5 M Tris-HCl, pH 8
- 20x SSPE (1 l): 175.3 g NaCl/27.6 g NaH_2PO_4áH_2O/7.4 g EDTA, pH 7.4
- Sterile glycerol

Equipment

- 37 °C water bath
- 15 °C waterbath
- 37 °C incubator
- 37 °C shaker incubator
- Refrigerated centrifuge, capable of at least 5000 g
- Microfuge
- Electrophoresis equipment and UV viewing/photography apparatus
- 4 °C refrigerator
- Micropipettes
- Eppendorf tubes

2.4.3.1
Preparation of Plant DNA for Cloning

Procedure

1. A pilot experiment is set by partially digesting the high-molecular-weight genomic DNA with *Sau*3A at a series of increasing concentrations up to 1 unit/µg DNA. The enzymatic reactions (0.5 µg DNA in 15 µl) are incubated at 37 °C for 1 h and terminated by adding agarose gel loading buffer. DNA is separated on a 0.4% agarose gel along with the high-molecular-weight DNA size marker. Electrophoresis is carried out at 1.5 V/cm for 16–24 h or until the bromophenol blue just migrates off the gel. Photograph the gel and determine the amount of enzyme needed to produce the maximum intensity of fluorescence in the desired size range.

2. Scale up digestion to contain 10–20 µg high-molecular-weight DNA, using the concentrations which yield most of the DNA between 30 and 50 kb.

3. The digests are combined and fractionated on a 0.4% agarose gel as in the pilot experiment.

4. Locate the region of the gel with the fragment size (30–50 kb) by comparing the size marker with a hand-held, long-wavelength ultraviolet lamp. Cut out the gel spanning the region.

5. DNA is electroeluted from the gel slices and purified according to Sambrook et al. (1989).

6. The DNA is resuspended in ddH_2O to a final concentration of 1 μg/μl.

Notes

- The most frequent cause of failure to make a good cosmid library is the problems associated with obtaining high-molecular-weight plant DNA (≥200 kb). There are published protocols for preparation of high-molecular-weight genomic plant DNA for cosmid cloning by Kislev and Rubenstein (1980), Steinmuller and Apel (1986), and ten Lohuis et al. (1993). Some protocols originally demonstrated for cosmid cloning of non-plant DNA can be adapted (Sambrook et al. 1989; Briley and Bidwell 1994).

- Partial digestion can also be achieved by limiting the reaction time. There are two reasons for choosing a particular restriction enzyme. First, the restriction enzyme should cleave DNA frequently, producing the most random collection of fragments. Normally, restriction enzymes that recognize 4-base sequences produce a more random collection of insert fragments than do enzymes that recognize 6-base sequences. Secondly, the particular restriction enzyme must produce fragments that can be ligated into the desired cosmid vector. In this case, the ends of *Sau*3A in the genomic DNA are compatible with *Bam*HI of SuperCos I (Fig. 2.4.1).

- Size selection of plant DNA is the most effective method to eliminate the problem of cloning non-contiguous fragments in a single cosmid. Size selection should be performed whenever possible because it will increase the cloning efficiency (cfu/μg insert DNA) 10- to 20-fold. Size fractionation may not be necessary when backfilling cloning strategy is used (Loftus et al. 1992), and high-efficiency in vitro packaging extract is available.

- The gel electrophoresis method used to fractionate partially digested plant DNA was modified from a procedure developed by Rackwitz et al. (1984). It proved to have better resolution and is easier to handle than electrophoresis on a 0.2% agarose gel. Alternatively, low-melting-point agarose in combination with regular melting temperature agarose can be used for easy separation and high efficiency purification of plant DNA (Steck 1994). PFGE (Pulsed-Field Gel Electrophoresis) or sucrose gradients are other methods commonly used for size fractionation if loss of plant DNA is not a limiting factor.

- Losses of DNA are inevitable through the DNA preparation steps. To ensure construction of a cosmid library of reasonable size (5×10^5 to 1×10^6 independent colonies), it is desirable to obtain at least 1 µg of size-fractionated plant genomic DNA.

2.4.3.2
Preparation of Vector DNA

Procedure

1. Vector DNA SuperCos I is digested (20 µg in 200 µl vol) to completion with *Xba*I according to the manufacturer's instructions.

2. Analyze 2 µl (0.2 µg) of the reaction by electrophoresis through a 0.8% agarose gel (in 1x TBE) to assess the extent of digestion by the restriction enzyme. Use as markers (1) undigested SuperCos I; and (2) SuperCos I digested with an excess of *Bam*HI.

3. When digestion is complete, add 1 unit of CIAP directly to the reaction and incubate for a further 1 h at 37 °C. The reaction is terminated by adding EDTA to a final concentration of 5 mM followed by heating to 75 °C for 10 min.

4. Extract the solution once with phenol:chloroform and once with chloroform alone. Dissolve DNA in 180 µl ddH_2O and add 20 µl of 10x *Bam*HI buffer. *Bam*HI (40 units) is added to the solution and the reaction is incubated at 37 °C for 4 h.

5. Analyze an aliquot by electrophoresis on a 0.8% agarose (in 1x TBE) gel to assess the extent of digestion.

6. When digestion is complete, extract the solution once with phenol:chloroform and once with chloroform alone. Precipitate the DNA with 2 vol of ethanol followed by washing with 70% ethanol. Redissolve the DNA in 20 µl ddH_2O

Notes

- It is not necessary to change buffer for CIAP reaction at step **3** because CIAP (New England Biolabs) is active in all four standard NEBuffers.
- Other cosmid vectors with single *cos* sites can be used with appropriate modifications on the preparation of two arms of vector DNA. Cosmid vectors pHC79 (Hohn and Collins 1980), and pWE15 (Wahl et al. 1987) are commercially available and widely used for construction of plant genomic cosmid libraries. There are, however, advantages of SuperCos I over other cosmid vectors. First, dual *cos* sites in SuperCos I eliminate the need to prepare two separate cosmid arms, and the internal phosphatased *Xba*I site prevents cosmid concatemerization. Secondly, bacteriophage T3 and T7 promoter sequences flanking the cloning site allow for the synthesis of end-specific probes for chromosome walking. Thirdly, the presence of unique *Not*I sites flanking the cloning site is useful for removal of the cloned DNA insert from the vector.

2.4.3.3
Ligation and Packaging

Procedure

1. Ligations are performed using 1 μg of vector arms and 1–3 μg of *Sau*3A digested and size-selected plant genomic DNA. Reactions are incubated with 3 Weiss units of T_4 DNA ligase at 15 °C for 15–24 h. To assay the efficiency of ligation, remove 2-μl samples before and after the ligation reaction and analyze these samples on a 0.4% agarose gel in the same way as in procedure 2.4.3.1.
2. Carry out positive control experiments to check that the efficiency of the packaging extracts exceeds 2×10^8 pfu/μg of unligated intact lambda DNA. Protocols outlined by Sternberg et al. (1977) and Scalenghe et al. (1981) can be followed for preparation of packaging extracts and packaging of bacteriophage λ and ligated plant DNAs in vitro.
3. Carry out another control experiment to check that the efficiency of the packaging extracts exceeds 1×10^5 colonies/μg ligated DNA. Use an aliquot (2 μl) of the ligated DNA (no more than 0.5 μg) for the packaging reaction. Two aliquots (1 μl and 10 μl) of packaged cosmids are mixed with 100 μl of SM and 200 μl of host cells (MN554 or ED8767) diluted in 10 mM MgSO4 to OD_{600} = 0.5.

Incubate the infected culture for 30 min at 37 °C. Then add 1 ml of LB and continue the incubation for a further 1 h at 37 °C.

4. Spread 0.05 ml and 0.2 ml of the bacterial cells onto LB plates containing 100 μg/ml ampicillin. Count the number of bacterial colonies after an overnight incubation at 37 °C. Store the unused portion of the ligation mixture at 4 °C, and keep the remaining packaged cosmids in SM containing a few drops of chloroform until the library is amplified or screened.

5. Pick a number of individual colonies and grow 3-ml overnight cultures in LB containing 100 μg/ml ampicillin. Isolate the cosmid DNA using alkaline lysis method (Sambrook et al. 1989). Digest the cosmid DNA with restriction enzymes and analyze the sizes of the resulting fragments by agarose gel electrophoresis.

6. If the efficiency of the packaging extracts and the average size of cosmid clones are satisfactory, package the remainder of the ligated DNA into bacteriophage λ particles.

7. After packaging is complete, add 500 μl SM per packaging reaction and 20 μl chloroform. Spin for 20 s to sediment debris after mixing the solution gently.

The supernatant solution contains the packaged cosmids and is ready for screening and amplification.

Notes

- A simple way to check the ligation of vector and insert DNA is to compare the mobility shifts of the vector DNA before and after ligation. Before ligation, the DNA should be resolved into three bands; the two vector arms and a smear of high-molecular-weight insert DNA. After successful ligation some of genomic DNA should have been converted to high-molecular-weight concatemers. Some of the vector DNA should remain unligated.

- It may be more efficient to purchase than to make your own packaging extracts if only one or two cosmid libraries are constructed at a given time. If the packaging extracts are purchased, the manufacturer's instructions should be followed for cosmid packaging, titration, library screening and amplification.

2.4.3.4
Amplification and Storage of a Cosmid Library

Procedure

1. Obtain a 50-ml overnight culture of *E. coli* MN554 in 0.2% maltose and 10 mM $MgSO_4$. Spin down the culture for 5 min at 5000 *g* at 4 °C. Resuspend the cells with 1/2 vol of 10 mM $MgSO_4$. Dilute the cells to OD_{600} = 0.5 with 10 mM $MgSO_4$.
2. In a series of sterile tubes, mix packaged recombinant cosmid DNA (1×10^4 cfu) with an equal volume of MN554 host cells at OD_{600} = 0.5. Incubate the mixture for 30 min at 37 °C. Add 4 vol of LB to each mixture and incubate for 1 h at 37 °C with shaking.
3. Spin the culture for 5 min at 5000 *g* at room temperature and resuspend the pellet in 0.6 ml LB medium.
4. Spread the cell mixture onto 150 mm LB/ampicillin plates. Incubate the plates at 37 °C until colonies are 0.2–0.3 mm in diameter (about 10–15 h).
5. Scrape the colonies from each plate into 5 ml of LB. Collect LB broth into a sterile tube and wash each plate with additional 3 ml LB.
6. Repeat with all the plates. Pool all the LB broth in a single tube.
7. Vortex the pooled cell suspension and add sterile glycerol to a final concentration of 20%. Mix well by vortexing, and dispense aliquots of the bacterial suspension into sterile vials (1.5 ml of each). Store the bacteria cells at −80 °C.

Notes

- It may be wise to titre the bacterial concentration after the bacteria are frozen at –80 °C. There is no need to titre the library at the time of each screening because the bacteria should remain viable without appreciable loss for at least 1 year.

2.4.4
Suggestions for Screening a Cosmid Library and Analyzing Recombinant Cosmid Clones

2.4.4.1
Screening the Cosmid Library

Cosmids are screened using the same methods that have been designed for plasmids with the following modifications. I recommend plating no more than 10 000 colonies per 138-mm filter because the copy number of recombinant cosmids in *E. coli* is relatively low. The bacteria containing recombinant cosmids should be allowed to grow larger than colonies of a regular plasmid library. To reduce the hybridization background resulting from non-specific binding of radioactive probe to cell debris, my experience is to rub the filters after fixation (at 80 °C under vacuum) in neutralization solution at different temperatures. After the filter is fully hydrated, hand-rub the colony surface successively in three generous volumes of buffer: (1) buffer at room temperature; (2) buffer at 50–60 °C; (3) rinse in the buffer at room temperature; and finally (4) rinse in 2x SSPE. Air-dry filter on bench top at room temperature.

2.4.4.2
Restriction Mapping of Recombinant Cosmids

Mapping of recombinant cosmids by restriction enzymes is more difficult than plasmid and λ clones because of the large size of the insert. Many restriction endonucleases cleave recombinant cosmid DNA into too many fragments to map using methods involving single and double digestions. The protocols designed for cosmid mapping include *cos*- (Rackwitz et al. 1985), rare-cutter-, and transcription-mappings (Table 2.4.1).

cos-mapping is based on linearization of circular cosmid DNA by phage λ terminase (Gold et al. 1983). The linearized cosmid is then partially digested with restriction enzymes, and either *cos* end is labelled by hybridization with radioactive oligos complementary to the cohesive end sequence. After gel electrophoresis and autoradiography, the restriction map can be directly determined from the "ladder" of partial digestion products.

Table 2.4.1. Comparison of methods for cosmid restriction mapping

Method of mapping[a]	Structural feature of the cosmid required for mapping	Examples of cosmid mappings and references	Advantages	Disadvantages
Cos mapping	*cos* site	pHC79, Liu and Rubenstein (1992a)	1. Versatile 2. Highly sensitive 3. Low background	Requires high quality of lambda terminase
Rare-cutter mapping	1. Rare-cutter site 2. Primer sequences flanking insert DNA, e.g. T3, T7 or SP6;	1. sCos I, Evans et al. (1989) 2. pWE vectors, Wahl et al. (1987)	1. Highly reproducible 2. Highly sensitive 3. Low background	Depends on absence of the rare-cutter site in insert DNA
Transcription mapping	T3, T7 or SP6 promoters flanking insert DNA	1. sCos I, Evans et al. (1989) 2. pWE vectors (Wahl et al. 1987)	1. Rapid 2. Independent of insert sequence 3. Sensitive	1. Premature termination of transcription 2. Degradation of transcripts 3. High background

[a]The descriptions of mapping methods are given in the text.

Rare-cutter-mapping is very similar to *cos*-mapping. The only significant difference is that linearization of cosmid is carried out by the restriction enzymes with recognition sequences least frequent in insert plant DNA such as *Not*I, *Sfi*I and *Sac*II. A set of fragments, which begin at the T7 or T3 promoter of the vector and end at the site of cleavage of the first enzymes, is generated after partial digestion of the linearized cosmid. These rare-cutter-terminated partial digestion products are then fractionated on an agarose gel and blotted to a solid support. The fragments can then be blotted with end-labelled oligonucleotide-sequencing primers specific for these promoters.

Transcription-mapping is designed for the vectors possessing T7 and T3 promoters flanking the insert DNA. The collection of fragments generated by partial digestion are transcribed in vitro under conditions that allow for the synthesis of transcripts larger than 10 kb. Each transcript produced should indicate the distance between the site of cleavage and the position of the first transcribed nucleotide. Fractionation of these transcripts according to size reveals the position of each cleavage site relative to the promoter. Because the two promoters in each vector are positioned to allow for transcription into the insert, the same set of fragments produced by partial digestion allows one to map from both ends of inserted sequence by merely setting up two transcription reactions. Each method has advantage(s) and disadvantage(s) (Table 2.4.1).

Choice of mapping methods depends on structure of the vector (e.g. availability of the rare-cutter and flanking promoters etc.) and complexity of the insert (e.g. presence of sites of rare-cutters in the vector, size of the fragments generated relative to the resolution of gel electrophoresis etc.). Sometimes it may be necessary to use a combination of different methods to verify a restriction map. The most universal mapping method is *cos*-mapping because every cosmid vector should have at least one *cos*-site and theoretically the probability of the presence of a *cos* sequence in a plant genome is about 1×10^{-7} ($=1/4^{12}$). In addition, good-quality λ terminase with high efficiency of *cos* cleavage is now commercially available from several sources (BRL Life Technologies Inc.,Gaithersburg, MD; Epicentre technologies, Madison, WI; Takara Biochemical Inc., Berkeley CA).

Acknowledgments. I would like to thank Dr. S.B. Gelvin for critical reading of the manuscript and helpful discussions.

References

Arumuganathan K, Earle ED (1991) Nuclear DNA content of some important plant species. Plant Mol Biol Rep 9:208–218
Briley GP, Bidwell CA (1994) Use of agarose block DNA to make cosmid libraries. Biotechniques 17:278–280
Clarke L, Carbon J (1976) A colony bank containing synthetic Col E1 hybrid plasmids representative of the entire *E. coli* genome. Cell 9:91–99
Evans GA, Lewis KA (1989) Physical mapping of complex genomes by cosmid multiplex analysis. Proc Natl Acad Sci USA 86:5030–5034
Evans GA, Lewis K, Rothenberg BE (1989) High efficiency vectors for cosmid microcloning and genomic analysis. Gene 79:9–20
Geraghty D (1985) Structure and organization of the zein multigene family. PhD Thesis, University of Minnesota, St. Paul
Gold M, Becker A, Parris W (1983) The bacteriophage lambda terminase enzyme. Methods Enzymol 100:183–191
Hohn B, Collins J (1980) A small cosmid for effecient cloning of large DNA fragments. Gene 11:291–298
Kislev N, Rubenstein I (1980) Utility of ethidium bromide in the extraction from whole plants of high molecular weight maize DNA. Plant Physiol 66:1140–1143
Liu C-N, Rubenstein I (1992a) Genomic organization of an α-zein gene cluster in maize. Mol Gen Genet 231:304–312
Liu C-N, Rubenstein I (1992b) Molecular characterization of two types of 22 kilodalton α-zein genes in a gene cluster in maize. Mol Gen Genet 234:244–253
Liu C-N, Rubenstein I (1993) Transcriptional characterization of an α-zein gene cluster in maize. Plant Mol Biol 22:323–336
Loening UE (1967) The fractionation of high-molecular-weight ribonucleic acid by polyacrylamide-gel electrophoresis. Biochem J 102:251–257
Loftus MG, Foster LM, Ross IK (1992) A rapid method for cosmid cloning. BioTechniques 12:172–175
Rackwitz H-R, Zehetner G, Frischauf A-M, Lehrach H (1984) Rapid restriction mapping of DNA cloned in lambda phage vectors. Gene 30:195-200
Rackwitz HR, Zehetner G, Murialdo H, Delius H, Chai JH, Poustka A, Frischauf A, Lehrach H (1985) Analysis of cosmids using linearization by phage lambda terminase. Gene 40:259–266
Sambrook J, Fritsch EF, Maniatis T (1989) Molecular cloning: a Laboratory Manual, 2nd edn. Cold Spring Harbor Laboratory Press, Cold Spring Harbor
Scalenghe F, Turco E, Edström JE, Pirrotta V, Melli M (1981) Microdisection and cloning of DNA from a specific region of *Drosophila melanogaster* polytene chromosomes. Chromosoma 82:205–216
Steck TR (1994) Use of low-melt agarose for the efficient isolation of large DNA fragments. BioTechniques 17:676–678
Steinmüller K, Apel K (1986) A simple and efficient procedure for isolating plant chromatin which is suitable for studies of DNase I-sensitive domains and hypersensitive sites. Plant Mol Biol 7:87–94

Sternberg N, Tiemeier D, Enquist L (1977) In vitro packaging of a λ *Dam* vector containing *Eco*RI DNA fragments of *Escherichia coli* and phage P1. Gene 1:255–280

ten Lohuis MR, Meyer A, Meyer P (1993) An improved method for the isolation of high molecular weight >300 kb plant DNA in liquid phase. BioTechniques 14:890–892

Wahl GM, Lewis KA, Ruiz JC, Rothenberg B, Zhao J, Evans GA (1987) Cosmid vectors for rapid genomic walking, restriction mapping, and gene transfer. Proc Natl Acad Sci USA 84:2160–2164

2.5
Yeast Artificial Chromosome (YAC) Libraries

E. Matallana, J. Simpson, and P.A. Guzmán

2.5.1
Introduction

The molecular genetic analysis of biological systems has been an important tool in unraveling the mechanisms which govern life. This type of analysis has been developed very successfully in model organisms with small genomes that have been subjected to extensive genetic investigation. In higher eukaryotes, mammals and plants, the access to genome manipulation has been impeded mostly because of technical obstacles. In recent years, new technical advances in physical mapping and molecular cloning have provided a way to bridge the gap between classical genetic analysis and molecular cloning in organisms with large genomes.

One of these techniques is pulsed-field gel electrophoresis (PFGE) (Carle and Olson 1984; Schwartz and Cantor 1984). This technique allows the manipulation of DNA molecules over two orders of magnitude larger than in conventional agarose gel electrophoresis. PFGE has proven very useful in separating chromosomes from microorganisms. In higher organisms, chromosomes are too large to be separated. However, PFGE has been used in the long-range mapping of genes by looking for the co-migration of probe sequences in a particular restriction fragment and in the construction of physical maps.

Another technical advance is the molecular cloning of large DNA segments into the budding yeast *Saccharomyces cerevisiae* (Schlessinger 1990; Anand 1992). Until recently, the capacity of cloning vectors for accepting exogenous DNA was about 40–50 Kb for bacteriophage lambda and cosmids. Segments of DNA as large as 1 Mbp have now been cloned into yeast artificial chromosome (YAC) vectors, improving the previous cloning capacity by at least one order of magnitude. Artificial chromosomes were adapted for use as cloning vectors by Burke et al. (1987) and in a few years many interesting applications of this system have been developed. They are especially suitable for cloning complete gene units and for the assembly of chromosome-sized contigs in a single map that unifies genetic and physical data. YACs are large enough and representative enough to fit together relatively easily into long-range contiguous DNA, and each YAC is small enough to be a convenient starting point for sub-cloning and further analysis down to the sequence level.

Although cloning in YACs is possible for any genome, and YAC libraries are now available for some plant species (Ecker 1990; Ward and Jen 1990; Grill and Somerville 1991; Edwards et al. 1992; Martin et al.1992), the most significant successes have come from mammalian genome research. There are two major reasons for this: the existence of important mammalian genome mapping projects based on the use of YACs for physical linking purposes (Foote et al. 1992; Bellann-Chantelot et al. 1992) and the relative ease of introducing YAC clones into mammal cells by using appropriate selectable markers in the vector, enabling the functional analysis of large gene units (Elicieri et al. 1991; Jakobovits et al. 1993; Schedl et al. 1993).

Selectable markers for plant cells could also be introduced into YAC vectors, but no experimental data exists so far. Such transfer experiments would require a protoplasting step in order to remove the cell wall, and standard protocols should be then effective. The major uses of plant YAC libraries to date have been for map-based cloning of genetically mapped genes and for physical mapping purposes (Hwang et al. 1991; Bancroft et al. 1992; Matallana et al. 1992; Schmidt and Dean 1992, 1993; Chang et al. 1993; Martin et al. 1993).

2.5.1.1
YAC Vectors and Cloning Strategies

Yeast artificial chromosome vectors have the essential elements which permit the in vitro construction of linear molecules that can be propagated in yeast as artificial chromosomes. The original vector, pYAC4, (Fig. 2.5.1) was designed in such a way that cleavage with *Bam*HI and *Eco*RI generates two arms which can be ligated to the exogenous DNA molecules. One arm supplies a yeast origin of replication (ARS1), a centromere (CEN4), a selectable marker (TRP1) and a telomeric sequence (TEL), the other arm contains a selectable marker (URA3) and a telomeric sequence (TEL). The telomeric sequences are derived from *Tetrahymena* and act as starting points for the addition of yeast telomeres. The vector can also be propagated in *Escherichia coli* as it contains the pBR322 origin of replication and the ampicillin resistance selectable marker.

The basic cloning strategy is depicted in Fig. 2.5.1. Vector DNA is cleaved with the telomere-releasing enzyme, which in pYAC vectors is *Bam*HI and then cleaved with the cloning site enzyme, which in pYAC4 is *Eco*RI. Large-sized DNA from the source organism is digested partially with the same cloning site enzyme and ligated to the vector arms to construct lineal artificial chromosomes containing exogenous DNA inserts. Since the original description of YAC cloning, many modifications have been introduced in both YAC vector design and cloning strategies. All these modifications claim to increase the versatility of vectors and to improve the insert size and the cloning efficiency in yeast. Vector modifications have also been introduced in order to facilitate the subcloning or analysis of inserts. Some of the modifications include new restriction sites that are unique in the vector arms as optional cloning sites, new selectable markers to promote the retention or incorporation of DNA into yeast or other eukaryotic cells, and the addition of T3 and T7 promoters to permit RNA synthesis from the end of inserts. A particular example of a vector system which allows for an easier and more versatile cloning strategy is the pJS97–pJS98 vector pair (McCormick et al. 1989). Each of these two independent plasmids provides one arm of the YAC and both include polylinkers and T7 promoters.

YAC molecules are stably and faithfully propagated in yeast cells as natural chromosomes due to the presence in the vector of all the structural elements essential to maintain them as single-copy molecules. However, although desirable, this feature causes detection and sensitivity problems which can largely be circumvented by using an amplifiable YAC vector. Such vectors contain a conditional centromere constructed by placing the centromeric sequences downstream of a regulated yeast promoter (Smith et al. 1990, 1993). By growing yeast cells under inducing conditions, the centromere function can be disrupted by high-level transcription from the promoter towards the centromeric element and therefore the copy number of the YAC clone increases. When the promoter is switched off, the centromeric function is restored and the copy number drops within a few generations. Some useful applications of amplifiable YACs include signal enhancement in colony hybridization or Southern blots, direct restriction mapping of the amplified chromosome, direct fingerprinting, and direct probe preparation from the ends of cloned inserts by RNA polymerase transcription. A potential problem of this system is the high level of homologous recombination in yeast cells which could promote rearrangements between the multiple copies of the artificial chromosome.

Another significant modification refers not to the vector, but to the cloning strategy. The potential utility of YACs for long-range mapping purposes has stimulated many attempts to improve both the insert size and the efficiency of yeast transformation. A modification to the basic strategy is the introduction of a PFGE size fractionation step for the exogenous DNA after the partial restriction digest and after the ligation step (see Fig. 2.5.1). By using the appropriate PFGE conditions, DNA molecules below a specific size range can be selectively eliminated. As there is a bias for small molecules during yeast spheroplast transformation, the elimination of small molecules greatly improves the size of the clones in the resulting library; this size fractionation can be performed twice to further reduce the number of small molecules. A second PFGE size-fractionation step after the ligation is also highly recommended in order to eliminate the ligated vector arms and to select further for large YAC clones. In both cases, a careful selection of PFGE conditions may determine the separation range of the gel allowing for a very precise resolution of the appropriate molecule size.

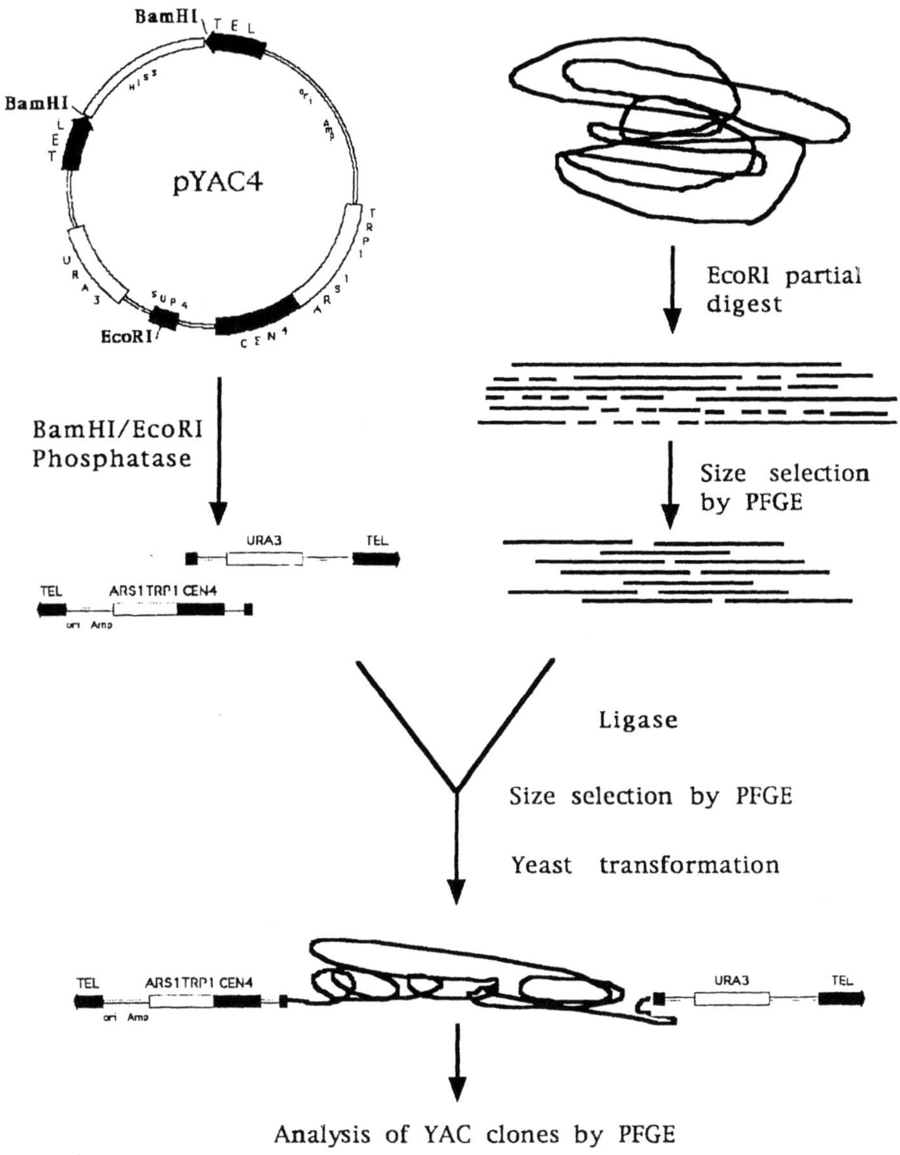

Fig. 2.5.1. Basic Scheme for YAC cloning. The vector is cleaved to generate the two chromosome arms containing all the sequences required for chromosome propagation in yeast cells: telomeres (*TEL*), centromere (*CEN4*), replication origin (*ARS1*) and selectable markers (*URA3* and *TRP1*). Vector arms are ligated to partially digested and size selected high-molecular-weight DNA from the source organism and the resulting artificial chromosomes are transformed into yeast spheroplasts (see text for details)

2.5.1.2
Megabase DNA from Plants

Important improvements in the cloning strategy have clear influences on the resulting YAC library. These improvements are completely useless if the DNA to be manipulated is not of megabase size. The isolation of megabase DNA is not usually a serious problem when animal cells are the source of the exogenous DNA. Cell cultures or tissue can be easily obtained for most purposes, and standard protocols to embed cells in agarose are protective and clean enough to yield a good-quality DNA. However, isolation of megabase DNA from plants is a much more difficult task due principally to the necessity of a protoplasting step in order to remove the cell wall. This process can result in a high loss of material and also the presence of many more contaminants in the purified DNA when compared with animal cells.

Several protocols have been described to obtain megabase DNA from protoplasts or nuclei from cultured cells or from plant tissue both in agarose or in liquid (Guzmán and Ecker 1988; Ganal and Tanksley 1989; Ecker 1990; Guzmán and Ecker 1990; Creusot et al. 1992; Hatano et al. 1992; Hernández et al. 1996). Each method has advantages and disadvantages and none can be indicated as the best. For each particular purpose and depending on the plant species or tissue used, several methods should be tested and optimized to obtain the highest yield and quality.

As the DNA integrity problem is present over the whole process of cloning large DNA in YACs, several agents have been tested for their capacity to protect DNA from physical damage. Among these, polyamines (spermine and spermidine) seem to be effective, and some protocols include these molecules in every step, from large DNA preparation to yeast transformation. The molecular effects of polyamines is unknown but they may stabilize DNA by ionic interactions when the appropriate DNA: polyamine ratio is achieved. A clearly detrimental effect on yeast viability has been observed for spermine, and the most recent protocols include only spermidine in steps previous to transformation.

2.5.2
Preparation of Megabase DNA from Plants

Several different methods for the preparation of megabase DNA from plants suitable for YAC cloning have been described. Generally, DNA is isolated by gentle lysis of protoplasts or nuclei obtained from leaf tissue or from cell suspension cultures which are embedded in agarose plugs or treated as a liquid suspension. Here, we will present general guidelines for isolating megabase DNA from leaf protoplasts, and an alternative protocol for isolation of plant nuclei. Depending on the plant species, the method and the source of DNA for isolation has to be worked out.

Reagents

- Washing solution: B5 salts, pH 5.6 (Gamborg 1968)
- Protoplasts isolation media: B5 salts, pH 5.6 (Gamborg 1968), 0.6 sucrose, 5 mM $CaCl_2$
- Enzyme mixture for protoplast isolation: 1% Cellulysin (Calbiochem, La Jolla, USA), 0.2% Pectolyase Y23 (Karlan Chemicals, Santa Rosa, USA) in protoplast isolation media.
- PMSF (phenylmethylsulfonylfluoride): 40 mg/ml in isopropanol
- NDS: 1% N-Lauroyl sarcosine/0.5 M EDTA/10 mM Tris-HCl (pH 8)
- TE: 10 mM Tris-HCl (pH 7.5)/1 mM EDTA
- 0.5 M EDTA
- 1x TBE: 0.1 m Tris-Base/0.1 M boric acid/0.2 mM EDTA; prepare 10x stock
- Proteinase K: 20 mg/ml in sterile distilled water
- Size markers: yeast chromosomes are used as size standards, they can be prepared as described in Section 4.6; they are also commercially available
- Diethyl ether
- Nuclei isolation buffer (NIB): 1M sucrose, 10mM Tris-HCl pH 7.2, 5mM $MgCl_2$, 2mM B-mercaptoethanol
- Nuclei wash buffer NWB: (NIB plus 0.5% Triton × 100)
- Azure C (optional)
- Miracloth
- Artists paintbrush

Equipment

- Pulsed-field gel electrophoresis apparatus, CHEF system from Bio-Rad

- Bench top centrifuge
- Water bath incubator
- Shaker incubator
- Inverted microscope
- Sorvall GSA rotor (or equivalent)
- Sorvall HB-4 rotor (or equivalent)
- Light microscope (optional)

Preparation of Protoplasts

1. Collect leaves form plants grown in the greenhouse or in a growth chamber before they reach the flowering stage.
2. Cut leaves and slice longitudinally into 1- to 2-mm strips, transfer the leaf tissue to a petri dish containing washing solution (B5 salts, pH 5.6).
3. Discard wash buffer and replace with enzyme solution consisting of 1% Cellulysin, 0.2% Pectolyase Y23 in protoplast isolation media. Leave cell wall digestion reaction in the dark with gentle shaking (25–40 rpm) at room temperature for 5 h to overnight. Check the release of protoplasts by observation, using an inverted microscope; stop when most of the protoplasts are released.
4. Filter the protoplasts through a 100-µm mesh nylon sieve and transfer to a 50-ml conical tube. Rinse the petri dish with medium and filter to recover remaining protoplasts.
5. Centrifuge the suspension at 200–500 *g* in a bench-top centrifuge for 10 min at room temperature.
6. Wash the pelleted protoplasts once with protoplast isolation media without enzymes and resuspend the pellet very gently in approximately one pellet-vol. Count the number of protoplasts in a small aliquot of the sample using a haemocytometer; adjust to a final concentration of $1–5 \times 10^7$/ml with protoplast isolation media.
7. Mix the protoplasts carefully with an equal volume of 1.6% low melting point agarose prepared in protoplast isolation media. Cast aliquots into plastic moulds and leave on ice for 15 min for the agarose to solidify.
8. Remove the plugs from the moulds and place them in a 15-ml tube containing NDS buffer plus 2 mg/ml of proteinase K; use enough solution to cover the plugs. Incubate at 50 °C for 36–48 h.

9. Wash the plugs twice in TE buffer containing 40 μg/ml of PMSF at 50 °C for 30 min each time and twice in TE buffer at room temperature for 30 min each time; for all washes use at least 5 ml of solution per plug.

10. Plugs are now ready for digestion or can be stored for several months in 100 mM EDTA at 4 °C for future use.

11. The integrity of the DNA is judged by PFGE. For routine use, standard 1% agarose gels in 0.5x TBE are made. The electrophoresis is set up depending on the type of PFGE apparatus. Initially, short runs of 80-s switch time, 200 V, 24 h at 13 °C can be used, then longer runs of 90 min switch time, 40 V, 168 h at 13 °C can be performed.

12. Stain gel with ethidium bromide for 30 min and de-stain for 15 min. The bulk of the DNA is visualized in the unresolved portion of DNA, which is focussed in the region just below the well.

Preparation of nuclei

1. Germinate and grow plant material in darkness for 10–15 days (or longer depending on the plant species) to produce etiolated seedlings.

If possible the following steps should be carried out at 4 °C.

2. Harvest aerial tissue, rinse in tap water and cut into small pieces with a scalpel.

3. Plasmolyze the tissue by immersion in ether for 7–10 minutes.

4. Homogenise the tissue in 4-ml/g of nuclei isolation buffer in a commercial kitchen blender.

5. Filter the homogenate through 4 layers of miracloth and centrifuge at 5K for 12 min in a Sorvall GSA rotor with the brake off.

6. Discard the supernatant and gently resuspend the pellet in 20 ml nuclei wash buffer, using an artists paintbrush.

7. Transfer samples to 50 ml centrifuge tubes and centrifuge at 5K for 12 min in a Sorvall HB-4 rotor with the brake off.

8. The washing steps (**6** and **7**) are repeated twice with NWB and finally with NIB.

9. Following the final centrifugation suspend the pellet in an equal volume of NIB using a cut-off micropipette tip.
10. At this stage aliquots may be taken to determine the purity and integrity of the nuclear preparation by staining with Azure C and observing under a light microscope.
11. Nuclei may now be processed as in steps **8–12** of the protoplast procedure.

2.5.3 Vector Preparation

The preparation of the vector is a very important step in the YAC cloning procedure. The cloning efficiency depends on the integrity of the cloning site in the vector arms: it is crucial to show non-self-ligation of the vector. We will describe the preparation of the YAC4 vector which has been the most commonly used; this method can be adapted for other cloning vectors.

Reagents

- Restriction enzymes *Bam*HI and *Eco*RI
- Reagents for CsCl-ethidium bromide gradients (see Sambrook et al. 1989)
- Alkaline phosphatase
- Polynucleotide kinase
- Nitrillotriacetic acid (NTA)
- Phenol/chloroform, and chloroform solutions (see Sambrook et al. 1989)
- Absolute ethanol

Equipment

- Microcentrifuge
- Thermoblock incubators
- Agarose gel electrophoresis apparatus and power supply

Procedure

1. Large-scale preparation of vector DNA. Isolate plasmid DNA by usual methods and band the plasmid twice in CsCl-ethidium bromide gradients (Sambrook et al. 1989).
2. Digest 500 μg of vector to completion with *Bam*HI. Check the digestion on an agarose gel; two restriction fragments of 9.6 and 1.7 kb are expected.

3. Treat with an excess of alkaline phosphatase, 1 unit per 50 μg of DNA for 30 min at 37 °C. Inactivate the phosphatase with 15 mM NTA (nitrilotriacetic acid) for 15 min at 65 °C. Extract twice with phenol/chloroform and once with chloroform, recover the DNA by ethanol precipitation and resuspend in TE.

4. Digest the vector DNA to completion with *Eco*RI. Check the digestion on an agarose gel: three restriction fragments of 5.9, 3.7 and 1.7 Kb are expected. Treat gently with alkaline phosphatase, 0.5–2 units per 500 μg of vector. Inactivate phosphatase as in step **3**, resuspend DNA in TE at 1 μg/μl.

5. Verify the integrity of the *Eco*RI cloning site by treating the vector DNA with polynucleotide kinase and ATP and then carrying out a ligation reaction (see Sambrook et al. 1989). Check the reaction in an agarose gel, the kinased vector should form linear concatemers or supercoiled molecules.

2.5.4 *Eco*RI Partial Digests of Genomic DNA and Size Selection of *Eco*RI Fragments by PFGE

*Eco*RI partial digestion is achieved by allowing *Eco*RI and *Eco*RI methylase enzymes to compete for *Eco*RI sites. The advantage of this competition is a more uniform digestion of the DNA sample. When the partial digestion is achieved by limiting enzyme concentration or incubation time, DNA is not uniformly digested because the restriction enzyme has difficulties in reaching the innermost molecules of the sample. The result is a DNA sample with some totally digested and some totally undigested molecules. In a competition experiment no limitation is imposed, either in enzyme concentration or in incubation time, and the two components of the competition system are enzymes with more or less equal possibilities of reaching all molecules in the sample; the result is a uniform partially digested sample. Initially, to titrate the partial digestion combine 1 unit of *Eco*RI with a wide range of concentrations of *Eco*RI methylase. Then, it is recommended to further titrate the enzyme concentration by combining a determine amount of *Eco*RI methylase with a wide range of concentrations of *Eco*RI endonuclease (Larin et al. 1991).

Reagents

- Restriction enzyme *Eco*RI
- EcoRI methylase
- T4 DNA ligase
- TE: 10 mM Tris-HCl (pH 7.5)/1 mM EDTA.
- 0.5 M EDTA
- S-adenosyl methionine
- Bovine serum albumin 50 mg/ml
- 1 x EcoRI digestion buffer: 1 x buffer is 100 mM NaCl/50 mM Tris-HCl (pH 7.5)/1 mM $MgCl_2$

Equipment

- Pulsed-field gel electrophoresis apparatus, CHEF system from Bio-Rad
- Waterbath incubator
- Thermoblock incubators

Procedure

1. Wash the plugs, that were previously treated with PMSF and stored in 100 mM EDTA, extensively with several changes of TE. Equilibrate four agarose plugs in a 50-ml Falcon tube for 1 h in 40 ml of TE on ice; repeat once.
2. Place half a plug in an Eppendorf tube and set up eight reactions as indicated: 40 µl of DNA (half plug), 10 µl of 10 x *Eco*RI digestion buffer, 80 µM of S-adenosyl methionine, 500 µg/ml of bovine serum albumin (BSA), and sterile distilled water to 100 µl.

   Add 1 unit of *Eco*RI and the following amounts of *Eco*RI methylase: A,1000 units; B, 500 units; C, 200 units; D, 100 units; E, 50 units. Controls: F, no *Eco*RI methylase + 1 unit of *Eco*RI; G, no *Eco*RI methylase + 20 units of *Eco*RI; H, no *Eco*RI methylase and no *Eco*RI.
3. Equilibrate reactions on ice for 2 h.
4. Transfer the tubes to 37 °C to start the reactions, incubate for 4 h.
5. Terminate reactions with the addition of EDTA to 50 mM, and incubate 30 min on ice.
6. Prepare a 1% agarose gel to analyze the restriction digests by PFGE. Electrophoresis is performed at 180 V for 16 h at 13 °C; use an 80-s switch time.

7. Stain the gel with ethidium bromide for 30 min and de-stain for 15 min. Photograph the gel and determine the best partial digestion. This is done empirically, select the concentration which shows a minor but visible digestion of the DNA.

8. Titrate the enzyme concentration again by combining a determined amount of *Eco*RI methylase, which usually is between 100 to 200 units, with a wide range of *Eco*RI. Equilibrate four plugs as in step **1**. Place half of a plug into an Eppendorf tube and set up eight reactions as indicated in step **2**.

9. Add 200 units of *Eco*RI methylase and the following amounts of *Eco*RI: A, 20 units; B, 10 units; C, 5 units; D, 1 unit; E, 0.5 units; F, 0.1 units; G, 0.05 units; H, 0.01 units; I, no enzymes control.

10. Follow steps **3** to **7**.

11. Scale up the best digestion reaction for ten agarose plugs.

12. Prepare a 1% low-melting-point agarose gel (SeaPlaque GTG, FMC) to size-fractionate the *Eco*RI partial digests.

13. Load the samples in one continuous slot along with yeast chromosomal size standards at both edges of the gel. Perform electrophoresis at 180 V for 16 h at 13 °C in 0.5% TBE using a 20-s switch time (for selection of 300-kb fragments).

14. Cut out the two edges of the gel containing the size markers and a fraction of the sample, and stain with ethidium bromide for 30 min and de-stain for 15 min; keep the central portion of the gel in the PFGE box. Photograph the gel with a ruler adjacent to it. Identify the unresolved portion of DNA which is focussed as a region just below the well.

15. Align the unstained portion of the gel with the fraction containing the size markers; excise the unresolved region from the unstained portion.

16. Place the strip of the gel which contains the size-selected *Eco*RI partial digests in a 50-ml tube and equilibrate the slice thoroughly with TE for at least 4 h; the strip can be stored in 50 mM EDTA.

2.5.5
Ligation Mixtures and Preparation of Sample for Transformation

Reagents

- 0.1 M ATP
- 0.1 M DTT
- TE: 10 mM Tris-HCl (pH 7.5)/1 mM EDTA
- 0.5 M EDTA
- Agarose

Equipment

- Pulsed-field gel electrophoresis apparatus, CHEF system from Bio-Rad
- Waterbath incubator
- Thermoblock incubators

Procedure

1. Equilibrate the strip of the gel which contains the partially digested DNA with three changes of 1 h each time in 25 mM NaCl/50 mM Tris-HCl (pH 7.5)/10 mM $MgCl_2$ and 3 mM spermidine trihydrochloride.

2. Cut out the strip in small slices and place about 500 µl of slices in an Eppendorf tube. Add an approximately 20-fold molar excess of vector DNA, melt at 65 °C for 10 min and then place at 37 °C. To the 500-µl reaction add 15–25 µg of vector DNA.

3. Calculate the actual volume of the sample; add ATP and DTT to 1 mM and 20 units of T4 DNA ligase. Mix very slowly with a blunt-ended micropipette tip and leave at 37 °C for 1 h.

4. Incubate ligation mixtures overnight at room temperature.

5. Prepare a 1% agarose gel with low melting point agarose for selection of the ligation products.

6. Melt the ligation mixtures at 65 °C for 10 min and place at 37 °C; load the molten agarose into the gel. Load the samples in one continuous slot along with yeast chromosomal size standards at both edges of the gel. Perform electrophoresis at 180 V for 16 h at 13 °C in 0.5% TBE using a 20-s switch time.

7. Cut the marker lanes from the gel and a small fraction of the region containing the resolved ligation products. Stain this portion of the gel with ethidium bromide for 30 min and de-stain for 15 min. Photograph the gel with a ruler adjacent to it.

8. Identify the unresolved portion of DNA which is focussed as a band just below the well. Excise this region from the ligation mixture lane of the unstained gel and place it in an Eppendorf tube. The strip can be stored in 50 mM EDTA.

9. Equilibrate about 200 µl of the strip in 10 mM Tris-HCl (pH 7.5)/ 1 mM EDTA/25 mM NaCl; at least three changes for 30 min each.

10. Melt the strip by placing the tube at 65 °C for 10 min and then place at 37 °C.

11. Add 1 unit of agarase per 100 µl of melted agarose, incubate for 1–2 h at 40 °C. Set up the agarase treatment while preparing spheroplast transformation.

2.5.6 Yeast Spheroplast Transformation

The procedure used for yeast transformation was developed by Burgers and Percival (1987). Before attempting the transformation with the selected ligation mixtures, it is essential to optimize the efficiency of transformation using plasmid YCP50 (Rose et al. 1987) and a YAC clone as controls. It is recommended to titrate each lot of lyticase by preparing yeast spheroplasts with a wide range of lyticase concentrations and checking the efficiency of transformation for each concentration.

Reagents

- 1% Bacto yeast extract, 2% Bactopeptone, 2% glucose
- Lyticase (Sigma, L8012); 10000 units/ml in 20 mM phosphate buffer (pH 7.5), 50% glycerol. Store at –20 °C.
- Sorbitol: 1 M
- SCE: 1 M Sorbitol, 10 mM EDTA, 0.1 M sodium citrate (pH 7.5), 30 mM β-mercaptoethanol. Add *β*-ME just before using.
- STC: 1 M Sorbitol, 10 mM Tris-HCl (pH 7.5), 10 mM $CaCl_2$
- PEG solution: 20% PEG 8000, 10 mM Tris-HCl (pH 7.5), 10 mM $CaCl_2$ PEG and Tris-HCl are autoclaved in 500 ml of water. Add $CaCl_2$ after autoclaving. A white precipitate may form, but does not affect the transformation.
- SOS: 1 M Sorbitol, 6.5 mM $CaCl_2$. 0.25% yeast extract, 0.5% Bactopeptone

- SORB plates: 0.67% Yeast Nitrogen Base (YNB) without amino acids, 0.9 M Sorbitol, 3% glucose, 2% agar. Autoclave agar separately in 500 ml of water; then add to autoclave mixture of YNB, sorbitol and glucose. Add supplements.
- TOP AGAR: 1 M Sorbitol, 2.5% agar, 0.67% Yeast Nitrogen Base without amino acids, 2% glucose. Autoclave solutions separately as above and then mix. Add supplements.
- Carrier DNA: calf thymus DNA sheared, phenol, phenol/chloroform extracted and ethanol-precipitated, resuspended at 5 mg/ml in sterile distilled water.
- 5% SDS
- Drop-out supplements: 0.2 g arginine, 0.1 g histidine, 0.6 g isoleucine, 0.6 g leucine, 0.4 g lysine, 0.1 g methionine, 0.6 g phenylalanine, 0.5 g threonine, 0.4 g tryptophan. Dissolve in 100 ml distilled water. Use 10 ml/l after autoclaving.
  (Note: For double -URA-TRP selection we leave out tryptophan)
- Tyrosine, adenine hemisulphate and uracil solutions: each is made to 1%, 1 g dissolved in 90 ml distilled water + 10 ml 1 M NaOH. To 1 l of medium add 5 ml of tyrosine, 1 ml of adenine, 4 ml of uracil; leave out uracil for -URA selection.
- SD plates: 0.67% Yeast Nitrogen Base without amino acids 2% dextrose, 2% agar. Add supplements as needed.

Equipment

- Water bath incubator
- Thermoblock incubators
- 30 °C oven incubator
- Bench-top centrifuge
- Microscope
- Spectrophotometer

Procedure

1. Inoculate 5 ml of YPD with a single yeast colony of strain AB1380 (*ade 2-1, can 1-100, lys2-1, trp1, ura 3, his 5,* [*psi*$^+$]). Grow cells overnight with vigorous shaking at 30 °C.

2. Inoculate two 50-ml cultures in YPD with 10 and 100 µl from the fresh overnight yeast culture. Incubate overnight with vigorous aeration, allowing one of the cultures to reach 3×10^7 cells/ml (OD_{600} 3–4); discard the other culture.

3. Harvest cells at 3000 rpm in a tabletop centrifuge for 5 min in a 50-ml Falcon tube.

4. Re-suspend cells in 20 ml of distilled sterile water and centrifuge as above. (At this point the cells may be resuspended by pipetting or gentle vortexing).
5. Wash once with 20 ml 1-M Sorbitol and pellet as above.
6. Resuspend cells in 20 ml SCE and add 1000 units of lyticase. Incubate at 30 °C with gentle agitation for about 20 min. The cells can be monitored for formation of spheroplasts by observing a sample in 5% SDS under the microscope, or by measuring the decrease in turbidity at 800 nm of a ten fold dilution of spheroplasts in water (80–90% spheroplasts should be obtained).
7. Pellet spheroplasts at 1000 rpm in a bench-top centrifuge for 4 min.
8. Wash spheroplasts twice, first with 20 ml 1-M sorbitol and then with 20 ml STC (the spheroplasts should be treated gently at this stage).
9. Resuspend the spheroplasts in 2 ml STC.
10. Mix extremely gently 5 μg of carrier DNA with agarase-treated ligation mixture. Carry out a positive and a negative control for transformation; use plasmid YCP50 as a positive control and carrier DNA alone as a negative control.
11. Add 10 μl of the above mixture to 100 μl of spheroplasts and incubate for 10 min at room temperature.
12. Add 1 ml of PEG solution. Mix gently by inverting the tube several times. Incubate for 10 min at room temperature.
13. Centrifuge for 4 min at 1000 rpm in a tabletop centrifuge. Remove as much of the supernatant (PEG) as possible.
14. Resuspend the spheroplasts in 150 μl of SOS medium. Incubate 30–40 min at 30 °C without shaking.
15. Add 8 ml of TOP AGAR inverting gently to mix. Plate on selective SORB plates. Incubate 3–5 days at 30 °C.
16. Examine plates every day for contaminants. Remove contaminants with a scalpel or with a Pasteur pipette.
17. Pick colonies from the transformation plates onto -TRP- URA SD plates.

2.5.7
Preparation of Yeast Artificial Chromosome Clones and Analysis by PFGE

Reagents

- YPD: Bacto yeast extract 1%/bactopeptone 2%/glucose 2%
- Zymolyase 20T: 10 mg/ml stock in 10 mM $NaPO_4$ buffer (pH 7.5), 50% glycerol. Store at –20 °C
- 0.5 M EDTA
- LET: 0.5 M EDTA/10 mM Tris-HCl pH 7.5/7.5% β-mercaptoethanol.
- NDS: 1% N-Lauroyl sarcosine/0.5 M EDTA/10 mM Tris-HCl (pH 8).
- YLB: 1% lithium dodecyl sulphate/0.1 M EDTA/10 mM Tris-HCl (pH 8).
- TE: 10 mM Tris-HCl (pH 7.5)/1 mM EDTA.
- Size markers: yeast chromosomes are used as size standards, they can be prepared as described in this Section, they are also commercially available.

Equipment

- Pulsed-field gel electrophoresis apparatus, CHEF system from Biorad
- Tabletop centrifuge
- Water bath incubator
- Shaker incubator

Procedure

1. Inoculate 20 colonies into separate test tubes with 5 ml of YPD medium and grow overnight at 30 °C.
2. Harvest cells by centrifuging for 5 min at 3000 rpm in a bench-top centrifuge. It is convenient to use 8-ml plastic disposable tubes; all the washes and manipulation of the plug can be readily done in the same tube.
3. Wash cells twice in 5 ml 50 mM EDTA, pH 8; centrifuge as above.
4. Resuspend cells in 150 ml 50 mM EDTA, pH 8.
5. Add 4 μl of zymolyase solution. Mix gently.
6. Add 250 ml of 1% low-melting-point agarose at 39 °C prepared in 125 mM EDTA. Mix gently, avoiding bubbles; let solidify 15 min at 4 °C.
7. Cover plugs with 1 ml of LET. Incubate overnight at 37 °C (Plugs can be also left incubating for 4–6 h.)

8. Change the LET to 0.5 ml of YLB buffer to cover the plugs. Incubate overnight at 50 °C.
9. Wash plugs several times with 50 mM EDTA. Store plugs in 50 mM EDTA at 4 °C.
10. Prepare a 1% agarose gel to run the yeast chromosome preparations. Perform electrophoresis at 60-s switch time, 160 V for about 36 h at 13 °C.
11. Stain the gel with ethidium bromide for 30 min and de-stain for 15 min. Photograph the gel, examine for the presence of new bands among the chromosomal ones.
12. Using standard procedures, transfer the DNA to a nylon membrane and hybridize with labelled pBR322 plasmid DNA to identify the artificial chromosomes.
13. Re-hybridize the membrane using total plant DNA as a probe to confirm the presence of plant DNA inserts.

2.5.8 Storage and Screening of YAC Libraries

Storage of a YAC library is not an easy task. For many purposes, the library is maintained as an ordered collection of individual clones. The most convenient way to preserve a YAC library is to transfer every individual clone to a well of 96-well microtitre plates and to store the plates at –80 °C. This transfer can be done from liquid cultures or from large colonies which were grown on agar plates by picking them with sterile toothpicks; the liquid media for storage should contain 20% glycerol. The viability of yeast clones stored in this way is high, but very sensitive to repetitive freezing and thawing. It is highly recommended to prepare work copies of the library in exactly the same way (but using complete media to increase the inocule and cell viability) and keep the master copy continually frozen. This kind of storage is the most appropriate if the library is going to be used for colony hybridization screening because it allows for an easy replication of every 96-well microtitre plate onto membranes by using a 96-pin replicating device which can be obtained from the Washington University School of Medicine Instrument Shop; St. Louis, Missouri 63110-1095. Filters are then grown on agar plates and colonies treated to fix the DNA onto the membrane for hybridization with radiolabelled probes (Matallana et al. 1992). Colony hybridization is the most common kind of YAC library screening, although other systems can be used. PCR on pooled clones (Green and Olson 1990) has become a very efficient screening technique which overcomes the difficulties associated with yeast colony hybridization.

Notes and troubleshooting

The conditions for isolating megabase DNA need to be worked-out for the desired plant species. Different procedures for isolation of protoplasts or preparation of nuclei have to be examined and compared to choose the one that gives DNA of larger size and greater integrity.

- It is difficult to control and obtain reproducible partial digests with plant DNA in agarose plugs; it is convenient to prepare various sets of plugs and to titrate for each one the conditions to achieve partial digests. If DNA in agarose plugs is consistently not being digested, the plug can be treated with agarase to release the DNA; the agarase-treated samples should be handled with extreme care to reduce shearing. Recently, success with a method for the isolation of plant megabase DNA by encapsulating plant protoplasts in microbeads has been reported (Wing et al. 1993). Severalfold increase in the surface area is obtained in the microbeads allowing more homogeneous access of restriction enzymes to the DNA, almost equivalent to that in an aqueous solution.
- The size selection of *Eco*RI fragments and the size selection of ligation mixture by PFGE are necessary in order to obtain YAC clones containing large inserts (>300 kb average insert size). Although YAC libraries have been constructed without any size selection or only with the selection of the ligation mixture by PFGE, the size of the clones in such libraries is small (> 150 kb average insert size).
- Always pipette plant protoplasts with wide-bore pipets
- Never pipette DNA, use blunt-ended tips to mix solutions containing DNA
- Use cover slips to cut the agarose plugs and agarose gel strips

References

Anand R (1992) Yeast artificial chromosomes (YACs) and the analysis of complex genomes. Trends Biotech 10:35–40

Burgers PMJ, Percival KJ (1987) Transformation of yeast spheroplasts without cell fussion. Anal Biochem 163:391–397

Burke DT, Carle GF, Olson MV (1987) Cloning a large DNA segments of exogenous DNA into yeast by means of artificial chromosome vectors. Science 236:806–812

Bancroft I, Westphal L, Schmidt R, Dean C (1992) PFGE-resolved RFLP analysis and long-range restriction mapping of the DNA of *Arabidopsis thaliana* using whole YAC clones as probes. Nucleic Acids Res 20:6201–6207

Bellann-Chantelot C, Lacroix B, Ougen P, Billault A, Beaufils S, Bertrand S, Georges I, Glibert F, Gros I Lucotte G, Susini L, Codani JJ, Gesnouin P, Pook S, Vaysseix G, Lu-Kuo J, Ried T, Ward D, Chumakov I, LePaslier D, Barillot E, Cohen D (1992) Mapping the whole human genome by fingerprinting yeast artificial chromosomes. Cell 70:1059–1068

Carle GF, Olson MV (1984) Separation of chromosomal DNA molecules from yeast by orthogonal-field-alternation gel electrophoresis. Nucleic Acids Res 12:5647–5664

Chang C, Kwok SF, Bleecker AB, Meyerowitz EM (1993) *Arabidopsis* ethylene-response gene *ETR1:* similarity of product to two-component regulators. Science 262:539–544

Creusot F, Sevignac M, Dron M (1992) Large-scale mapping using high-molecular weight DNA from *P. vulgaris*. Plant Mol Biol Rep 10:207–218

Ecker JR (1990) PFGE and YAC analysis of the *Arabidopsis* genome. Methods 1:186–194

Edwards KJ, Thompson H, Edwards D, de Saizieu A, Sparks C, Thompson JA, Greenland AJ, Eyers M, Schuch W (1992) Construction and characterization of a yeast artificial chromosome library containing three haploid maize genome equivalents. Plant Mol Biol 19:299–308

Elicieri B, Labella T, Hagino Y, Srivastava A, Schlessinger D, Pilia G, Palmieri G, D'Urso M (1991) Stable integration and expression in mouse cells of yeast artificial chromosomes harboring human genes. Proc Natl Acad Sci USA 88:2179–2183

Foote S, Vollrath D, Hilton A, Page DC (1992) The human Y chromosome: overlapping DNA clones spanning the euchromatic region. Science 258:60–66

Gamborg OL, Miller RA, Ojima K (1968) Nutrient requirements of suspension cultures of soybean root cells. Exp Cell Res 50:151–158

Ganal MW, Tanksley SD (1989) Analysis of tomato DNA by pulsed-field gel electrophoresis. Plant Mol Biol Rep 7:17–28

Grill E, Somerville C (1991) Construction and characterization of a yeast artificial chromosome library of *Arabidopsis* which is suitable for chromosome walking. Mol Gen Genet 226:484–490

Guzmán P, Ecker JR (1988) Development of large DNA methods for plants: molecular cloning of large DNA segments of *Arabidopsis* and carrot DNA into yeast. Nucleic Acids Res 16:11091–11105

Guzmán P, Ecker JR (1990) Development of large DNA methods for plants. In: Bennett A (ed) Horticultural biotechnology. Plant Biology series. AR Liss, New York, pp 95–107
Green ED, Olson MV (1990) Systematic screening of yeast artificial chromosome libraries by use of the polymerase chain reaction. Proc Natl Acad Sci USA 87:1213–1217
Hatano S, Yamaguchi J, Hirai A (1992) The preparation of high-molecular-weight DNA from rice and its analysis by pulse-field gel electrophoresis. Plant Sci 83:55–64
Hernández S, Sánchez M, Guzmán P, Simpson J (1996) Isolation of High molecular weight DNA from plant nuclei. Methods Mol Cel Biol (in press)
Hwang I, Kohchi T, Hauge BM, Goodman HM, Schmidt R, Cnops G, Dean C, Gibson S, Iba K, Lemieux B, Arondel V, Danhoff L, Somerville C (1991) Identification and map position of YAC clones comprising one-third of the *Arabidopsis* genome. Plant J 1:367–374
Jakobovits A, Moore AL, Green LL, Vergara GJ, Maynard-Currie CE, Austin HA, Klaphloz S (1993) Germ-line transmission and expression of a human-derived yeast artificial chromosome. Nature 362:255–258
Larin Z, Monaco AP, Lehrach H (1991) Yeast artificial chromosome libraries containing large inserts from mouse and human DNA. Proc Natl Acad Sci USA 88:4123–4127
Martin GB, Ganal MW, Tanksley SD (1992) Construction of a yeast artificial chromosome library of tomato and identification of cloned segments linked to two disease-resistance loci. Mol Gen Genet 233:25–32
Martin GB, Brommonschenkel SH, Chunwongse J, Frary A, Ganal MW, Spivey R, Wu T, Earle ED, Tanksley SD (1993) Map-based cloning of a protein kinase gene conferring disease resistance in tomato. Science 262:1432–1436
Matallana E, Bell CJ, Dunn PJ, Lu M, Ecker JR (1992) Genetic and physical linkage of the *Arabidopsis* genome: methods for anchoring yeast artificial chromosomes. In: Koncz C, Chua N-H, Schell J (eds) Methods in *Arabidopsis* research. World Scientific, Singapore, pp 144–169
McCormick M, Shero J, Chung M-C, Dan YW, Hieter P, Antonarakis S (1989) Construction of chromosome 21 specific yeast artificial chromosomes. Proc Natl Acad Sci USA 86:9991– 9995
Rose MD, Novick P, Thomas JH, Botstein D, Fink GR (1987) A *Saccharomyces cerevisiae* genomic plasmid bank based on a centromere-containing shuttle vector. Gene 60:237–243
Sambrook J, Fritsch EF, Maniatis T (1989) Molecular cloning: a laboratory manual, 2nd edn. Cold Spring Harbor Laboratory Press, Cold Spring Harbor
Schedl A, Montoliu L, Kelsey G, Schütz G (1993) A yeast artificial chromosome covering the tyrosinase gene confers copy number-dependent expression in transgenic mice. Nature 362:258–261
Schlessinger D (1990) Yeast artificial chromosomes: tools for mapping and analysis of complex genomes. TIG 6:248–258
Schmidt R, Dean C (1992) Physical mapping of the *Arabidopsis thaliana* genome. In: Davies KE, Tilghman SR (eds) Genome Analysis, vol 4. Strategies for physical mapping. Cold Spring Harbor Laboratory Press, Cold Spring Harbor, pp 71–98

Schmidt R, Dean C (1993) Towards construction of an overlapping YAC library of the *Arabidopsis thaliana* genome. BioEssays 15:63–69
Schwartz DC, Cantor CR (1984) Separation of yeast chromosome-sized DNAs by pulsed-field gradient gel electrophoresis. Cell 37:67–75
Smith DR, Smyth AP, Moir DT (1990) Amplification of large artificial chromosomes. Proc Natl Acad Sci USA 87:8242–8246
Smith DR, Smyth AP, Strauss WM, Moir DT (1993) Incorporation of copy-number control elements into yeast artificial chromosomes by targeted homologous recombination. Mamm Genome 4:141–147
Ward ER, Jen GC (1990) Isolation of single-copy-sequence clones from a yeast artificial chromosome library of randomly sheared *Arabidopsis thaliana* DNA. Plant Mol Biol 14:561–568
Wing RA, Rastogi VK, Zhang H-B, Paterson AH, Tanksley SD (1993) An improved method of plant megabase DNA isolation in agarose microbeads suitable for physical mapping and YAC cloning. Plant J 4:893–398

Chapter 3

Extraction of RNA, Cloning and Subtractive Hybridization

3.1 Isolation and Analysis of Messenger RNA from Plant Cells Cloning of cDNAs

P. LESSARD, V. DECROOCQ, and M. THOMAS

3.1.1 Introduction

The development of an organism depends on the co-ordinated expression of specific genes which lead to the determination of each cell type. In plants, besides the nucleus and mitochondria, present in all eukaryotes, there are also (chloro)plasts which contain their own genetic information.

Different types of RNAs are present in plant cells:

- Ribosomal RNAs (rRNAs), transcribed by RNA polymerases I and III, are the most important species, representing more than 70% of total RNA population. They are transcribed by the three genomes as large transcripts that are then cleaved into different classes. The different families of RNA assemble together with ribosomal proteins to form a ribosome, the site of protein synthesis.
- Transfer RNAs (tRNAs), synthesized by RNA polymerase III in the three cellular compartments, are also relatively abundant (15%) and are composed of 20 species implicated, after major post-transcriptional modifications, in translation processes.
- Heterogeneous RNAs (hnRNA) are the primary products of transcription by RNA polymerase II of nuclear genes. They are the precursors of messenger RNAs (mRNAs) and are located in the

nucleoplasm. One characteristic of most nuclear genes is the presence of non-coding sequences (or introns) interrupting the coding sequence. hnRNAs are faithful copies of the genes and therefore the intron sequences are also transcribed by the polymerase. Most hnRNAs are modified by the addition of a methylated cap at the 5′ end during transcription and by polyadenylation of the 3′ terminus immediately after the end of transcription. They will then be transferred to the cytoplasm and matured into mRNA [or poly $(A)^+$RNA] by splicing out of the introns (Fig. 3.1.1).

Regulation of gene expression can occur at different stages of RNA production. It is the rate of synthesis, processing and degradation of

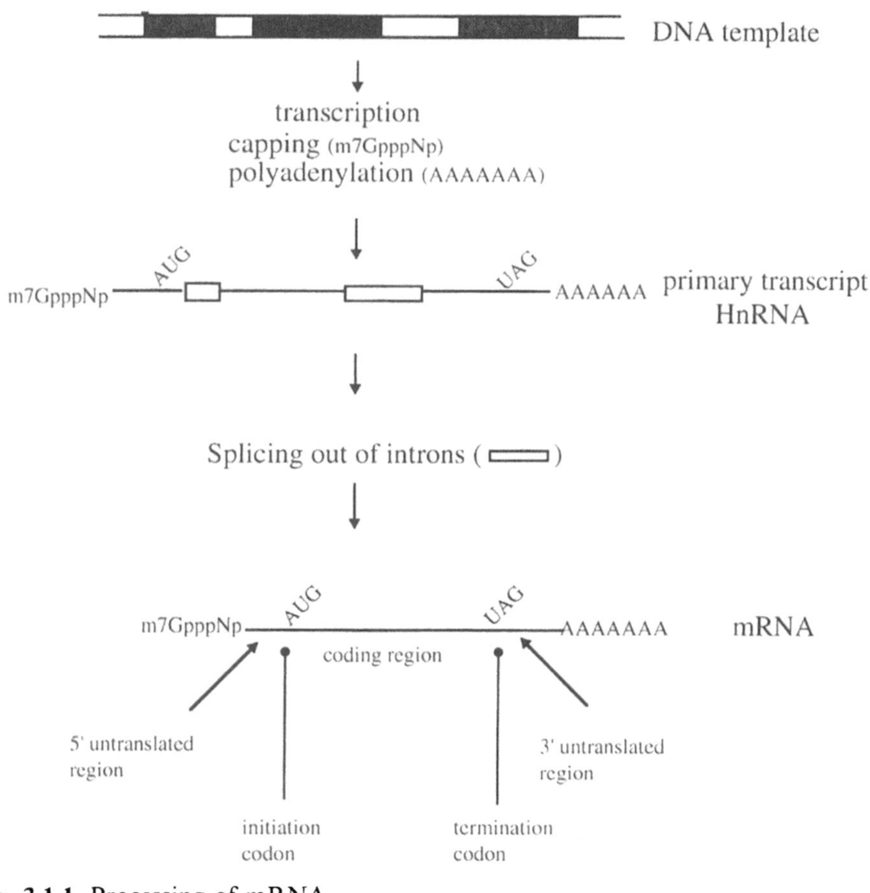

Fig. 3.1.1. Processing of mRNA

transcripts which determines the abundance of an mRNA during development in response to physiological and environmental cues.

Purification of intact RNAs is fundamental for the study of gene expression and for the isolation of new gene products via their cDNAs. Recombinant DNA techniques together with PCR-based technologies make it now possible to detect and analyze the transcription products of virtually any gene, even of those that are weakly expressed.

The following methods will be described in this chapter (Fig. 3.1.2):

- RNA and mRNA isolation and purification,
- determination of transcription rate of specific genes by in vitro transcription of isolated nuclei,
- qualitative and quantitative analysis of RNA by primer extension, northern and slot blotting, quantitative competitive RT-PCR,

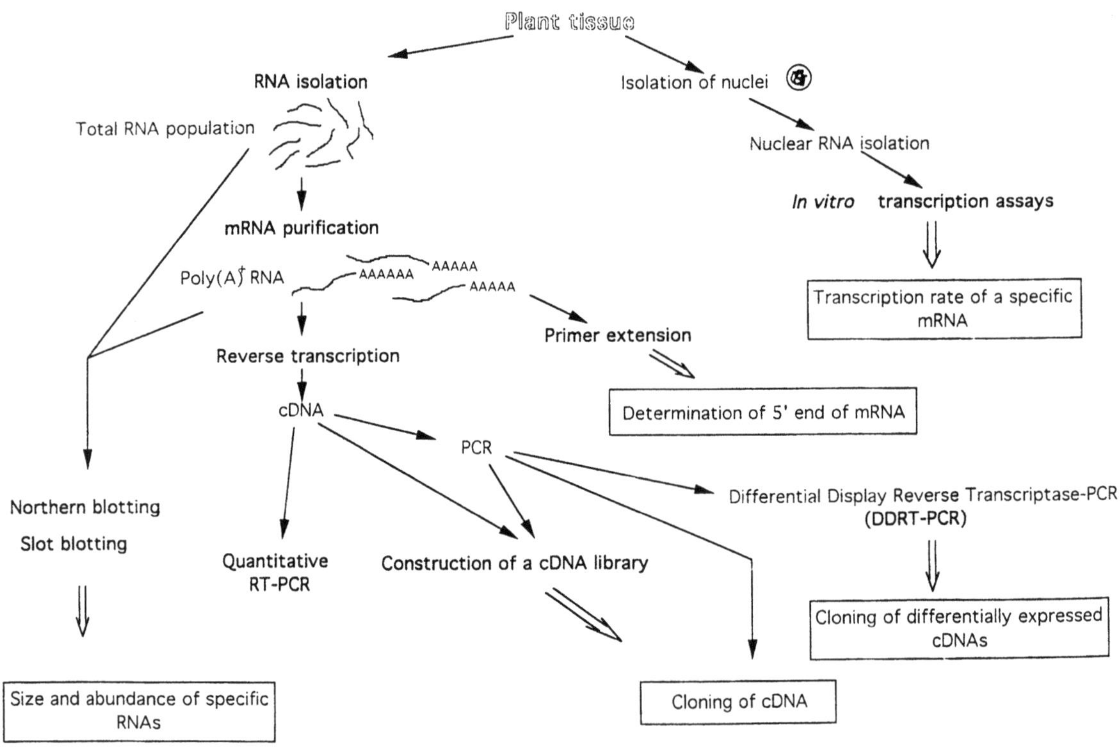

Fig. 3.1.2. The techniques described in this section

- isolation of cDNAs: construction of a representative cDNA library and corresponding screening.
- we will finally present a very recent technique for isolating differentially expressed cDNAs (DDRT-PCR).

Precautions To Be Taken When Working with RNA

Several specific precautions should be taken in order to avoid degradation of RNA by ribonucleases. RNAse is a very robust enzyme, not inhibited by classical autoclave treatment. The source of contamination by RNAses can be endogenous, some plant material such as roots being particularly nuclease-rich, or exogenous (glassware, buffers or manipulator skin). Human skin is a rich source of ribonucleases!

All the precautions listed below do not have to be used simultaneously, but using one or combining several can easily solve many degradation problems:

- wear gloves,
- before preparing solutions, ensure that all chemicals are RNAse-free,
- do not use spatulas to weigh chemicals,
- bake all glassware in a Pasteur oven at 180 °C for at least 2 h,
- treat solutions and buffers with diethyl pyrocarbonate (DEPC). DEPC is a very powerful and irreversible RNAse inhibitor (Fedorcsak and Ehrenberg 1966).

  Add 20 mM DEPC to each solution, mix carefully. After at least 1 h, autoclave at 120 °C for 20 min to remove residual DEPC.

  DEPC cannot be used to treat Tris-containing buffers. Use Tris powder from a new unopened bottle and dissolve in DEPC-treated sterile distilled water (DEPC-SDW).

  Be careful, DEPC is a suspected carcinogen and therefore should be handled with gloves under a chemicalhood.
- Competitive RNAse inhibitors can be added to specific reactions (cDNA synthesis, in vitro transcription experiments etc.).

3.1.2 RNA Isolation

3.1.2.1 *Isolation of Total Cellular RNA*

Several successful methods have been described for plant cellular RNA extraction. Total RNA comprises mitochondrial and chloroplastic RNAs, rRNAs, tRNAs, hnRNAs and mRNAs. The extraction buffer usually contains a strong detergent such as SDS and an organic denaturant such as guanidinium isothiocyanate, guanidine hydrochloride or phenol, which inhibit RNAses and facilitate removal of non-nucleic acid components.

The most currently used techniques are described below.

- CsCl cushion centrifugation method (Sambrook et al. 1989). After extraction, the RNA is recovered by isopycnic centrifugation through a CsCl cushion. The RNA pellet is then dissolved, ethanol-precipitated and re-dissolved in sterile distilled water. However, RNA pellets after CsCl centrifugation are often very difficult to re-dissolve, especially for material rich in polysaccharides.

- Guanidine hydrochloride method (Logemann et al. 1987). After extraction in guanidine hydrochloride buffer and centrifugation, proteins are removed from the RNA containing supernatant by phenol/chloroform/isoamyl alcohol. RNA is then recovered by ethanol precipitation. We found it necessary to add a DNAse treatment of the preparation, the RNA being often contaminated with genomic DNA.

We will describe in detail the procedure we routinely use in the lab which has proven efficient for most materials.

This whole cell RNA isolation procedure relies on guanidinium isothiocyanate, a strong protein denaturant together with the reductant β-mercaptoethanol. The simultaneous use of these two reagents followed by an ethanol precipitation and phenol/chloroform extractions leads to an efficient inactivation of nucleases. The procedure described below is a modification of the method first described by Chirgwin et al. (1979). We have used this technique with success for various organs including stems, roots, leaves, flowers or cell cultures. However, in our hands, this method was not suitable for extraction of

RNA from seeds or other starch rich material. RNA extracted by this procedure can be used for northern (Sects. 3.1.4.1/2) and dot blot (Sect. 3.1.4.3) analysis, cDNA synthesis (Sect. 3.1.7) as well as for in vitro translation experiments. However, for the latter application, best results are obtained using poly(A)$^+$ RNA (Sect. 3.1.3).

Reagents

All solutions, except RNA buffer 1 should be DEPC-treated and then autoclaved:

- RNA buffer 1: 25 mM Tris-HCl pH 8/ 4M guanidinium isothiocyanate/100 mM β-mercaptoethanol (add extemporarily)/ DEPC-SDW. Filter the buffer using Whatmann 1-M paper and keep at room temperature.
- RNA buffer 2: 50 mM Tris-HCl pH 8/10 mM EDTA/100 mM NaCl/0.2% (p/v) SDS. Keep at 4 °C.
- DEPC treated SDW (DEPC-SDW)
- 70 and 100% ethanol
- 3 M sodium acetate pH 5
- 1 M magnesium chloride
- 8 M lithium chloride
- 4 M NaCl
- 10% (w/v) SDS
- Acetic acid
- Molecular grade phenol saturated with 0.1 M Tris-HCl pH 8
- Chloroform

Equipment

- Balance
- 1.5-ml Eppendorf tubes
- Micropipettes (1–20, 1–200 and 200 µl-1 ml, 1–5 ml capacity plus tips)
- Ice
- Liquid nitrogen
- Mortar and pestle
- –20 °C freezer
- 15- and 50-ml disposable centrifuge tubes
- Vortex
- Dounce homogenizer
- Vacuum desiccator
- pH indicator
- 4 °C refrigerator
- Centrifuge

Procedure

1. Weigh out 2 to 10 g of fresh tissue and freeze in liquid nitrogen.
2. Grind the material to fine powder using mortar and pestle with liquid nitrogen.
3. Transfer the powder to a clean mortar containing 10 ml of RNA buffer 1, at room temperature, mix powder and buffer for several minutes.
4. Transfer the mixture to 50-ml centrifuge tube and centrifuge at 10 000 *g* for 20 min at 4 °C.
5. Add 0.03 vol 3 M sodium acetate pH 5 and 0.75 vol 100% ethanol to the supernatant. Mix by inversion and place at –20 °C for at least 2 h.
6. Centrifuge at 10 000 *g* for 20 min at 4 °C. Discard supernatant and add 10 ml of RNA buffer 2 to the pellet.
7. Transfer pellet and buffer in a Dounce homogenizer. Make the mixture as homogeneous as possible and transfer to 50-ml centrifuge tube.
8. Add 1 vol phenol/chloroform (1/1). Vortex for 1 min and centrifuge at 10 000 *g* for 15 min at 20 °C.
9. Transfer the aqueous phase (upper phase) in a clean 50-ml tube and add 1 vol chloroform. Vortex for a few seconds and centrifuge at 10 000 *g* for 10 min at 20 °C.
10. Transfer the aqueous phase to a clean tube. Adjust the pH to 5 with one or two drops of acetic acid (check the pH with pH indicator).
11. Add 1/20 vol 4 M NaCl and 0.6 vol isopropanol. Mix by inversion and place at –20 °C for 2 h.
12. Centrifuge at 10 000 *g* for 20 min at 4 °C. Discard supernatant.
13. Drain pellet by leaving the tube upside down for 5 min on filter paper and resuspend in 4 ml DEPC-SDW.
14. Add 5 μl 1 M magnesium chloride and 1 ml 8 M lithium chloride and place at 4 °C overnight.
15. Centrifuge at 10 000 *g* for 20 min at 4 °C. Discard supernatant.

16. Drain pellet and resuspend in 1 to 5 ml DEPC-SDW depending on the solubility of the pellet.

17. Add 1/10 vol 3 M sodium acetate and 2.5 vol 100% ethanol. Place at –20 °C for 2 h.

18. Centrifuge at 10 000 g for 20 min at 4 °C. Discard supernatant, drain pellet briefly, wash pellet in 1 ml 70% ethanol and dry it in a vacuum desiccator for 10 min.

19. Resuspend in 200 µl DEPC-SDW and transfer to an Eppendorf tube. Store at –20 or –70 °C.

Notes

- The above method can be modified for use with 0.5 to 20 g of material.
- The yield of total RNA extracted by this method varies from 0.1 to 2 mg/g fresh weight material, depending on the plant tissue.
- For large amounts of material, difficulties may occur in resuspending nucleic acid pellets at steps 13, 16 or 19. Add 0.1% SDS to DEPC-SDW, break the pellet with a pipette tip and place on ice for 20 min.
- RNA concentration can be determined by measuring the OD_{260} of an aliquot of the preparation. A solution containing 40 µg/ml will give a $OD_{260} = 1$.
  A ratio OD_{260}/OD_{280} between 1.8 and 2 is a sign of a clean RNA preparation with little protein contamination.
- The quality of the RNA preparation can be checked by loading 1 to 5 µg of RNA on a 1.5% agarose gel containing 0.5 µg/ml of ethidium bromide in 1x TAE electrophoresis buffer, prepared in a RNAse-free way (Sect. 3.1.1). No high molecular weight band (> 20 kb) should be visible (sign of DNA contamination). Several distinct bands should be seen under UV illumination corresponding to rRNAs and tRNAs (in the case of photosynthetic tissue, more bands are visible corresponding to chloroplastic rRNAs).
- Avoid thawing and freezing the RNA preparation too often. Keep stocks at –70 °C.

3.1.2.2
RNA Isolation from Small Amounts of Material

This protocol, adapted from Logemann et al. (1987), is performed with small amounts of tissue in Eppendorf tubes. The RNA isolated using this method is often contaminated with DNA, therefore a purification step using DNAse treatment is recommended (and necessary if the RNA is to be used in Reverse Transcriptase-PCR experiments; see Sect. 3.1.7.1).

Reagents

All solutions should be DEPC-treated and then autoclaved.
- Extraction buffer: 8 M guanidine hydrochloride/20 mM Mes (4-morpholineethansulphonic acid)/20 mM EDTA/50 mM β-mercaptoethanol. Adjust pH to 7.
- phenol/chloroform/isoamylalcohol (25/24/1)
- DNAse buffer x 10: 60 mM $MgCl_2$/400 mM Tris HCl, pH 8
- RNAse-free DNAse (e.g. DNase FPLCpure, Pharmacia)
- 1 M acetic acid
- 70 and 100% ethanol
- 3 M sodium acetate
- DEPC-SDW
- Sea sand

Equipment

- 1.5-ml Eppendorf tubes
- Micropipettes (1–20, 1–200 and 200 μl-1 ml, 1–5 ml capacity plus tips)
- Ice
- –20 °C freezer
- Microfuge
- Vacuum desiccator
- 70 °C water bath
- Vortex

Procedure

RNA Extraction.

1. Grind 2 cm^2 of material with sea sand in 400 μl extraction buffer.
2. Add 1 vol (500 μl) of phenol/chloroform/isoamyl alcohol.
3. Vortex 1 min.

4. Centrifuge at 12 000 *g* for 15 min at room temperature.
5. Add 0.2 vol 1 M acetic acid.
6. Add 0.7 vol 100% ethanol. Allow to precipitate for 2 h at –20 °C.
7. Centrifuge at 12 000 *g* for 20 min at 4 °C. Discard the supernatant. Dry the pellet for 5 min at room temperature.
8. Resuspend the pellet in 100 µl 3 M sodium acetate.
9. Centrifuge at 12 000 *g* for 15 min at 4 °C. Discard the supernatant.
10. Resuspend the pellet in 100 µl 3 M sodium acetate.
11. Centrifuge at 12 000 *g* for 15 min at 4 °C. Discard the supernatant.
12. Wash pellet with 70% ethanol.
13. Resuspend in 50 µl DEPC-SDW.

DNAse Treatment of RNA

1. In an Eppendorf tube, mix 40 µl: 10 µg of RNA, 4 µl of 10x DNAse buffer, 200 units of RNAse-free DNAse.
   Incubate for 10 min at 37 °C.
2. Inactivate DNAse by incubating for 10 min at +70 °C.
3. Add 60 µl DEPC-SDW, 1/10 vol 3 M sodium acetate and 2.5 vol 100% ethanol. Allow to precipitate for 2 h at –20 °C.
4. Centrifuge at 12 000 *g* for 20 min at 4 °C.
5. Discard supernatant, dry pellet and resuspend RNA in 15 µl DEPC-SDW.

Notes

- RNA concentration can be determined by measuring the OD_{260} as described in the previous section. Determination of quality of the RNA sample is also described in this section.

3.1.2.3
Isolation of RNA from Recalcitrant Plant Tissue

The presence of secondary compounds such as polysaccharides, anthocyanins, phenolic compounds, latex etc. often leads to very low yields and poor quality of RNA preparations when using standard RNA extraction procedures. Successful methods have been described, preventing oxidation and leading to differential precipitation of RNA from polysaccharides and phenolic compounds (Bahloul and Burkard 1993; Ainsworth 1994; Schultz et al. 1994). Nevertheless, it may be necessary to modify these protocols to adapt them to each type of recalcitrant tissue.

3.1.2.4
Isolation of RNA from Organelles

Extracting RNA from organelles necessitates first a reliable method for isolation of pure organelles (mitochondria or chloroplasts) (see Chap. 5). After lysis, total organelle RNA can be purified after ultracentrifugation through a CsCl cushion (Sambrook et al. 1989) or by LiCl differential sedimentation (Sect. 3.1.2.1, steps **14** to **19**). Mitochondrial tRNA, rRNA and mRNA can be separated following a procedure reported by Marechal-Drouard et al. (1995).

3.1.3
Purification of mRNAs

Most mRNAs from eukaryotic cells carry a poly(A) tail at their 3′ end. Poly(A) is added after transcription to hnRNA by a poly(A) polymerase, this step being necessary for the passage from nucleus to cytoplasm and the maturation of mRNA. Poly(A)$^+$ RNA can be purified from total or polysomal RNA.

The following procedure was first described by Aviv and Leder (1972) and adapted by Tobin and Klein (1975).

Reagents All solutions should be DEPC-treated and then autoclaved except for column and TE buffers.

- DEPC-SDW
- Column buffer: 50 mM Tris-HCl, pH 7.5/400 mM NaCl/1 mM EDTA/0.5% (w/v) SDS
- TE buffer: 10 mM Tris-HCl, pH 7.4/1 mM EDTA
- 0.1 M NaOH
- 4 M NaCl
- 3 M sodium acetate

- 70 and 100% ethanol
- Sulfochromic acid
- Oligo(dT)cellulose (Sigma)

Equipment

- Pasteur pipettes or syringes
- DEPC-treated and siliconized glass wool
- 1.5-ml Eppendorf tubes
- Micropipettes (1–20, 1–200 and 200 µl-1 ml, 1–5 ml capacity plus tips)
- Ice
- Spectrophotometer cuvettes
- Spectrophotometer
- Pasteur oven
- pH indicator
- –20 °C freezer
- Microfuge
- 65 °C water bath
- Vacuum desiccator

Procedure

1. Rehydrate 0.2 to 0.5 g of oligo(dT) cellulose in 10 ml of 0.1 M sodium hydroxide.

2. Plug a Pasteur pipette or a syringe with sterile glass wool. Preplugged disposable columns are also commercially available.

3. Pour 0.1 ml (packed volume) of oligo(dT)cellulose in the column (for up to 5 mg of total RNA).

4. Wash the column with 2 to 5 ml of column buffer until the pH of the eluate is 8 (check with pH indicator).

5. Dilute RNA preparation with 1 ml of TE buffer, heat at 65 °C for 5 min and allow to cool on ice.

6. Add 1/10 vol 4 M NaCl and load on the column.

7. Collect the eluate in a clean Eppendorf tube, heat for 5 min at 65 °C and load again on the column.

8. Collect the eluate in a clean Eppendorf tube [poly(A)$^-$RNA].

9. Wash the column with 2 ml of column buffer.

10. Elute poly(A)$^+$RNA by 100 μl fractions with DEPC-SDW [5 to 10 fractions are enough to elute all poly(A)$^+$RNA fixed on the column].

11. Group the fractions and directly estimate the amount of mRNA purified by measuring the absorbance at 260 nm.

12. Add 1/10 vol 3 M sodium acetate and 2.5 vol 100% ethanol. Place at –20 °C for at least 2 h.

13. Centrifuge at 12 000 *g* for 20 min at 4 °C. Discard supernatant.

14. Wash pellet in 1 ml 70% ethanol and dry it in a vacuum desiccator for 10 min.

15. Resuspend in DEPC-SDW at a concentration of 1 μg/ml. Keep at –20 or –70 °C.

Notes

- Binding capacities of oligo(dT)cellulose can vary with the supplier.
- Before measuring the absorbance, the quartz cuvettes should be treated with sulfochromic acid for 1 h and then carefully rinsed with DEPC-SDW. A solution containing 40 μg/ml will give an OD_{260} = 1.
- The yield obtained should be 1 to 3% of total RNA. A higher yield is often observed coming from a slight contamination of mRNA with poly(A)$^-$RNA. This problem can be solved by submitting the RNA to a second round of purification (after step **11**, proceed again from steps **5** to **11**).
- If the final eluate looks cloudy after staying on ice, it could be a sign of SDS contamination. Leave the preparation on ice for 10 min and centrifuge at 12 000 *g* for 5 min. Transfer the RNA-containing supernatant to a clean Eppendorf tube.
- When working with small quantities, other commercialized procedures, such as the use of oligo(dT) Dynabeads (Dynal UK), although more expensive, have proven to be very efficient. We have also performed direct purification of mRNA from as little as 6×10^4 cells with the Quickmicroprep-mRNA purification kit (Pharmacia).

3.1.4
RNA Analysis

A large array of techniques are now available in order to determine the size of a specific mRNA molecule, quantitate it among total RNA, estimate transcription rate or map its 5′ and 3′ ends. Some of the methods commonly used will be described in this section.

3.1.4.1
Northern Blotting: RNA Electrophoresis and Transfer

Northern blotting experiments allows the determination of size and relative abundance of specific RNA molecules in total or poly(A)$^+$ RNA preparations. RNA is separated by electrophoresis in denaturing conditions, transferred to a nitrocellulose or nylon membrane and hybridized with a ^{32}P-labelled DNA probe (Alwine et al. 1977). Several denaturing agents can be successfully used such as glyoxal and dimethyl sulphoxyde (McMaster and Carmichael 1977) or formaldehyde. We use the formaldehyde procedure derived from Lehrach et al. (1977) and Seed (1982).

Reagents

- 10x Mops buffer: 200 mM Mops-NaOH/50 mM sodium acetate/ 10 mM EDTA. Adjust to pH 7 with NaOH and protect from light
- 20x SSC: 3 M sodium chloride/0.3 M sodium citrate
- Staining buffer: 0.2% (w/v) methylene blue/0.4 M sodium acetate pH 5.2
- DEPC-SDW
- Formamide
- Formaldehyde (usually purchased as a 37% solution in water)
- RNA molecular weight markers ranging from 0.24 to 9 kb (Gibco-BRL)
- 1.2% denaturing gel (for 100 ml): 1.2 g of agarose/10 ml 10x Mops buffer/DEPC-SDW to 74 ml. Heat to dissolve agarose and let it cool down to 70 °C. Add 16 ml formaldehyde, mix and pour rapidly.
- Total or poly(A)$^+$RNA
- RNA mix: 1x Mops buffer/50% (v/v) formamide/6% (v/v) formaldehyde
- 0.1 M NaOH

Equipment

- Nitrocellulose or nylon membrane
- 1.5-ml Eppendorf tubes
- Micropipettes (1–20, 1–200 and 200 µl-1 ml, 1–5 ml capacity plus tips)
- Ice
- Gel electrophoresis tank and power pack
- 80 °C oven
- 65 °C water bath

Procedure

1. Wash gel electrophoresis tank and comb in 0.1 M NaOH for 5 min and rinse carefully with DEPC-SDW.
2. Prepare RNA samples; 10 to 30 µg of total RNA or 0.1 to 3 µg of mRNA are diluted in 20 to 30 µl of RNA mix. Heat the samples at 65 °C for 5 min and cool on ice.
3. Before loading the samples, pre-run the gel for 5 min. Separate RNA by denaturing agarose gel (1.2%) electrophoresis in 1x Mops buffer at 100 V for 2 to 3 h. Run 5 µg of marker RNAs in parallel.
4. Wash the gel in DEPC-SDW for 5–10 min.
5. Cut the marker lane. Place it in the staining solution for 5 min, wash it in DEPC-SDW for several hours (Herrin and Schmidt 1988).
6. Transfer the RNA onto a nylon membrane in 10x SSC buffer using the procedure described by Southern (1975) for 3 to 15 h (Sect. 1.2).
7. Wash briefly (1–2 min) the membrane in 2x SSC and air-dry.
8. Fix RNA on the membrane according to the supplier's instructions.

Notes

- Formaldehyde is toxic. It should be handled with gloves and the gels should be poured and run in a chemicalhood.
- There is no need to add loading buffer to the RNA samples, as formamide confers an appropriate density to the sample.
- We have obtained better separation by laying a clean glass plate on top of the gel during electrophoresis.
- RNA markers can also be stained with ethidium bromide (0.5 µg/ml in 0.1 M ammonium acetate) for 20 min.
- Concerning both transfer, RNA fixation and hybridization, we advise following the manufacturer's instructions for each type of

membrane (nitrocellulose or different types of nylon membranes). Changing these conditions may result in very high background problems [we routinely use Hybond N^+ membranes (Amersham)].

- For capillary blotting, a short period of transfer (3 h) is often preferred.
- Transfer can also be performed by electroblotting.

3.1.4.2
Northern Hybridization

Reagents

- 20x SSC: 3 M NaCl/0.3 M sodium citrate
- 20x SSPE: 3 M NaCl/0.18 M NaH_2PO_4/20 mM EDTA/Adjust to pH 7 with NaOH
- DEPC-SDW
- Prehybridization and hybridization buffer: 50% (v/v) formamide/ 5x SSPE/5x Denhardt's/1.5% (w/v) SDS/50 µg/ml denatured salmon sperm DNA

Equipment

- Hybridization oven (Hybaid oven)
- 1.5-ml Eppendorf tubes
- Micropipettes (1–20, 1–200 and 200 µl–1 ml, 1–5 ml capacity plus tips)
- 100 °C water bath
- Ice
- X-ray films
- Autoradiographic cassettes with intensifying screens
- Material for developing autoradiographs

Procedure

1. Wet the membrane in 2x SSC buffer.
2. Prehybridize the membrane in a tube under rotation in the hybridization oven at 42 °C for 2 to 15 h in 20 ml of prehybridization buffer.
3. Denature the probe at 100 °C for 10 min and cool on ice (for labelling of the probe, see Chap. 1.3).
4. Add the labelled probe to the prehybridization buffer (the volume of buffer should be reduced as much as possible, e.g. 5 ml of buffer for a 10 cm^2 membrane). Incubate at 42 °C for 15 h.

5. Discard the hybridization buffer.

6. Wash the filter twice with 25 ml of 2x SSC/0.5% SDS at room temperature for 15 min.

7. Wash the filter twice with 25 ml of 0.5x SSC/0.2% SDS at 62 °C for 20 min.

8. Never allow the membrane to dry out. Cover it with cling film or insert it in a plastic bag.

9. Put an X-ray film on membrane inside an autoradiographic cassette with two intensifying screens. Place at -70 °C for 2 h to 8 days.

Notes and troubleshooting

- Best results are obtained using 20 to 50 ng of probe (for a 10 cm^2 membrane) labelled to 5×10^8 to 5×10^9 dpm/μg of DNA.
- Some positively charged nylon membranes may give a very high background. You can lower this background by:
  - increasing the time of prehybridization,
  - increasing the amount of SDS (up to 7%) in the prehybridization and hybridization buffers,
  - filtering the solutions.
- Re-use of the filter: strip off the probe by placing the membrane for 10 min in 100 ml of 0.1x SSC/0.1% SDS previously heated at 100 °C. Perform this step at least twice.
- Although relative quantification can be obtained using northern blotting, electrophoresis and transfer of RNA may lead to differences in the amounts of RNA really accessible to the probe. Nevertheless, this problem can be solve by re-hybridizing the filter with a probe corresponding to a constitutively expressed mRNA.

3.1.4.3
Dot/Slot Blotting

Dot and slot hybridization experiments consist of immobilizing total or poly$(A)^+$RNA on a solid support (nitrocellulose or nylon membrane) and then hybridizing with a radiolabelled probe. Specific mRNA amounts are then quantified by densitometry of the corresponding autoradiographs. Filtration manifolds are available commercially (Minifold II, Schleicher and Schuell). The procedure described is adapted from Brown et al. (1983) and Chapman et al. (1983).

Reagents

- 0.1 M NaOH
- 10x Mops buffer: 200 mM Mops-NaOH/50 mM sodium acetate/10 mM EDTA. Adjust to pH 7 with NaOH and protect from light.
- 20x SSC: 3 M sodium chloride/0.3 M sodium citrate
- DEPC-SDW
- Formamide
- Formaldehyde (usually purchased as a 37% solution in water)
- Total or poly(A)$^+$RNA
- RNA mix: 2x SSC/50% formamide/6% formaldehyde/1x Mops buffer

Equipment

- 65 °C water bath
- Filtration manifold
- Whatmann 3-M filter paper
- Nitrocellulose or nylon membrane
- 1.5-ml Eppendorf tubes
- Micropipettes (1–20, 1–200 and 200 µl-1 ml, 1–5 ml capacity plus tips)
- Ice
- Vacuum
- X-ray films
- Autoradiographic cassettes with intensifying screens
- Material for developing autoradiographs

Procedure

1. Place the manifold in 0.1 M NaOH for 5 min and then thoroughly rinse it in DEPC-SDW.
2. Cut a sheet of nylon or nitrocellulose the same size as the manifold and wet it in DEPC-SDW. Place it in 10x SSC for 10 min.
3. Cut two sheets of Whatmann 3 M filter paper and wet them in 20x SSC.
4. Assemble the manifold following the instructions of the manufacturer.

5. Prepare RNA samples: dilute 10 to 50 μg of total RNA or 0.1 to 10 μg of poly(A)$^+$ RNA in RNA mix. Incubate at 65 °C for 10 min and cool rapidly on ice.
6. Wash the wells with 0.5 ml of 10x SSC.
7. Load the RNA samples. Apply gentle suction to the manifold until the RNA samples pass through the membrane.
8. Wash the wells with 2 vol 10x SSC.
9. Continue vacuum suction for 5 min after the buffers have passed through the wells.
10. Disassemble the manifold. The membrane can then be used in hybridization experiments as described in Section 3.1.4.2

Notes

- Dot blotting can also be performed by spotting RNA samples directly on nitrocellulose or nylon membranes. However, the quantification is more difficult due to the fact that the size of the spots is highly variable.

3.1.5 Run-Off Transcription

Transcription of isolated nuclei helps to characterize in vitro the transcriptional state of the nuclei at the time of isolation. This type of transcription, classically called run-off allows in vitro elongation of RNA molecules initiated in vivo. Transcription is performed in the presence of three cold nucleotides, the fourth one being brought as $\alpha(^{32}P)$-UTP. Radioactivity is then incorporated in high-molecular-weight polyribonucleotides. Transcription of a specific gene is shown by hybridization of the in vitro-labelled RNA with a DNA probe fixed in excess on a nylon membrane (Marzluff and Huang 1984).

The rationale of this approach is based on the generally accepted fact that isolated nuclei continue to effect transcription as they did in vivo at the moment they were fixed and extracted. Re-initiation of new transcripts does not take place under the assay conditions (Marzluff and Huang 1984). Therefore, in vitro-observed transcriptional rates reflect the actual and present capacity of nuclei to produce a given set of specific RNAs.

3.1.5.1
Isolation of Nuclei

The method of Luthe and Quatrano (1980) has been successfully used for isolating nuclei with a generally good yield of purity and free of contaminating ribonucleases. The principal improvement brought about by this technique compared to conventional ones, based on purification through sucrose cushions, is the centrifugation step through a discontinuous gradient of percoll. However, for some plants (e.g. sorghum), we found this method inadequate, giving low nuclei yield. The procedure we describe below is a modification of the chromatin isolation technique of Steinmüller et al. (1986) with minor amendments (Thomas et al. 1990).

Reagents

- Isolation buffer: 20 mM Tris-HCl pH 7.8/250 mM sucrose/5 mM $MgCl_2$ ($6H_2O$)/5 mM KCl/40% (v/v) glycerol/0.25% (v/v) Triton X-100/0.1% (v/v) β-mercaptoethanol
- Liquid nitrogen
- 30-mesh nylon filter
- Hoechst 33342 (5 mg/ml stock solution)

Equipment

- Mortar and pestle
- Balance
- Refrigerated low-speed centrifuge
- 50-ml disposable polyethylene centrifuge tubes (Falcon)
- −80 °C freezer
- Ice

Procedure

1. Weight out 20 g plant tissue and freeze in liquid nitrogen.
2. Grind to a powder using mortar and pestle with liquid nitrogen.
3. Resuspend in 45 ml of cold isolation buffer on ice.
4. Filter the first homogenate through a 30-mesh nylon filter.
5. Centrifuge at 4000 *g* for 30 min at 4 °C.
6. Discard the supernatant and carefully resuspend the pellet in 30 ml of cold isolation buffer.
7. Centrifuge at 4000 *g* for 20 min at 4 °C.

8. Steps **6** and **7** have to be carried out twice.

9. Resuspend the last pellet in 2 ml of buffer.

10. Dispatch in 500-µl aliquots, freeze rapidly in liquid nitrogen and keep at –80 °C.

11. Check the purification by microscopic observation after diluting ten fold in isolation buffer (Zeiss, Axiophot, Plan Néofluar 40x).

12. Mix equal volumes of nuclei preparation with a 5 mg/ml Hoechst 33342 solution. Count the nuclei in a Mallassez cell (excitation 365 nm, filter BP365-FT395-LP397).

Notes

- All the steps of the nuclei extraction procedure should be performed on ice.
- The nuclei prepared by this way can be kept for months without loss of transcriptional activity.
- The yield should be of about 5×10^5 nuclei/g fresh weight.

3.1.5.2
Transcription of Isolated Nuclei

Reagents

- Nuclei (1 to 3×10^7 nuclei per assay)
- Transcription buffer: 50 mM Tris-HCl, pH 8/5 mM $MgCl_2$/12.5% (v/v) glycerol/0.35 mg/ml dithiothreitol/50 mM $(NH_4)_2SO4$ /2 mM $MnCl_2$/0.5 mM (ATP, GTP, CTP)
- TE buffer: 10 mM Tris-HCl pH 7.4/1 mM EDTA
- DEPC-SDW
- 100% ethanol (cold)
- Sephadex G100 equilibrated in TE buffer
- Carrier tRNAs
- Ribonuclease inhibitor (Promega Biotech)
- DNAse1
- Proteinase K
- Chloroform
- Molecular grade phenol saturated with 0.1 M Tris-HCl pH 8
- UTP

Equipment

- Water bath at 30 °C
- Centrifuge

- 15-ml disposable polyethylene centrifuge tubes (Falcon)
- Microfuge
- 1.5-ml Eppendorf tubes
- Micropipettes (1–20, 1–200 and 200 μl–1 ml, 1–5 ml capacity plus tips)
- Ice
- Syringe
- Siliconized glass wool
- Vortex
- Vacuum desiccator

Procedure

1. Aliquot 1 to 3×10^7 nuclei in an Eppendorf tube.
2. Briefly spin (30 s at 6000 *g*) to pellet the nuclei.
3. Carefully resuspend the nuclei in 1.2 ml of transcription buffer and transfer to 15 ml centrifuge tubes.
4. Add 125–175 μCi of (α-^{32}P)UTP (400 Ci/mmol) and 100 units of ribonuclease inhibitor (Promega Biotech). Incubate at 30 °C for 40 min.
5. Stop transcription reaction by the addition of 100 μg/ml of DNAseI in the presence of 500 μg/ml tRNA and 120 mM cold UTP. Incubate at room temperature for 5 min.
6. Add 200 μg/ml of proteinase K and incubate for 15 min at room temperature.
7. Add 1 vol phenol:chloroform (1:1) mix. Vortex for a few seconds.
8. Centrifuge at 10 000 *g* for 10 min at 15 °C.
9. Transfer the aqueous phase to a new tube.
10. Add 2 vol cold 100% ethanol. Incubate at least 2 h at –20 °C.
11. Centrifuge at 20 000 *g* for 20 min at 4 °C. Discard the supernatant.
12. Dry the pellet and resuspend in 50 μl of DEPC-SDW.
13. Place the RNA on top of a 0.6-ml Sephadex G100 column equilibrated in TE buffer (this step can be performed in a 1-ml syringe plugged with glass wool).
14. Elute the purified RNA by adding 250 μl of TE buffer onto the column.

15. The eluate of the column can be stored at –80 °C or used in hybridization experiments (see below).

3.1.5.3
Hybridization of Cloned Probes to In Vitro-Labelled Transcripts

Reagents

- Hybridization buffer: 50 mM sodium phosphate pH 6.5/50% (v/v) formamide/4x SSC/0.2% SDS (w/v)/5x Denhardt's/120 µg/ml *E. coli* tRNA /100 µg/ml poly(A).
- 10% (w/v) SDS
- 20x SSC: 3 M sodium chloride/0.3 M sodium citrate
- Plasmid DNA corresponding to the mRNA analyzed

Equipment

- 42 and 62 °C water bath
- X-ray films
- Autoradiographic cassettes with intensifying screens
- Material for developing autoradiographs
- Hybridization oven (Hybaid)

Procedure

1. Plasmid DNA containing the sequence of interest (0.5 to 2 µg) is restricted in order to separate the insert, from the vector, electrophoresed on a 1% agarose gel and Southern blotted (see Sect. 1.2).
2. Prehybridize the membrane for 2 h at 42 °C in the hybridization buffer.
3. Add the in vitro-labelled transcripts and incubate for 10 to 24 h at 42 °C.
4. Wash filter once with 2x SSC/0.2% SDS for 5 min at room temperature. Wash filter twice with 1x SSC/0.2% SDS for 20 min at 62 °C. Treat the filter with 20 µg/ml RNAse A in 2x SSC for 20 min at room temperature.
5. Expose filter overnight at –80 °C to X-ray film using an intensifying screen.

Notes

- The RNAse A treatment is performed in order to lower background by eliminating all the RNA fragments not engaged in a double-stranded structure (Chappell et Hahlbrock 1984).

3.1.6
Mapping of the 5′ and 3′ Termini of Specific mRNA

Among the different techniques employed to determine 5′ and 3′ termini of a specific mRNA, the two most commonly used are mapping of RNA using nuclease S1 and primer extension:

- *Mapping of RNA using nuclease S1.* This technique allows precise mapping of 5′ and 3′ termini of mRNA on genomic DNA template and was first described by Berk and Sharp, 1977. ^{32}P-labelled genomic fragments are hybridized to mRNAs under conditions favourizing DNA/RNA hybrid formation. Single-strand nucleic acids will be digested during incubation with S1 nuclease while the hybrids are protected and further analyzed by gel electrophoresis.

- *Primer extension.* This technique allows mapping of the 5′ termini of mRNA. One specific oligonucleotide is labelled at its 5′ terminus and hybridized to the mRNAs. cDNA is synthesized by reverse transcriptase using the oligonucleotide as primer and mRNA as a template. The DNA/RNA hybrid is then denatured and labelled DNA is electrophoresed on a denaturing polyacrylamide gel. Its size is determined by comparison with a sequence reaction of the genomic DNA using the same oligonucleotide as a primer and run on the same gel.

Reagents

- 10x T4 kinase buffer: 500 mM Tris-HCl, pH 7.5/100 mM $MgCl_2$/50 mM dithiothreitol/1 mM EDTA/1 mM spermidine
- Hybridization buffer: 50 mM Tris-HCl, pH 8/500 mM NaCl/ 5 mM EDTA
- Reverse transcriptase buffer: 50 mM Tris-HCl, pH 8.3/ 10 mM $MgCl_2$/70 mM KCl
- Loading buffer: 0.03% (w/v) bromophenol blue/20 mM EDTA/ 98% formamide
- TE buffer: 10 mM Tris-HCl, pH 7.4/1 mM EDTA
- DEPC-SDW
- 3 M sodium acetate
- 100% ethanol
- $\gamma(^{32}P)$-ATP
- T4 kinase
- Reverse transcriptase
- dNTPs solution: 10 mM of each dATP, dCTP, dGTP, dTTP

- 5 μg mRNA
- 25 to 40-mer oligonucleotide
- RNAsin
- Carrier tRNA

Equipment

- 1.5-ml Eppendorf tubes
- Micropipettes (1–20, 1–200 and 200 μl-1 ml, 1–5 ml capacity plus tips)
- Ice
- 30, 37 and 70 °C water bath
- –20 °C freezer
- Microfuge
- Vacuum desiccator
- X-ray films
- Autoradiographic cassettes
- Material for developing autoradiographs

Procedure

1. Dilute 10 pmol of dephosphorylated oligonucleotide in 20 μl of labelling solution containing 2 μl of 10x T4 kinase buffer, 15 μl $\gamma(^{32}P)$-ATP (150 μCi) and 2 units T4 kinase. Incubate at 37 °C for 30 min.

   Add 50 μg carrier tRNA, 1/10 vol 3 M sodium acetate and 2.5 vol 100% ethanol. Place at –20 °C for 2 h.

   Centrifuge at 12 000 *g* for 20 min at 4 °C. Dry the pellet and resuspend in 200 μl of hybridization buffer.

2. Dilute 5 μg mRNA in 30 μl of labelled oligonucleotides in hybridization buffer.

   Incubate at 70 °C for 5 min and then overnight at an adequate temperature for annealing of the mRNA and oligonucleotides.

3. Add 1/10 vol 3 M sodium acetate and 2.5 vol 100% ethanol. Place at –20 °C for 2 h.

   Centrifuge at 12 000 *g* for 20 min at 4 °C. Dry the pellet and resuspend in 3 μl of reverse transcriptase buffer.

4. Add 6 μl of 500 mM dithiothreitol, 6 μl of dNTP solution, 40 units of reverse transcriptase, 80 units of RNAs in, 5 μl DEPC-SDW. Incubate at 30 °C for 1 hour.

5. Add 1/10 vol 3 M sodium acetate and 2.5 vol 100% ethanol. Place at –20 °C for 2 h.

Centrifuge at 12 000 *g* for 20 min at 4 °C. Dry the pellet and resuspend in 4 µl of TE buffer and 6 µl of loading buffer.

6. Perform a sequence reaction of the genomic DNA (Sect. 4.1) using the same oligonucleotide as a primer.
7. Load both reactions on a denaturing 6% polyacrylamide gel (see Sect. 4.1).
8. After electrophoresis, fix and dry the gel. Establish an autoradiographic image of the gel.

Notes

- To calculate the annealing temperature, use the formula:
  T_m = 81.5 + 0.41 [%(G+C)] + 16.6 log(Na) - 0.65 (%formamide) -600/n
  n = length of the fragment in nucleotides
- The oligonucleotide, a 25- to 40-mer, should be located within 100 nucleotides of the 5′ terminus of mRNA.

3.1.7 Production of cDNA

3.1.7.1 *Reverse Transcription*

A cDNA is a strand of DNA complementary to a RNA. cDNA synthesis is performed with an RNA dependent-DNA polymerase usually called Reverse Transcriptase (RT). Two RT enzymes, isolated from AMV (Avian Myeloblastosis Virus) and M-MLV (Moloney Murine Leukemia Virus) are generally used in reverse transcription experiments at respectively 30 and 300 units per reaction (Sects. 3.1.7, 3.1.9 and 3.1.10). One enzyme showing higher activity has been recently commercialized (Sect. 3.1.9.2).

Reagents

- 10x RT buffer: 500 mM Tris-HCl pH 8.3/80 mM $MgCl_2$/300 mM KCl
- DEPC-SDW
- primer oligo (dT) (Pharmacia)
- DNAse-free RNAse (Pharmacia)

- 0.1 M Dithiothreitol (DTT)
- 20 mM dNTP (Pharmacia)
- RNAse Block (40 units/µl, Stratagene)
- AMV reverse transcriptase (Pharmacia).

Equipment

- 42, 65 and 92 °C water bath
- 1.5-ml Eppendorf tubes
- Micropipettes (1–20, 1–200 and 200 µl-1 ml, 1–5 ml capacity plus tips)
- Ice

Procedure

1. Mix 1 µg DNA-free RNA with 25 ng of primer oligo (dT).
2. Add DEPC-SDW up to 14 µl.
3. Incubate at 65 °C for 15 min.
4. Allow to cool for 10 min at room temperature.
5. Add 2 µl of 0.1 M DTT (10 mM final), 1 µl of 20 mM dNTP (1 mM final), 2 µl of 10x RT buffer, 1 µl of RNAse Block, 0.5 µl (19 units) of AMV reverse transcriptase. Incubate at 42 °C for 60 min.
6. Inactivate reverse transcriptase by incubating the mixture at 92 °C for 4 min.

Note

- Reverse transcription can be performed using M-MLV reverse transcriptase. In this case, inactivate the enzyme by incubating at 65 °C for 5 min.

3.1.7.2
Association of Reverse Transcription with PCR

Association of reverse transcription with PCR allows rapid isolation of partial cDNAs that can be used as probes to screen cDNA or genomic libraries.

This may also be used as a rapid, very sensitive but nevertheless not quantitative method to check the expression of a specific gene in a given tissue, even in the case of very low expression levels.

After the reverse transcription reaction, RNA is removed with a DNAse-free RNAse. An aliquot of the cDNA is then used as template in a PCR reaction in order to amplify a fragment of the targeted cDNA.

Reagents

- SDW
- DNAse-free RNAse (Pharmacia)
- 100 mM $MgCl_2$
- dNTP (20 mM stock solution, Pharmacia)
- Oligonucleotide primers for the PCR
- Taq polymerase (Appligene)

Equipment

- Ice
- 70 °C water bath
- Thermocycler
- 1.5-ml Eppendorf tubes
- Micropipettes (1–20, 1–200 and 200 µl-1 ml, 1–5 ml capacity plus tips)

Procedure

1. Eliminate RNA by adding 1 µl of DNAse-free RNAse at 2.5 mg/ml. Incubate at 70 °C for 10 min.
2. Amplification

   PCR mix contains: 1 to 5 µl of the cDNA (obtained as described in Sect. 3.1.7), 0.2 mM of dNTP, 1µM of each primer, 2.5 units of Taq polymerase (Appligene).

   PCR program: will depend on the characteristics of the primers used (see corresponding note)

Notes

- The RNAse treatment prior to PCR is a necessary step. The generation of unexpected PCR products due to the presence of RNA has been reported (Brown et al. 1993).
- Ensure that RNAseA is DNAse-free: dissolve RNAse A at a concentration of 10 mg/ml in 10 mM Tris-HCl (pH 7.5)/15 mM NaCl. Heat at 100 °C for 15 min and allow to cool slowly to room temperature. Aliquot and store at –20 °C.
- The PCR program will differ depending on the primers pairs used (Rychlik et al. 1990; see also Sect. 2.1).
- To identify amplified fragments derived from potential genomic DNA contamination, you should choose, whenever possible, primers which anneal either side of an intron.

3.1.8
Quantification of Specific mRNA (Quantitative RT-PCR)

Competitive RT-PCR is a highly sensitive method to quantify transcripts of very weakly expressed genes (Siebert and Larrick 1992). The limit of detection of a Northern Blot is usually around 10^5 copies of an RNA molecule, whereas the PCR allows the detection of 10^3–10^4 copies. Nevertheless, this technique is far more expensive and difficult to run than a Northern blot experiment!

After reverse transcription of RNA, part of the cDNA is co-amplified with a competitor DNA by PCR. The competitor can be used as an internal standard since it is amplified by the same primers as the cDNA to be quantified.

PCR products are then run on an agarose gel. The intensity of the bands corresponding to the specific amplification of the cDNA of interest and of the competitor are compared.

In theory, in order to be amplified with the same efficiency, an ideal competitor should use the same primers, have the same length and the same sequence.

To distinguish easily competitor and template after electrophoresis, we have decided to use a "home-made" competitor presenting a small difference in size (30 to 50 pb), sufficient to separate the two bands on an agarose gel, but negligible with respect to the efficiency of amplification by PCR with a consistent polymerization step (Förster 1994).

Reagents

- Primers: 0.55 µg/µl stock solution
  Primers 1 and 2: 20-mer, GC% = 50
  Primer 3: 40-mer, GC% = 50
- PCR template: 1 µl of DNA obtained by RT (Sect. 3.1.7.1) or 1 ng of cDNA cloned in a plasmid vector
- dNTP: 20 mM stock solution
- SDW

Equipment

- 1.5-ml Eppendorf tubes
- Micropipettes (1–20, 1–200 and 200 µl-1 ml, 1–5 ml capacity plus tips)
- Ice
- Thermocycler
- Electrophoresis tank and power pack
- Eppendorf tubes

- Taq polymerase
- Oil for PCR (the liquid paraffin found at the chemist's works fine)

Procedure

Construction of the Competitor (Fig. 3.1.3)

1. Amplify the cDNA fragment to quantify.
PCR in 100 µl: 1µl of DNA obtained by RT (Sect. 3.1.7.1) or 1 ng of cDNA cloned in a plasmid vector, 1 µl of dNTP (0.2 mM final), 10 µl of 10x buffer, 1 µl of primers 1 and 2 (1 µM final), 0.5 µl (2.5 units) of Taq polymerase.
PCR program: 94 °C for 2 min
35 cycles: 94 °C for 1 min; 55 °C for 2 min; 72 °C for 1.5 min
72 °C for 2 min.

2. Run PCR products on an agarose gel (Sect. 1.2).

3. Excise the band of interest from the gel and elute DNA (Sect. 1.3).

4. Re-amplify the DNA fragment with the 40-mer primer 3 and primer 1.
PCR conditions: same conditions as step 1 except for:
- PCR template: 1 ng of DNA from the step **3**
- PCR primers: primer 1 and 3
- PCR program: same as step **1**.

5. Run PCR products on an agarose gel.

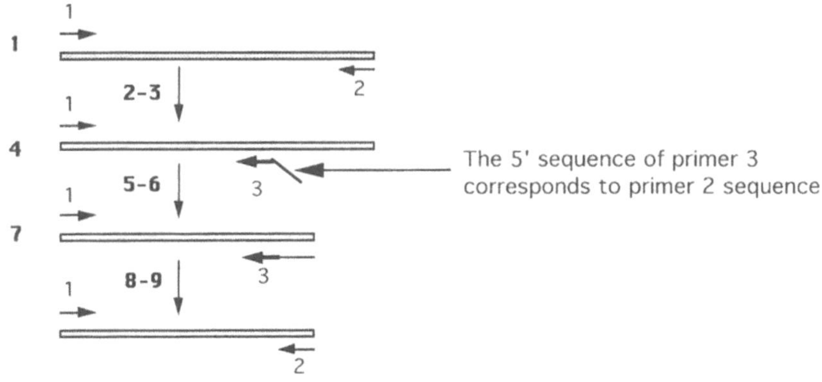

Fig. 3.1.3. Construction of the competitor for quantitative competitive RT-PCR

6. Excise the band of interest for the gel and elute DNA from the gel.

7. Amplify the competitor.
PCR conditions: same conditions as step 1 except for:
PCR template: 1 ng of DNA from point **6**.
PCR primers: primer 1 and 2.
PCR program: same as step **1**.

8. Run PCR products on an agarose gel.

9. Excise the band containing the competitor from the gel and elute DNA.

10. Determine the concentration of the competitor DNA (Sect. 1.1).

Quantitative competitive RT-PCR (Fig. 3.1.4)

11. Make serial dilutions of the competitor to a final concentration of 10^4 to 10^5 copies per µl.

12. Amplify cDNA and competitor.
PCR conditions: same as step **1** except: PCR DNA template: 1 µl of DNA obtained by RT (Sect. 3.1.7.1) + competitor

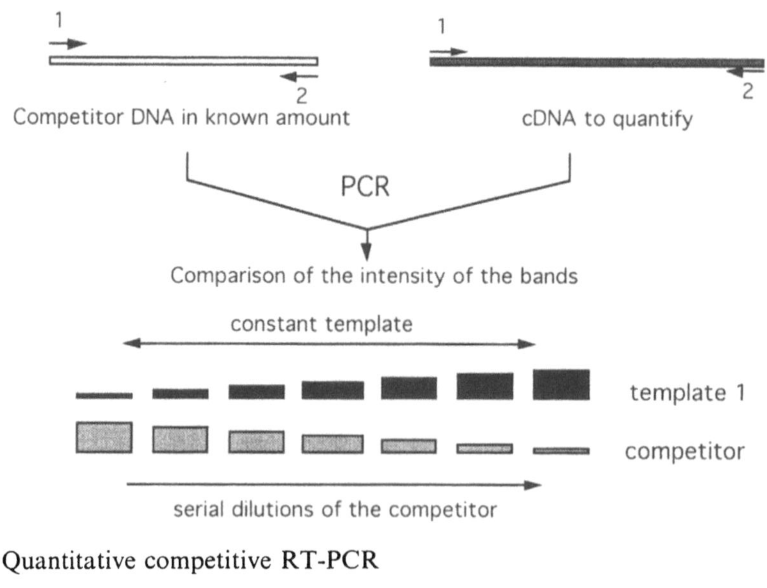

Fig. 3.1.4. Quantitative competitive RT-PCR

PCR program: 94 °C for 2 min
25 to 30 cycles: 94 °C for 1 min; 55 °C for 2 min; 72 °C for 1.5 min 72 °C for 2 min.

13. Run PCR products on an agarose gel (Sect. 1.2).

Notes

- We have used this protocol successfully to compare the expression of a given gene in two different populations of RNAs. In this protocol, the efficiency of reverse transcription step is not measured, so we have used this technique only to compare the expression of a low expressed gene in RNAs from two different plants, but NOT to quantify the number of transcripts in a plant compared to the number of copies of the competitor.
  To prevent differences between the two reverse transcription reactions, we usually isolate the two RNAs at the same time and the reverse transcription experiments are performed using the same stock of reaction mixture.
- To estimate the absolute number of RNA molecules, the efficiency of the reverse transcription reaction can be measured or equalized (Kolls et al. 1993).
- The PCR conditions must be adapted to the primers and the sequence to amplify.
  We use the formula from Rychlik et al. (1990):

  $$T°\text{annealing} = 0.3\ T_m\ \text{primers} + 0.7\ T_m\ \text{products} - 14.9$$
- Check that the competitor is actually derived from the DNA to quantify by direct sequencing or hybridization before starting the competitive reactions.
- For quantitative competitive RT-PCR, the number of cycles should be kept to a minimum to avoid reaching the plateau and must be optimized for each case.
- As for all PCR reactions, the magnesium concentration may need to be optimized for each primer pair used.

3.1.9
Cloning of cDNAs

3.1.9.1
Construction of a cDNA Library: Introduction

In recent years, the construction of cDNA libraries has become more widely accessible and several well-documented kits have been commercialized (we have been satisfied with cDNA synthesis kits from Stratagene and Pharmacia). Therefore, successful construction of a cDNA library will almost uniquely rely on the availability of a good quality mRNA preparation (Sect. 3.1.2). For a full description of the vector and strategies available, we refer the readers to Sambrook et al. (1989).

In this section, we will first outline the steps of construction of an oriented cDNA library in a Lambda vector (e.g. λ-ZAPII vector, Stratagene) and further describe a PCR-assisted method for dealing with minute amounts of material.

A good cDNA library should

- contain a number of independent clones in the library large enough to represent transcripts present in the original mRNA preparation,
- be constituted by nearly full-length copies of the mRNAs from which it was derived.

Main steps of the construction of an oriented cDNA library in a Lambda vector are described in the following (see Fig. 3.1.5).

First Step: first strand cDNA synthesis. This reaction requires an RNA template, a primer, a reverse transcriptase enzyme, an appropriate buffer and deoxyribonucleotides. The template is usually poly(A)$^+$ RNA, as this lowers any possible genomic and/or rRNA contamination in the final cDNA product. The initiation of the first strand synthesis depends on the hybridization to the mRNA of, either a poly(dT) primer carrying a restriction site (R1) for directional cloning, or random hexanucleotides. As full-length cDNAs can only be produced from non-degraded RNA templates, the quality of the starting RNA material will be critical.

Step 2: second strand replacement. The RNA-DNA hybrid resulting from the first strand synthesis is converted into double stranded cDNA by the combined action of RNAse H and DNA polymerase I. The mRNA template is nicked at multiple sites by the RNAse and

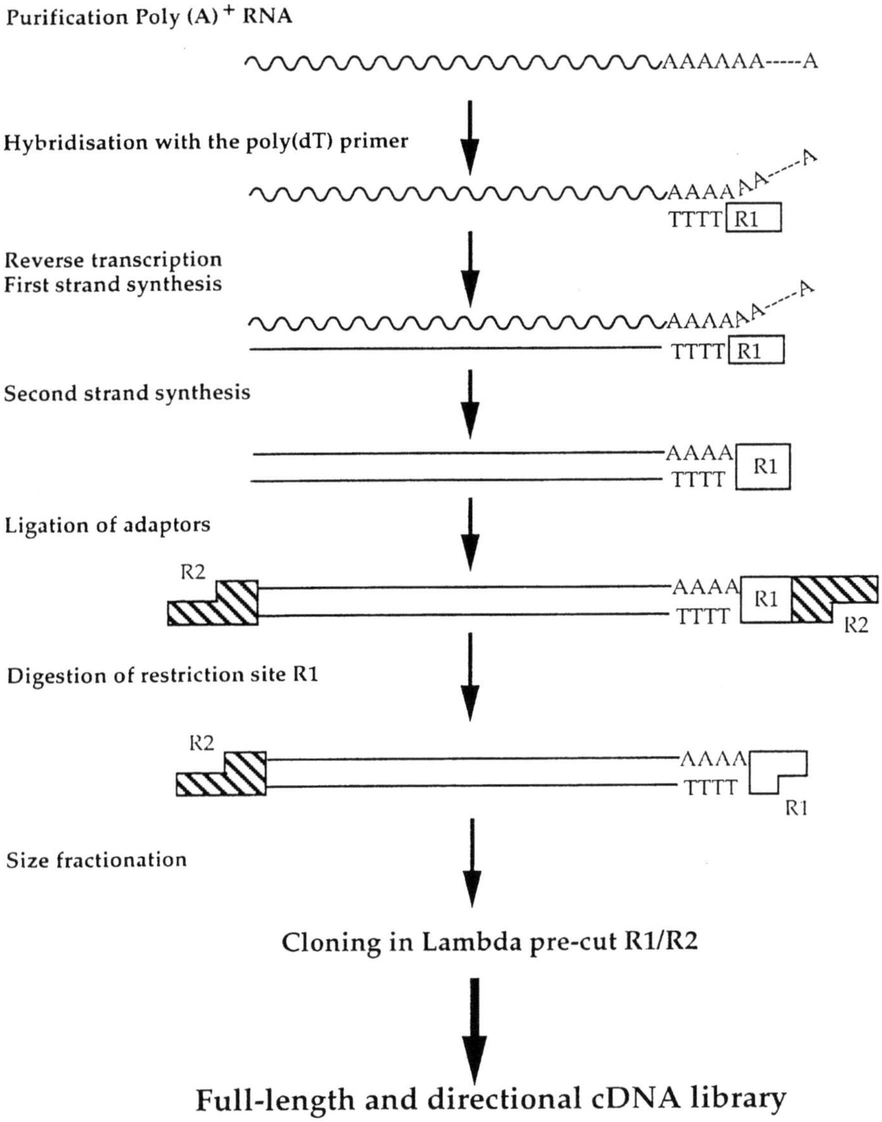

Fig. 3.1.5. Construction of a full-length directional cDNA library starting from limited amounts of plant material

DNA PolI is replacing it with DNA by nick translation. DNA polymerase I may be replaced by Klenow fragment which appears to be more efficient (Ray and Ray 1991).

Step 3: ligation of the adapters. The uneven termini of the double-stranded cDNA are blunt-ended with T4 DNA polymerase, and adaptors are ligated on. The adaptor is a short, duplex oligomer synthesized so that it has one blunt end and one dephosphorylated cohesive end (R2). Blunt-end ligation of the adapters to the cDNA will produce a new molecule with cohesive ends, providing compatible ends for cloning.

In the case of directional cloning, the restriction site introduced by the primer for reverse transcription will be digested after the ligation of the adaptors, which will carry a site for a different enzyme. To avoid partial cDNA clones resulting from internal digestion, one of the two strands, usually the first strand, may be methylated during synthesis. These two cohesive ends, one provided by the primer for reverse transcription located at the 3′ end of the cDNA and the other by the adaptors at the 5′ end, allow for an oriented insertion of cDNA molecules in the cloning vector.

Step 4: size selection of ds cDNA. Size fractionation of cDNA on Sephacryl S-400 columns following adaptor addition and digestion of the RT primer is important for several reasons:

- to eliminate excess of adaptors and primer fragments released from the cDNA by digestion, as these compete in the subsequent ligation between cDNA and the cloning vector,
- to eliminate partial cDNA molecules (i.e. < 500 bp) due to degradation of the mRNAs or the failure of RT to synthesize complete first strands.

Step 5: ligation to a λ phage cloning vector and packaging. In most cases, size-fractionated cDNAs are ligated to compatible lambda (λ) - derived vector arms, ideally in a molar ratio of insert to vector between 1:1 and 3:1 to avoid religation of the vector arms or formation of concatamers.

The phage vector might be either a λ ZAP-derived vector or a λgt10 or a λgt11-derived vector. The most convenient, the λ ZAP-derived vectors, allow in vivo excision of the phagemid and blue/white colour selection.

Recombinant phages are then packaged using commercially available extracts using conditions recommended by the manufacturer (we have used Gigapack II packaging extract from Stratagene). Plating and amplification of the cDNA library must be followed as detailed by the manufacturers (Stratagene). Unamplified libraries are more representative of the original mRNA population but the amount of phage obtained is usually low (around 10^6 to 10^7 pfu in total). Moreover, the titre of this library will diminish considerably with time. Amplification is often recommended and allows the production of large volumes (20 to 50 ml) of high-titre library stocks (10^9 to 10^{10} pfu/ml).

The library should then be aliquoted and stored at 4 °C (add 0.3% of chloroform) or –70 °C (add 70 μl of 100% DMSO to 1-ml aliquots). For more detail on lambda cloning, see Section 2.3.

3.1.9.2
Construction of cDNA Libraries Following In Vitro Amplification of cDNA

In recent years, interest in the isolation of new genes expressed in a particular cell type or in a narrow developmental stage has necessitated technical advances in the construction of cDNA libraries from cells available in limited amounts. Until recently, small quantities of RNA could be copied faithfully into cDNA but not cloned efficiently. To overcome this problem, the polymerase chain reaction technique, which is independent of abundance, has been used to amplify minute amounts of cDNA. In order to isolate sufficient amounts of RNA to act as a template for first strand synthesis, scaled-down versions of RNA extraction methods can be used in reaction volumes of 100–150 μl. Alternatively, direct purification of mRNA from as little as 6×10^4 cells was performed using the Quickmicroprep-mRNA purification kit (Pharmacia).

Current strategies of cDNA synthesis and cloning starting from few cells are based on in vitro amplification of cDNA after homopolymeric tailing at the 3′ end of the single-stranded cDNA (Belyavsky et al. 1989; Domec et al. 1990; Dresselhaus et al. 1994). We prefer to use a method based on the ligation of linkers to the 3′ and 5′ termini of the double-stranded cDNA (Don et al. 1993). The linkers are used as primer sites for a single primer which initiates PCR amplification on both strands. This also provides restriction sites for the cloning of the amplified cDNA copies into vector arms.

A PCR-assisted construction of a directional cDNA library in a λ vector is describe here (Fig. 3.1.6).

Reagents

- SuperScript reverse transcriptase (Life Technologies)
- NotI-RT primers (RT_N)
- UniAmp linkers (UA_S) (Clontech).

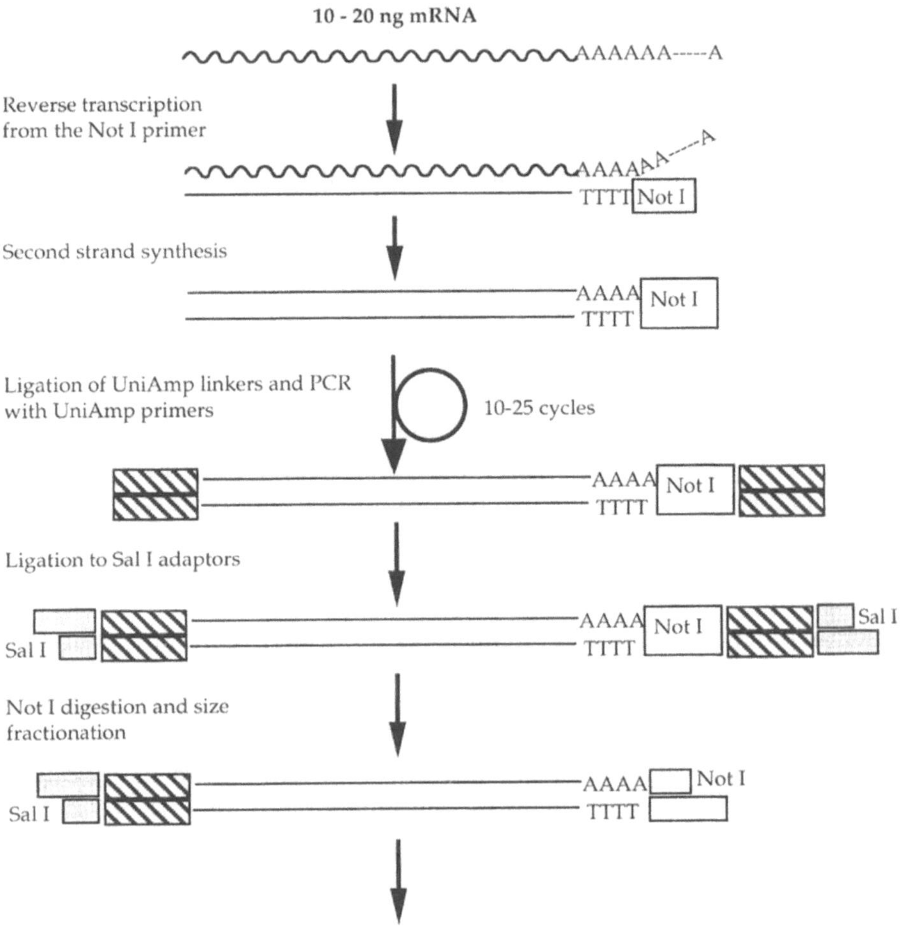

Fig. 3.1.6. PCR-assisted construction of a directional cDNA library in a λ vector

- UniAmp primers (Clontech)
- 20 mM dNTP mix
- 1 M Tris-HCl pH 8.3
- 1 M $MgCl_2$
- Taq Polymerase (AmpliTaq DNA Polymerase from Perkin Elmer Cetus)
- 3 M sodium acetate
- 70 and 100% ethanol
- Klenow enzyme

- Sephacryl S200 microspin column (Pharmacia)
- Sephacryl S300 cDNA spun column (Pharmacia)
- SalI-adapters (Life Technologies)
- NotI restriction enzyme
- Sephacryl S400 column (Stratagene)
- λgt22A vector (Life Technologies)

Equipment

- 16 and 37 °C water bath
- Thermocycler (Thermocycler)
- 1.5- and 0.5-ml microfuge tubes
- Microfuge
- Micropipettes (1–20, 1–200 and 200 µl-1 ml, 1–5 ml capacity plus tips)
- Ice

Procedure

1. Convert 10–20 ng mRNA to the first strand of cDNA with SuperScript reverse transcriptase and NotI-RT primers (RT_N) using the conditions recommended by the manufacturer. Perform second-strand synthesis and blunt ending.
2. Ligate half of the blunt-ended ds-cDNA to the UniAmp linkers (UA_S).
3. Amplify ds-cDNA by PCR:
   Total volume of 25 µl containing: 0.5 to 10 ng of cDNA, 0.5 µM UniAmp primers, 0.2 mM each dNTP, 10 mM Tris-HCl pH 8.3, 3 mM $MgCl_2$, 1.25 Units of Taq Polymerase.
   Amplification program: 10–25 cycles of 1 min at 94 °C; 2 min at 60 °C; 3 min at 72 °C.
4. Add 1/10 vol 3 M sodium acetate and 2.5 vol 100% ethanol to the amplified cDNA. Place at –20 °C for 2 h. Centrifuge at 12 000 *g* for 20 min at 4 °C, wash the pellet with 70% ethanol, air dry and resuspend in SDW.
5. Treat with Klenow enzyme to generate blunt ends.
6. Pass the amplified cDNA successively through a Sephacryl S200 *microspin* column and a Sephacryl S300 *cDNA spin* column.
7. Ligate the purified cDNA to the SalI-adapters.
8. Digest using NotI restriction enzyme.

9. Fractionate on Sephacryl S400 column to select for cDNAs > 500 bp. The full-length cDNAs are by now ready for directional cloning in the SalI-NotI arms of the vector λgt22A.

Screening of a cDNA library. This has already been described fully in the chapter on lambda genomic cloning (Sect. 2.3). Refer to this section for all further steps in phage library generation, screening, phage isolation and storage.

Notes

- The SuperScript RT is reported to provide higher yield of full-length cDNA since it lacks the intrinsic RNAse activity.
- The two-step protocol (point **6**) was developed in order to remove most of the unligated UA_S linkers and excess of UniAmp primers.
- Since the available starting material can be limiting, the efficiency of each step is very important. To improve recovery of RNA and cDNA from ethanol precipitation, a carrier substance can be used (tRNA, rRNA, poly I, glycogen). PCR amplification requires many precautions to prevent contamination and a control experiment should always be performed in parallel.
- In our case, after fractionation on Sephacryl S400 column, the peak fraction ranged from 0.4 to 3 kb in size.
- The number of PCR cycles should be kept to the minimum needed to obtain sufficient material (100 ng), to avoid misrepresentation of rare transcripts.

3.1.10 A New Way of Cloning Differentially Expressed cDNAs: Differential-Display Reverse Transcriptase-Polymerase Chain Reaction (DDRT-PCR)

The Differential Display Reverse Transcriptase Polymerase Chain Reaction (DDRT-PCR) is a promising method, initially published by Liang and Pardee in 1992, allowing the analysis of differential gene expression.

Two populations of mRNA extracted from two types of tissues (roots and leaves for example) or from the same tissue at different stages of development, differ by the presence or absence of some messengers and/or by a large difference in copy number of some messenger RNAs. These differences are often hidden in a background of constitutive transcripts such as those generated by "housekeeping" genes, often making the differential screening of cDNA libraries an potentially haphazard process.

DDRT-PCR technology allows the visualization and isolation of such differentially expressed mRNAs. It has been successfully used in the isolation of a mammal gene involved in cell proliferation (Liang et al. 1994), and has proven efficient in plant research (Oh 1995; Sharma and Davis 1995; Song et al. 1995).

The aim of this section is an introduction to this recent and powerful technique and to problems which. may arise in its use. For a further technique to detect differentially expressed RNAs, see the next section (3.2) on subtractive hybridization.

Principle (Fig. 3.1.7). Different populations of total RNA are reverse transcribed with primers derived from the usual oligo(dT).

A fraction of the newly synthesised cDNA is used as a template for the PCR amplification of the 3′ termini in the presence of radioactive nucleotides.

The products are separated by migration on a polyacrylamide gel and visualized by autoradiography.

The different populations of mRNA will give different patterns of bands.

Polymorphic bands will be cloned and sequenced and may be further used to screen a cDNA or genomic library.

Choosing the primers. Most of the protocols published so far differ in the sequences of the primers used for the reverse transcription and the PCR.

The major difficulties are to avoid non reproducible results and to display the largest number of distinct PCR fragments without saturating the gel.

With the aim of subdividing the total cDNA into equal fractions, particular oligo d(T) primers have been used in the reverse transcription reaction. These primers hybridize to the 3′ end of the mRNA [poly(A)$^+$ tail] plus one or two additional bases.

A primer such as $^{5'}T_{11}AT^{3'}$ will anchor mRNA terminating by $^{5'}$AT-polyA$^{3'}$ which represent 1/12th of the total mRNAs. This supposes that 12 independent reverse transcription reactions must be run in order to amplify all the mRNA population.

Recently, the use of degenerate primers such as $T_{12}MG$, $T_{12}MC$, $T_{12}MT$, $T_{12}MA$ (where M is a degenerate nucleotide: A, C or G) to

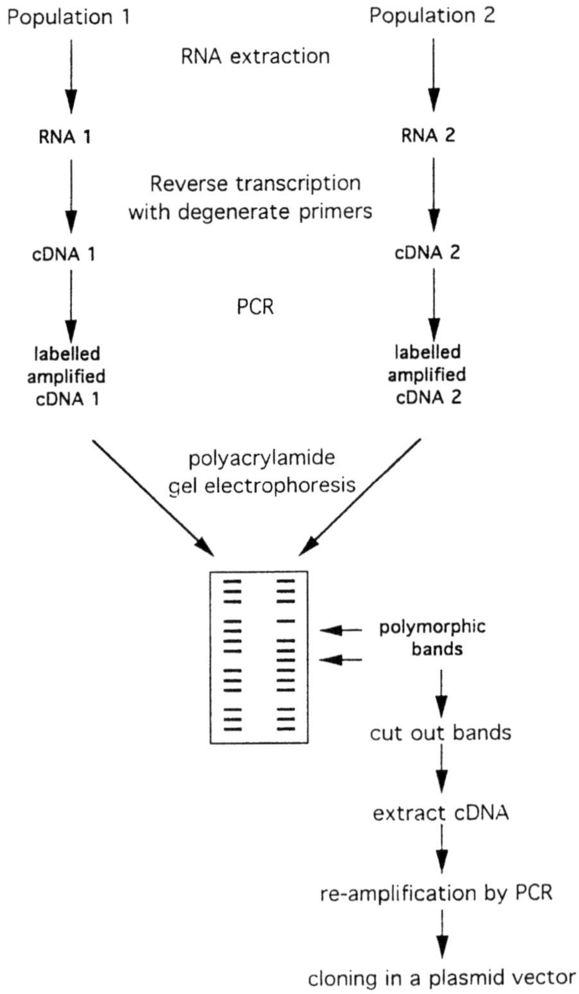

Fig. 3.1.7. Isolating differentially expressed RNAs by DDRT-PCR

initiate reverse transcription allowed the subdivision of the cDNA pool into only four fractions (Liang et al. 1994). Furthermore, Oh et al. (1995) have shown that the use of non-degenerate primers for the PCR reaction only, may lead to a better pattern of products.

The second primer used for amplification is a 10-mer random primer with a GC content of about 50%.

Reagents

- 200 ng DNA-free RNA
- Four primers ($T_{12}MG$, $T_{12}MC$, $T_{12}MT$, $T_{12}MA$, 50 μM stock solution)
- DEPC-SDW
- 10x RT buffer (use conditions preconized by the supplier)
- Taq DNA polymerase (Perkin-Elmer Cetus)
- 10x Taq buffer (use conditions preconized by the supplier)
- M-MLV reverse transcriptase
- dNTP mixes (10 μM, 0.4 mM and 20 mM stock solutions)
- 1 M $MgCl_2$
- 3 M sodium acetate
- 70 and 100% ethanol
- α^{32}P-dCTP (or α^{35}S-dATP or α^{33}P-dCTP)
- 10-mer random primer

Equipment

- 37, 65, 95 and 100 °C water bath
- Ice
- 1.5 microfuge tubes
- Microfuge
- Micropipettes (1–20, 1–200 and 200 μl-1 ml, 1–5 ml capacity plus tips)
- Thermocycler
- Vacuum desiccator
- Cutter
- 3 M Whatmann filter paper
- –20 °C freezer
- X-rays films
- Autoradiographic cassettes with intensifying screens
- Material for developing autoradiographs

Procedure

The reverse transcription and PCR conditions are derived from Bauer et al. (1993). Isolate total RNA (Sect. 3.1.2) and treat them with DNAse.

Reverse Transcription of RNA

1. For each of the four reactions:
   Mix 200 ng of DNA-free RNA with 1 μl of a 50 μM solution of

one of the four primers ($T_{12}MG$, $T_{12}MC$, $T_{12}MT$, $T_{12}MA$) (2,5 μM final). Add DEPC-SDW to a final volume of 10 μl.

2. Denature the template and primers by incubation at 65 °C for 10 min and let them cool at room temperature for 10 min.
3. Spin the tubes briefly in a microfuge.
4. Add 2 μl of 10x RT buffer, 1 μl of a 0.4 mM dNTP solution, 300 units of M-MLV reverse transcriptase, DEPC-SDW up to 20 μl. Incubate at 37 °C for 60 min.
5. Inactivate reverse transcriptase by incubation at 95 °C for 5 min.

PCR

6. The resulting cDNAs are diluted 2x prior to PCR.
7. PCR is performed in a final volume of 20 μl containing: 2 μl of cDNA, 2 μl 10x Taq buffer, 0.8 μl (2 μM) of one $T_{12}MN$ primer, 0.5 μM of a 10-mer random primer, 4 μl (2 μM) dNTP (10 μM stock, it is a low concentration compared to normal PCR using 20 μM), $MgCl_2$ if necessary (see note), 5 μCi of α^{32}P-dCTP, 1 unit of Taq polymerase.
   Amplification program
   95 °C for 2 min
   40 cycles: 95 °C for 0.5 min; 42 °C for 1 min; 72 °C for 0.5 min
   72 °C for 3 min
8. Dry down the PCR mixture using a "speed-vac". Redissolve the pellet in 5 μl SDW.
9. Load on a 6% denaturing polyacrylamide gel (Sect. 4.2).
10. After electrophoresis, dry the gel on 3 M Whatman paper (do not fix it!) and expose to an autoradiographic film overnight.

Recovery of Bands of Interest and Re-Amplification

11. Superpose carefully the autoradiograph on the gel. Excise the bands of interest (polymorphic bands) and transfer them into Eppendorf tubes.
    Incubate the bands in 100 μl SDW.

12. Boil the gel slice for 10 min.

13. Eliminate filter paper using a quick spin at high speed.

14. Transfer the supernatant to a new Eppendorf tube.

15. Add 1/10 vol 3 M sodium acetate, 2.5 vol ethanol and 20 µg of glycogen as carrier. Place at –20 °C for at least 2 h.

16. Centrifuge at 12 000 *g* for 20 min at 4 °C. Discard supernatant and wash the pellet with 100 ml of 70% ethanol. Dry the pellet under vacuum.

17. Redissolve the DNA in 10 µl SDW.

18. Perform a second PCR with 5 µl of DNA, using the same primers and PCR conditions except for the reaction volume (50 µl) and dNTP concentration (use 0.5 µl of a 20 mM stock solution).

19. PCR products can then be loaded on an agarose gel (Sect. 1.2) and cloned (Sect. 2.1).

Notes and troubleshooting

- Steps **15** and **16** can be replaced by concentrating the DNA using speed-vacuum centrifugation.
- Because there is no way to check the efficiency of the reverse transcription reaction, the use of very clean RNAs is essential.
- Ensure that the gel and autoradiograph are aligned properly (use fluorescent or labelled marks).
- The use of PCR involves taking precautions to avoid contamination:
  – use clean water (traces of DNA can still be in bottles even after washing and autoclaving)
  – use filter tips to avoid DNA cross-contamination from the pipettes.
- If the primer chosen does not give differential pattern, try another one.
- Double-check PCR reactions to avoid working with false positive bands.

- A single band on the gel may result from the co-migration of different PCR fragments of the same size. We advise you to perform a direct sequencing of the PCR fragments (an unreadable sequence would be the outcome of a mixed template; Reeves et al. 1995). Another solution is the cloning of the PCR products in a vector before sequencing.

- The use of total RNA can lead to artefacts, some of the selected PCR fragments can result from a non-specific amplification of ribosomal RNA. Before sequencing, hybridize the colonies with a rDNA probe.

- The radioelement used for labelling can be either $\alpha(^{35}S)$-dATP, $\alpha(^{32}P)$-dCTP or $\alpha(^{33}P)$-dCTP. The latter gives better resolution and necessitates minimal physical protection (Tokuyama and Takeda 1995; Trentmann et al. 1995). $\alpha(^{35}S)$dATP may lead to very high contamination levels of the thermocycler.

- Many differentially expressed PCR products can be isolated from the same gel.

- Very small amounts of plant material are needed (200 ng of total RNA per reverse transcription); Zimmermann and Schultz (1994) have related the use of DDRT-PCR starting with 50 mouse two-cell embryos, 25 eight-cell embryos or 10 blastocysts.

Acknowledgements. The authors thank M.W. Bianchi for critical reading of the manuscript. PL is supported by an MRT fellowship (Ministère de la Recherche et de la Technologie).

References

Ainsworth (1994) Isolation of RNA from floral tissue of *Rumex acetosa* (Sorrel). Plant Mol Biol Rep 12:198–203

Alwine JC, Kemp DJ, Stark GR (1977) Method for detection of specific RNAs in agarose gels by transfer to diazobenzylomethyl-paper and hybridization with DNA probes. Proc Natl Acad Sci USA 74:5350

Aviv H, Leder P (1972) Purification of biologically active globin messenger RNA by chromatography on oligo(dT)cellulose. Proc Natl Acad Sci USA 69:1408–1412

Bahloul M, Burkard G (1993) An improved method for the isolation of total RNA from spruce tissues. Plant Mol Biol Rep 11:212–215

Bauer D, Müller H, Reich J, Riedel H, Ahrenkiel V, Warthoe P, Strauss M (1993) Identification of differentially expressed mRNA species by an improved display technique (DDRT-PCR). Nucleic Acids Res 21:4272–4280

Belyavsky A, Vinogradova T, Rajewsky K (1989) PCR-based library construction: general cDNA libraries at the level of few cells. Nucleic Acids Res 17: 2919–2933

Berk AJ, Sarp PA (1977) Sizing and mapping of early adenovirus mRNAs by gel electrophoresis of S1 endonuclease-digested hybrids. Cell 12:721

Brown PC, Tlsty TD, Schimkert RT (1983) Enhancement of methotrexate resistance and dihydrofolate reductase gene amplification by treatment of mouse 3T6 cells with hydroxyurea. Mol Cell Biol 3:1097

Brown JWS, Simpson GG, Clark G, Lyon J, Kumar A, Simpson C (1993) Detection of antisense transcripts in transgenic plants by RT-PCR. Plant J 4: 883–885

Chapman AB, Costello MA, Lee F, Ringold GM (1983) Amplification and hormone regulated expression of a mouse mammary tumor virus-Eco gpt fusion plamid in mouse 3T6 cells. Mol Cell Biol 3:1421

Chappell J, Hahlbrock K (1984) Transcription of plant defense genes in response to UV light or fungal elicitor. Nature 311:76–78

Chirgwin JM, Przybyla AE, McDonald RJ, Rutter MJ (1979) Isolation of biologicaly active ribonucleic acid from sources enriched in ribonucleases. Biochemistry 18:5294–5299

Domec C, Garbay B, Fournier M, Bonnet J (1990) cDNA library construction from small amounts of unfractionated RNA: association of cDNA synthesis with Polymerase Chain Reaction Amplification. Anal Biochem 188:422–426

Don RH, Cox PT, Mattick JS (1993) A "one-tube reaction" for synthesis and amplification of total cDNA from small numbers of cells. Nucleic Acids Res 21:783

Dresselhaus T, Lörz H, Kranz E (1994) Representative cDNA libraries from few plant cells. Plant J 5:605–610

Fedorcsak I, Ehrenberg L (1966) Effects of diethyl pyrocarbonate and methyl methane sulfonate on nucleic acids and nucleases. Acta Chem Scand 20:107

Förster E (1994) An improved general method to generate internal standards for competitive PCR. Biotechniques 16:18–20

Herrin DL, Schmidt GW (1988) Rapid, reversible staining of northern blots prior to hybridization. Biotechniques 6:196

Kolls J, Deininger P, Cohen JC, Larson J (1993) cDNA equalization for reverse trancription-polymerase chain reaction quantification. Anal Biochem 208:264–269

Lehrach H, Diamond D, Wozney JM, Boedtker H (1977) RNA molecular weight determination by gel electrophoresis under denaturing conditions, a critical reexamination. Biochemistry 16:4743

Liang P, Pardee AB (1992) Differential display of eucaryotic messenger RNA by means of the polymerase chain reaction. Science 257:967–971

Liang P, Zhu WM, Zhang XY, Guo Z, Oconnell RP, Averboukh L, Wang FL, Pardee AB (1994) Differential display using one-base anchored oligo-dT primers. Nucleic Acids Res 22:5763–5764

Logemann J, Schell J, Willmitzer L (1987) Improved method for the isolation of RNA from plant tissues. Anal Biochem 163:16–20
Luthe DS, Quatrano RS (1980) Transcription in isolated wheat nuclei, I and II. Plant Physiol 98:503–518
McMaster GK, Carmichael GG (1977) Analysis of single- and double-stranded nucleic acids on polyacrylamide and agarose gels by using glyoxal and acridine orange. Proc Natl Acad Sci USA 74:4835
Marechal-Drouard L, Small I, Weil J-H, Dietrich A (1995) Transfer RNA import into plant mitochondria. Methods Enzymol 260:310–327
Marzluff WF, Huang RCC (1984) Transcription of RNA in isolated nuclei. In: Hames BD, Higgins SJ (eds) Transcription and translation. A practical approach. IRL Press, Oxford, Washington DC, pp 89–129
Oh BJ, Balint DE, Giovannoni JJ (1995) A modified procedure for PCR-based differential display and demonstration of use in plants for isolation of genes related to fruit ripening. Plant Mol Biol Rep 13:70–81
Ray BK, Ray A (1991) Synthesis of a highly efficient cDNA library. Nucleic Acids Res 19:4559
Reeves SA, Rubio MP, Louis DN (1995) General method for PCR amplification and direct sequencing of mRNA differential display products. Biotechniques 18:18–20
Rychlik W, Spencer WJ, Rhoads RE (1990) Optimization of the annealing temperature for DNA amplification in vitro. Nucleic Acids Res 18:6409–6412
Sambrook J, Fritsch EF, Maniatis T (1989) Molecular cloning: a laboratory manual, 2nd edn. Cold Spring Harbor Laboratory Press. Cold Spring Harbor, pp F.4-F.5
Schultz DJ, Craig R, Cox-Foster DL, Mumma RO, Medford JI (1994) RNA isolation from recalcitrant plant tissue. Plant Mol Biol Rep 12:310–316
Seed B (1982) Attachment of nucleic acids to nitrocellulose and diammonium-substituted supports. In: Setlow JK, Hollaender A (eds) Genetics engineering: principles and methods, vol.4. Plenum, New York, p91
Sharma YK, Davis KR (1995) Isolation of a novel *Arabidopsis* ozone-induced cDNA by differential display. Plant Mol Biol 29:91–98
Siebert PD, Larrick W (1992) Competitive PCR. Nature 359:557–558
Song P, Yamamoto E, Allen RR (1995) Improved procedure for differential display of transcripts from cotton tissues. Plant Mol Biol Rep 13:174–181
Southern E (1975) Detection of specific sequences among DNA fragments separated by gel electrophoresis. J Mol Biol 98:503–518
Steinmüller K, Batschauer A, Apel K (1986) Tissue-specific and light-dependent changes in chromatin organization in barley. Eur J Biochem 158:519–525
Thomas M, Cretin C, Vidal J, Keryer E, Gadal P (1990) Light regulation of phosphoenolpyruvate carboxylase mRNA in leaves of C4 plants: evidence for phytochrome control on transcription and circadian rythmicity. Plant Sci 69:65–78
Tobin EM, Klein AU (1975) Isolation and translation of plant messenger RNA. Plant Physiol 56:88–92
Tokuyama Y, Takeda J (1995) Use of ^{33}P-labeled primer increases the sensitivity and specificity of mRNA differential display. Biotechniques 18:424–425
Trentmann SM, Van der Knaap E, Kende H (1995) Alternatives to ^{35}S as a label for differential display of eucaryotic messenger RNA. Science 267:1186

Zimmermann JW and Schultz RM (1994) Analysis of gene expression in the preimplantation mouse embryo: use of mRNA differential display. Proc Natl Acad Sci USA 91:5456–5460, (Correction) 91:9664 (1994)

3.2 Subtractive Hybridization of Different mRNA Populations

H.C.C. FOOTE, G. BRADY, G.J. THORLBY, and C.H. FRANKLIN

3.2.1 Introduction

The study of eukaryotic gene structure and expression is dependent on the isolation or cloning of the genes concerned. The first stage in the cloning of a particular gene is often the isolation of a DNA copy of the messenger RNA (mRNA) encoded by the gene. This copy, known as a complementary or cDNA clone, may be identified as follows. A cDNA library is constructed which represents the mRNA population of the appropriate tissue. The clone of interest is then identified by screening the library with a labelled probe. The probe used will vary according to the information available concerning the gene of interest. Firstly a synthetic oligonucleotide or antibody probe may be utilized. However, if neither is available, the library may be screened with probes representing differentially expressed mRNA (Huyn et al. 1985).

The effectiveness of differential screening depends on the abundance of the target gene in its native tissue. It is generally accepted that a level of at least 0.1% of total mRNA, which represents highly abundant transcripts (Timblin et al. 1990), is required for successful isolation by differential screening. Consequently, few low abundance differentially expressed messages are identified, and also the number of unique clones isolated will be limited before re-isolation of previously identified transcripts predominates (Hodge et al. 1992).

The production of subtracted cDNA probes is believed to permit the identification of moderately abundant species, representing down to approximately 0.01% of mRNA transcripts (Sambrook et al. 1989). In this procedure, a cDNA probe prepared from one pool of mRNA is depleted of common sequences present in a second pool of mRNA by subtractive hybridization (Timberlake 1980). A number of more complex subtractive hybridization protocols have now been developed, but typically a "tracer" cDNA pool is hybridized two or three times in succession to a 20-fold excess of a second pool of "driver" mRNA. The cDNA:mRNA hybrids formed are then removed, originally by chromatography on hydroxyapatite (HAP). The remaining un-hybridized cDNA pool which is enriched for sequences specific to the original tracer mRNA is then used to probe a cDNA library or may itself be used in the construction of an enriched or subtracted cDNA library (e.g. Timblin et al. 1990).

There are a number of limitations to this type of procedure, some of which have now been overcome in a variety of ways. First, the use of mRNA at the subtraction stage requires extra care due to its instability, especially at the high temperatures used in HAP chromatography (e.g. 60 °C). Second, the HAP chromatography procedure itself is complicated and involves dilution of the un-hybridized sample (Hara et al. 1991). Third, a large amount of driver mRNA is required which may be difficult to obtain if the tissue required is in limited supply (Lebeau et al. 1991). Finally, the amount of tracer cDNA left after successive rounds of subtraction may be severely restricted. The latter two difficulties have led to the application of polymerase chain reaction (PCR) technology in subtractive hybridization, firstly in the production of cDNA pools from small amounts of starting tissue (Belavysky et al. 1989) and also in the amplification of subtracted tracer cDNA (Duguid and Dinauer 1989). The use of PCR technology in itself does, however, bring further difficulties as it is apparent that not all DNA sequences are amplified with the same efficiency (Elrich 1989). The preferential amplification of smaller molecules may lead to the under-representation of larger species in PCR produced in vitro libraries (McConlogue et al. 1988), and will also decrease the efficiency of subtraction. A final complication that exists in many protocols is that of the ligation of linker molecules to the subtracted cDNA before PCR amplification or cloning into a vector. Although probably not a major limiting step, this stage can be circumvented.

The simple and efficient PCR-based subtractive hybridization protocol described here overcomes all of the problems mentioned above. The protocol has been used to isolate stigma-specific sequences from the field poppy, *Papaver rhoeas* L., by subtracting sequences common to leaf (driver) and stigma tissues from stigma (tracer) sequences.

The starting point for this subtractive hybridization protocol is that of the production of "representative" cDNA from small amounts of total RNA (Brady et al. 1990). In this procedure, synthesis of first strand cDNA is primed with oligo(dT) and the resulting strands are tailed with poly(dA). The cDNA is then amplified by PCR using a single oligo(dT) containing stabilizer sequence. By limiting the length of the first strand synthesis, efficient amplification is achieved independent of the size of the original mRNA transcripts. Brady et al. (1990) went on to demonstrate that the relative abundance of sequences was preserved during amplification.

The complete subtractive hybridization procedure is performed as follows. First, two pools of cDNA are produced as above, using an independent PCR primer for each tissue type. Each PCR primer contains a unique restriction site to facilitate subsequent cloning. During PCR, the driver pool is biotinylated by the inclusion of biotinylated nucleotide (Bio-11-dUTP) in the reaction mixture (Lo et al. 1988). A 20-fold excess of biotinylated driver cDNA is then hybridized (following denaturation of both cDNA pools) with 200 ng of tracer cDNA. Biotinylated hybrids representing re-hybridized driver sequences and sequences common to both driver and tracer are then removed by the addition of streptavidin (which has a high affinity for biotin) followed by phenol chloroform extraction (Sive and St. John 1988). The remaining subtracted tracer cDNA is then re-amplified using the tracer-specific primer before being subjected to further hybridization and subtraction cycles. After the final subtraction the remaining subtracted tracer is re-amplified and may be tested for tracer tissue specificity.

3.2.2 RNA Preparation

Extract total RNA by standard procedures and check its integrity by (1) gel electrophoresis, and one of; (2) in vitro translation or (3) the ability to synthesize large first strand cDNA molecules (Sambrook et al. 1989). Dilute the total RNA to 120 ng/ml in diethyl pyrocarbonate (DEPC)-treated sterile distilled water (see Sect. 3.1 for details).

3.2.3 Production of Tailed First Strand cDNA

Reagents

- Driver RNA
- Tracer RNA
- Reverse transcriptase

- Reverse transcriptase buffer
- Human placental (or alternative) ribonuclease inhibitor
- Oligo d(T) 19–24
- Ultrapure deoxyribonucleotides (dNTPs: dATP, dCTP, dGTP, dTTP) e.g. Pharmacia
- Terminal transferase
- Terminal transferase buffer

Equipment

- Microfuge tubes (1.5 ml)
- Water bath (65 °C, 22 and 37 °C)
- Ice bucket

Procedure

Perform the following reactions using both RNA samples separately:

Limited First Strand cDNA Synthesis Reaction

2.0 μl diluted total RNA (120 ng/ml)
2.0 μl reverse transcriptase buffer (5x)
0.5 μl Human Placental RNAse Inhibitor (25 000 units/ml)
1.0 μl oligo dT 19–24 (1000 ng/ml)
2.0 μl dNTPs mix (dATP, dCTP, dGTP and dTTP each at 10 μM)
0.5 μl DEPC-treated water

1. Incubate this mixture at 65 °C for 1 min, then 22 °C for 2 min.
2. Place the tube on ice, then add 2 μl of reverse transcriptase (200 units/μl). Incubate at 37 °C for 10 min, then heat-denature the enzyme by incubating at 65 °C for 10 min.
3. Finally place the tube on ice. If the tailing reaction is not performed immediately store the first strand mixture at –70 °C.

Tailing Reaction

10 μl first strand reaction
4.0 μl terminal transferase buffer (5x)
4.0 μl dATP (1 mM)
10 units terminal transferase
Make up to 20 μl with sterile distilled water

1. Prepare the tailing reaction mixture listed above then incubate at 37 °C for 15 min.

2. Heat denature the enzyme by incubation at 65 °C for 10 min, before placing the tube on ice.

Notes and troubleshooting

- The use of a low concentration of primers and nucleotides, in addition to the short incubation with reverse transcriptase, serves to limit the length of the first strand cDNA. The low concentration of d(T) primer also minimizes its use as a template in subsequent tailing and amplification reactions.

3.2.4 Amplification of Representative cDNA Pools

Reagents

- Taq polymerase
- 10x PCR buffer:
  100 mM Tris/HCl (pH 8.3)
  500 mM KCl
  15 mM $MgCl_2$
  0.01% gelatin (Sigma)
  0.5% Tween (Sigma)
  0.5% Nonidet (Sigma)
- EcoRI PCR primer:
  5′-ATG/TGC/TCC/AGG/CCG/CTC/TGG/ACA/AAA/TAT/GAA/TTC/T(24)-3′
- Not I primer:
  5′-GCGGCCGC-T(17)-3′
- Bio-11-dUTP (Sigma)
- Mineral oil

Equipment

- Spectrophotometer measuring OD_{260}
- PCR machine
- Microfuge tubes (0.5 ml)
- Gel electrophoresis tank and power pack
- 1x TBE: 0.9 M Tris-borate, 0.02 M Na_2EDTA
- 0.8% agarose gel in 1x TBE containing 0.15 μg/ml ethidium bromide
- Ethidium bromide
- UV transilluminator
- Sephacryl S-300 Spin Column (Pharmacia) or similar

Procedure The two-tailed first strand cDNA pools are now used as templates in a PCR reaction to generate two separate pools of double-stranded cDNA using a different PCR primer for each pool. The driver cDNA is biotinylated at this stage. Prepare the following PCR reaction mixtures and amplify using the same program for each as shown.

Biotinylated Driver

1. Dilute 5 µl of the first-strand driver cDNA to 100 µl to give final concentrations of 1x PCR buffer, 0.2 mM dATP, 0.2 mM dCTP, 0.2 mM dGTP, 0.15 mM dTTP, 0.05 mM Bio-11-dUTP, 2 OD_{260} units/ml Not I primer, and 5 units of Taq polymerase/100µl.
2. Overlay with 80 µl of mineral oil.

Non-Biotinylated Tracer

1. Dilute 5 µl of the first-strand tracer cDNA to 100 µl to give final concentrations of 1x PCR buffer, 0.2 mM each dNTP (dATP, dCTP, dGTP and dTTP), 2 OD_{260} units/ml EcoRI primer, and 5 units of Taq polymerase/100 µl.
2. Overlay with 80 µl of mineral oil.

Amplification

25 cycles:	94 °C, 15 s
	42 °C, 30 s
	72 °C, 3 min
1 cycle:	72 °C, 10 min

1. Following amplification, remove mineral oil by extraction with diethyl ether.
2. Five or 10 µl aliquots may then be analyzed by gel electrophoresis through a 1.0% agarose gel (as Sect. 8.2). A smear of cDNA (usually approximately 10 µg per reaction) should be produced with most of the material ranging in size between 300–700 base pairs (see Fig. 3.2.2a).

3. Before proceeding with the hybridization, pass the reaction mixture through a Sephacryl S-300 Spin Column (or similar) to remove unincorporated biotin, primers and nucleotides.

Notes and troubleshooting

- In addition to providing unique restriction sites for subsequent cloning, the stabilizer (non-oligo d(T)) sequence of the two PCR primers serves to enhance hybridization specificity. The use of pure oligo d(T) primer in PCR results in products which increase in length as amplification progresses presumably due to the ability of homogenous d(T) primer to hybridize to any position in the target poly d(A) sequences.
- Oligonucleotide (primer) samples may vary in their ability to amplify cDNA. For reasons that remain unclear, certain oligonucleotides that have failed to amplify cDNA may be "rescued" by a simple one-step de-ionization using, for example, mixed bed resin TMD-8 (Sigma; Brady and Iscove 1993). The final concentration of the PCR oligonucleotides may also be varied between 1–4 OD_{260} units/ml.
- It may be necessary to optimize $MgCl_2$ concentration in the PCR reaction. Final concentrations of 15–50 mM $MgCl_2$ in increments of 5 mM may be tried for the 10x PCR buffer. We have also found that certain commercially supplied Taq polymerase buffers also work well, for instance Taq XL 10x reaction buffer (NBL) (H. Foote, unpubl. results).
- It is important to remove unincorporated biotin before proceeding with hybridization, as this could compete for streptavidin and thus decrease the efficiency of subtraction. Likewise, carry over of driver primer could lead to amplification of any remaining driver cDNA during re-amplification of the subtracted tracer. This cDNA purification stage should thus be performed carefully.
- It is best in the long term (longer than 2 months) to store amplified cDNA at –70 °C due endonuclease activity inherent to Taq polymerase. For particularly precious samples it is recommended to store cDNA, as dried down ethanol precipitates which can be resuspended and re-amplified at a later date.
- A 0.5 mM solution of Bio-11-dUTP should be prepared and stored dark at –20 °C. It is also easiest to prepare appropriate mixed stocks of deoxyribonucleotides rather than adding each separately.

3.2.5 Hybridization

Reagents

- tRNA
- 10x hybridization buffer:
  100 mM EPPS (Sigma, E-9502) (pH 8.25),
  10 mM Na_2EDTA, 1% SDS
- 3 M sodium acetate (NaOAC)
- 100% ethanol (EtOH)
- 5 M NaCl
- Mineral oil

Equipment

- Spectrophotometer measuring OD_{260}
- Microfuge tubes (1.5 ml)
- Microcentrifuge
- –70 °C freezer
- 68 °C water bath

Procedure

1. Place 200 ng of purified tracer cDNA in a 1.5-ml microfuge tube, then add 4 µg of purified biotinylated driver cDNA and 5 µg of tRNA.

2. Precipitate by adding a 1/10 vol NaOAc, 2 vol EtOH and incubating at –70 °C for 30 min.

3. Microcentrifuge for 10 min at 12 000 g. Wash pellet in ice-cold 70% ethanol. Air-dry the pellet, then resuspend in 4 µl of 5/4x hybridization buffer.

4. Add 1 µl of 5 M NaCl then 80 µl of mineral oil as an overlay. Boil for 5 min, then incubate the tube at 68 °C for 48 h to allow for hybridization.

Notes and troubleshooting

- A 1-M stock of EPPS should be prepared and titrated to pH 8.25 with concentrated NaOH, then autoclaved. The 10x hybridization buffer should be prepared from the above stock and filter-sterilized with a 0.22-micron filter.
- It is not advisable to vacuum dry the precipitated cDNA, as the pellet then becomes difficult to resuspend.

3.2.6 Subtraction

Reagents

- Extraction buffer: 50 mM EPPS (pH 8.25), 500 mM NaCl, 2 mM Na_2EDTA
- 4 mg/ml streptavidin (Gibco, BRL) in 50 mM EPPS (pH 8.25)
- Phenol/chloroform (1:1)
- Diethyl ether

Equipment

- Microfuge tubes (1.5 ml)
- Microcentrifuge

Procedure

1. To the hybridization reaction add 100 µl of extraction buffer. Centrifuge briefly, then remove excess mineral oil.
2. The subtraction process is performed as follows. Add 4 µl streptavidin (4 mg/ml in 50 mM EPPS (pH 8.25), Gibco/BRL), incubate the tube for 2 min at room temperature, then extract the reaction with 100 µl of phenol/chloroform. Repeat the subtraction process twice more.
3. Perform an extra phenol/chloroform extraction to remove trace amounts of streptavidin, then extract the aqueous phase once with diethyl ether.

Notes and troubleshooting

- The phenol/chloroform extraction partitions the streptavidin-biotin-driver complexes at the phenolic interface, leaving the non-hybridized subtracted tracer cDNA in the aqueous phase. Care should thus be taken to avoid this interface in order to achieve efficient subtraction. The disadvantage of losing a small amount of tracer cDNA is overcome by re-amplifying prior to further hybridization cycles.
- Prior to embarking on a succession of hybridization/subtraction cycles, a pilot experiment, using radiolabelled tracer cDNA, should be performed to give an indication of the efficacy of subtraction. A radioactively labelled pool of stigma tracer cDNA is most easily produced simply by adding 10µCi of (α-^{32}P)-dCTP to a standard cDNA amplification PCR mixture. By monitoring the amount of radioactivity in the phenol chloroform extractions and in the re-

maining tracer cDNA, and by running a parallel hybridization/subtraction experiment in which no driver cDNA is added, the specific removal of labelled tracer cDNA due to subtraction can be monitored.

3.2.7 Re-Amplification

After subtraction, the subtracted tracer cDNA is re-amplified using the tracer-specific primer. Reagents and equipment are as for Section 3.2.4.

Procedure Set up the following PCR reaction and amplify using the program shown below.

1. Dilute 5 µl of the subtracted cDNA to 100 µl to give final concentrations of 1x PCR buffer, 0.2 mM each dNTP (dATP, dCTP, dGTP and dTTP), 2 OD_{260} units/ml EcoRI primer, and 5 units of Taq polymerase/100 µl.
2. Overlay with 80 µl of mineral oil.
3. PCR re-amplification programme:
   25 cycles: 94 °C, 15 s
   49 °C, 30 s
   72 °C, 3 min

   1 cycle: 72 °C, 10 min

Notes and troubleshooting

- A PCR primer containing only the stabilizer fragment may be prepared and used here in order to eliminate the amplification of remaining driver cDNA through cross-priming to the poly d(A) region of the driver cDNA. We did not find this to be an essential precaution, however.

3.2.8 Further Hybridization/Subtraction Cycles

Following purification (Sephacryl S-300 spin column) of PCR amplified subtracted tracer cDNA, estimate the DNA concentration at OD_{260} using a spectrophotometer, then repeat the procedures in Sections 4, 5 and 6 using 200 ng of purified amplified subtracted cDNA in place of purified tracer cDNA. In our experience, three or four rounds of hybridization and subtraction have provided satisfactory enrichment levels.

3.2.9 Analysis of Subtracted cDNA

In order to test the success of the subtractive hybridization procedure, the amplified products of successive rounds of subtraction may be analyzed in various ways. We find the most satisfactory to be the production of a slot-blot and Southern blots containing original tracer cDNA and the amplified products of each successive round of subtraction. The filters produced may then be probed with labelled driver cDNA and a tracer-specific gene if available. The following are methods used in our laboratory; other standard procedures may also be used.

Slot-Blot Production

Reagents

- 20x SSC: 3 M NaCl, 0.3 M Na_3citrate
- Denaturation buffer: 0.5 M NaOH, 1.5 M NaCl
- Neutralization buffer: 1.5 M NaCl, 0.5 M Tris/HCl (pH 7.4)

Equipment

- Slot-blot apparatus
- DNA binding membrane (e.g. Hybond N, Amersham)
- Whatman 3-MM chromatography paper

Procedure

1. Prepare a fivefold dilution series in water using the original tracer cDNA and the amplified products of each successive subtraction to give samples containing 1, 1/5, 1/25 and 1/125 µg of cDNA (see Fig. 3.2.1).
2. Heat cDNA samples to 95 °C then place on ice.
3. Add an equal volume of 20x SSC to each sample.
4. Transfer samples onto a pre-wetted (in 10x SSC) DNA binding membrane using the slot-blot apparatus according to the manufacturer's instructions.
5. Place the membrane on a wad of Whatman 3-MM paper soaked in denaturation solution for 5 min.

Fig. 3.2.1. Slot-blot of subtracted tracer cDNA probed with total driver cDNA. *Horizontal numbers 1–1/125* refer to a fivefold dilution series, where 1 = 1 μg. *Vertical numbers 0–3* refer to successive rounds of subtractive hybridization, where 0 = non-subtracted original tracer cDNA

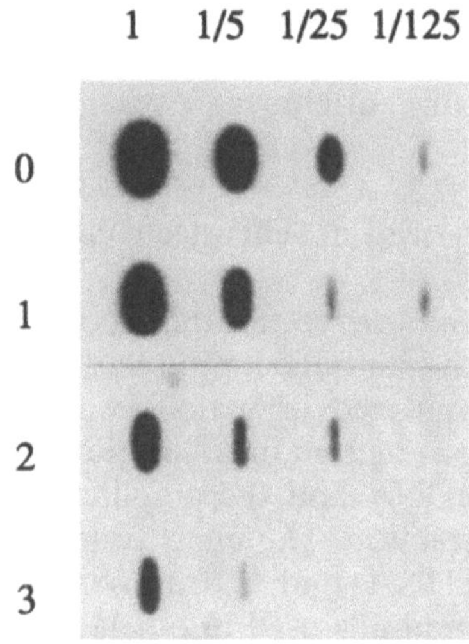

6. Transfer to a wad of Whatman 3-MM paper soaked in neutralization solution for 1 min.
7. Air-dry the membrane on a dry sheet of Whatman 3-MM paper before fixing the DNA to the membrane according to the manufacturer's instructions.

Southern Blot Production

Detailed instructions for southern blotting have already been given in Section 1.2. Prepare a duplicated Southern blot using (approximately 5 μg/lane) original tracer and driver cDNAs and the product of the final subtraction (see Fig. 3.2.2a).

Probe Labelling

Label 200 ng of the original amplified driver cDNA or a tracer specific gene using standard procedures. See Section 1.3 for further details.

Hybridization of Radiolabelled Probe

Reagents

- 100x Denhardt's solution: 2% (w/v) Ficoll
  2% (w/v) polyvinyl-pyrrolidone

	2% (w/v) bovine serum albumin (fraction V)
– Pre-hybridization buffer:	50% formamide, 25x Denhardt's solution, 50 mM $NaPO_4$ buffer, 1% glycine, 500 ng/ml denatured salmon sperm DNA
– Hybridization buffer:	50% formamide, 5x SSC, 1x Denhardt's solution, 20 mM $NaPO_4$ buffer, 100 ng/ml denatured salmon sperm DNA, 10% dextran sulphate

Equipment

- Hybridization oven (e.g. Hybaid)
- Tubes or plastic lunch boxes
- Temperature-controllable water bath

Procedure

The slot-blot and/or Southern membranes prepared as described above are hybridized in the same way as follows:

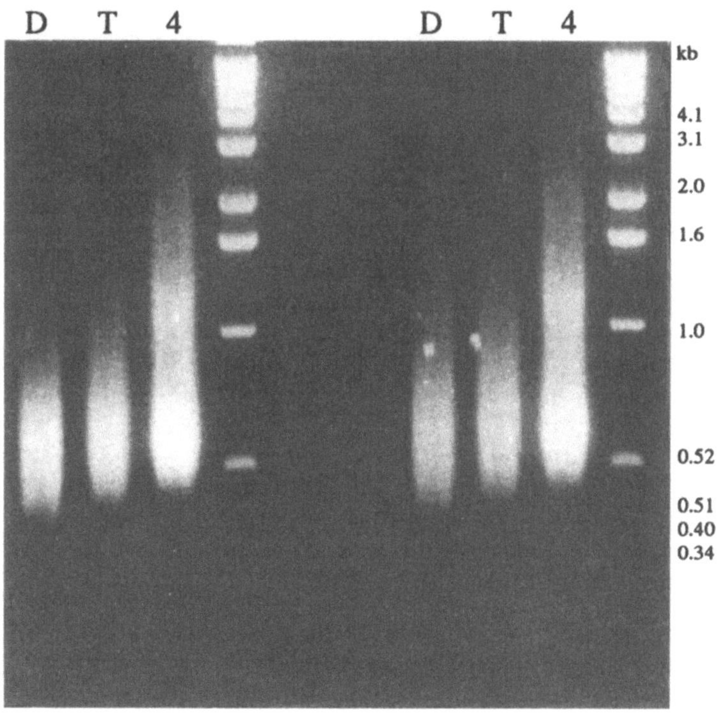

Fig. 3.2.2a. Photograph of gel containing original driver cDNA (*D*), original tracer cDNA (*T*) and the product of the fourth subtraction (*4*). Lanes contain 6 µg of each cDNA sample

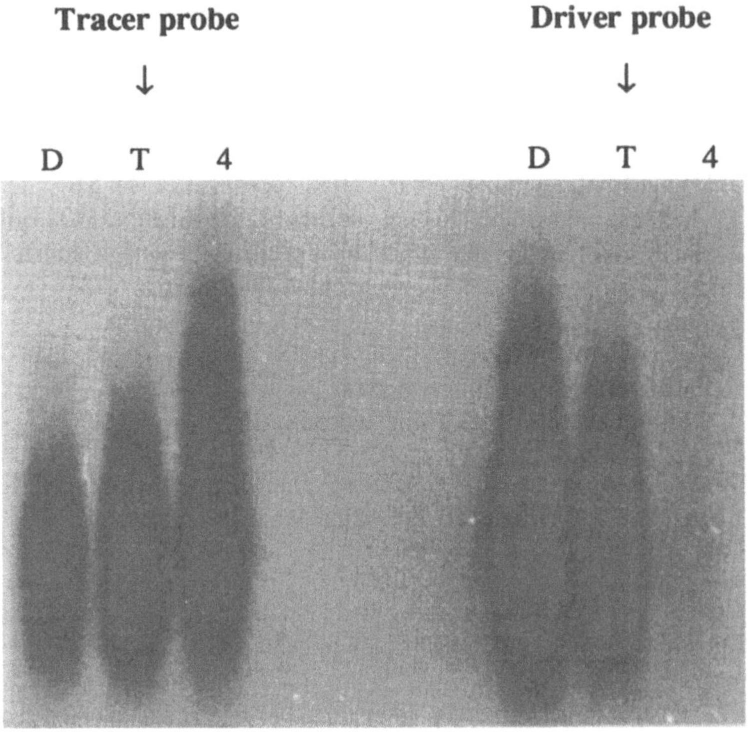

Fig. 3.2.2b. Southern hybridization of subtracted cDNA (gel in **a**), using total driver and total tracer cDNA probes separately. Lanes contain 6 μg each of original driver cDNA (*D*), original tracer cDNA (*T*) and the product of the fourth subtraction (4)

1. Incubate filters in 30 ml of pre-hybridization buffer at 42 °C for 4 h in a hybridization oven (or lunch box shaking in a water-bath).

2. Decant off this buffer and replace it with 15 ml of hybridization buffer. Add denatured radioactive probe DNA to the hybridization buffer, then incubate in the same way at the same temperature for approximately 16 h.

3. Decant off radioactive hybridization buffer then wash the filters in a plastic lunch box placed in a waterbath. Use decreasing concentrations of SSC with 0.1% SDS at increasing temperatures to give successively more stringent washes as necessary to remove background hybridization. For example 2x SSC, 0.1% SDS at 37 °C for 20 min followed by 0.5x SSC, 0.1% SDS at 50 °C for 20 min, followed by 0.1x SSC, 0.1% SDS at 65 °C for 20 min, may be used with monitoring of background radioactivity between washes using a Geiger counter.

4. Expose the filters to X-ray film until the desired signals are produced.

Typical results

- *Analysis of subtraction products using total driver and total tracer cDNA as probes.* The autoradiograph in Fig. 3.2.1, shows a slot-blot, produced as described in Section 1.2 probed with total driver cDNA. With each successive round of hybridization/subtraction, a decrease in hybridized signal is apparent. This corresponds to a decrease in driver homologous sequences due to the subtraction of sequences common to both driver and tracer from original tracer cDNA. To confirm that the remaining subtracted material is of tracer tissue origin, it is desirable to differentially screen the product of the fourth subtraction along with appropriate controls, using total driver and total tracer cDNAs as probes.

  The photograph in Fig. 3.2.2a, shows a gel produced as described in Section 1.2 and demonstrates the equal relative loading of approximately 6 µg per lane. Once blotted, the filter produced was cut in two and hybridized separately under the same conditions; the first half with a total driver cDNA probe, and the second identical half with a total tracer cDNA probe. Attempts were made to label each probe to the same specific activity. The two filters were then exposed together to the same X-ray film. The resulting autoradiograph is shown in Fig. 3.2.2b. The driver and tracer lanes were included as internal controls and demonstrate that driver probe to driver cDNA gives a stronger signal than driver probe to tracer cDNA, and that tracer probe to tracer cDNA gives a stronger signal than tracer probe to driver cDNA. More interestingly, however, the autoradiograph shows that the tracer probe to the fourth subtraction cDNA gives a far stronger signal than does the driver probe to the fourth subtraction cDNA. This clearly demonstrates the removal of driver sequences by the subtractive hybridization and the powerful enrichment for tracer-specific sequences.

- *Analysis of the subtraction products using a tracer specific probe.* The autoradiograph in Fig. 3.2.3, shows the same slot-blot as that in Fig. 3.2.1, following stripping and reprobing with a tracer-specific gene. In this case, the clearly stigma (tracer) specific *Papaver rhoeas* self-incompatibility (S-) gene was used (Foote et al. 1994). With

Fig. 3.2.3. Slot-blot of subtracted tracer cDNA probed with the (tracer-specific) S_1-cDNA sequence. *Horizontal numbers 1–1/125* refer to a fivefold dilution series, where 1 = 1 µg. *Vertical numbers 0–3* refer to successive rounds of subtractive hybridization, where 0 = non-subtracted original tracer cDNA

Fig. 3.2.4. Southern hybridization of subtracted cDNA material (gel in Fig. 3.2.2a) using the (tracer-specific) S_1-cDNA sequence as a probe. Lanes contain 6g each of original driver cDNA (*D*), original tracer cDNA (*T*) and the product of the fourth subtraction (*4*)

successive hybridization/subtraction cycles, the signal produced increases in intensity corresponding with an enrichment for tracer specific sequences by the removal of sequences common to driver and tracer. Figure 3.2.4, shows the autoradiograph produced after the stripping of the same filter Southern filter as shown in Fig. 3.2.2b. and the re-probing of (one half of) it with the tracer specific gene. This demonstrates the very strong enrichment for this tracer specific sequence in the amplified product of the fourth subtraction in comparison to the original tracer cDNA, where a much weaker signal is seen, and to the driver cDNA, where no signal whatsoever is detected.

3.2.10 Cloning to Form the Subtracted Library

Reagents

- EcoRI restriction enzyme
- EcoRI restriction enzyme buffer (supplied with enzyme)
- TE buffer: 10 mM Tris/HCl, 1mM Na_2EDTA
- Dephosphorylated vector
- Phage packaging extract and *E. coli* cells as necessary, depending on vector choice

Equipment

- Microfuge tubes (1.5 ml)
- Temperature controllable water bath
- Sephacryl S-300 Spin Column (Pharmacia) or similar

Procedure

The subtracted cDNA may now be cloned to form a subtracted library. Before this it is advisable to perform an additional polymerisation reaction to maximize the proportion of double-stranded, clonable material.

Additional Polymerization

1. Dilute 20 µl of amplified subtracted cDNA to 100 µl to give final concentrations of 1x PCR buffer, 0.2 mM each dNTP (dATP, dCTP, dGTP and dTTP), and 2 OD_{260} units/ml EcoRI primer. Overlay with 80 µl of mineral oil.
2. Boil the mixture for 2 min, then incubate for 4 min at 42 °C.
3. Add 5 units of Taq polymerase and incubate at 72 °C for 30 min.
4. Extract with ether to remove mineral oil then ethanol precipitate by adding 1/10 vol 3 M NaOAc and 2 vol cold ethanol. Place the tube at –20 °C for 30 min then centrifuge at 12 000 *g*.
5. Wash the pellet twice in 70% EtOH, then vacuum-dry.

Restriction Digestion

1. Resuspend the pellet in 100 μl of 1x restriction digest buffer then add 100 units of EcoRI and incubate for 4 h to overnight.
2. Separate the liberated "tag" fragment from the subtracted cDNA by passing the reaction twice through a suitable separation column (e.g. Sephacryl S-300 spin column).
3. Ethanol precipitate, wash and dry (as Sect. 10.1) the cDNA once more and finally resuspend in 50 μl of TE buffer.

Aliquots of the cDNA are now ready to be cloned into a suitable commercially available de-phosphorylated vector using standard procedures (Sambrook et al. 1989, or the vector supplier's instructions). Further information regarding the optimization of ligation reactions, and transformation into competent bacteria or packaging into lambda phage particles is given in Brady and Iscove (1993).

Notes and roubleshooting

- When the cDNA is completely digested, there will be a slight reduction in the overall size of the cDNA as viewed by electrophoresis. The "tag" fragment, which represents the 36-base pair stabilizer fragment of the amplification primer, should also become visible. This fragment possesses EcoRI ends and will therefore compete with the cDNA during vector ligation if not removed by the sizing column.
- The cloning of subtracted cDNA prepared as described above routinely produces libraries containing approximately 10^6 clones/μg cDNA for bacterial vectors, and 10^9 clones/μg cDNA for lambda vectors. See Section 2.3 for lambda cloning protocols.

3.2.11 Concluding Comments

- There are a number of ways in which the level of enrichment for a specific gene may be improved. First, a tissue with a more similar mRNA profile may be used to provide the driver. However, care must be taken in ensuring that the target gene is not simply expressed at a lower level in the driver tissue, as the large excess of driver to tracer may result in the subtraction of the target gene itself.

- Another way of increasing the enrichment of low abundance sequences is to use a self-subtracted driver following successive subtractions with the original driver. Highly abundant mRNA species, for example those encoding chlorophyll a/b binding protein, the small subunit of Rubisco, together represent a large proportion of the mRNA population in plants (Broglie et al. 1981). By self-subtracting through just one round, the most highly abundant sequences will be removed. Thus in using self-subtracted driver, low abundance common sequences still present in the subtracted tracer should be removed, therefore further enriching for low abundance target sequences.
- At the technical level, many of the steps may be further optimized. The hybridization is probably the stage in the protocol with the greatest potential for improvement. One possible approach would be to carry the hybridization to higher Cot values (see Hames and Higgins 1988), which can be accomplished by increasing either the hybridization time or nucleic acid concentration. However, this approach may be limited by nucleic acid solubility and the increased possibility of thermal damage to the cDNA.
- The subtraction stage could also be further optimized. For instance, the use of photoactivatable biotin in biotinylation of the driver (Sieve and St. John 1988) has been shown to be more efficient than incorporation of biotin during PCR (G. Brady, unpubl. results) and thus improves the efficiency of subtraction. Most of the other stages in this protocol appear to work efficiently.

Acknowledgements. This protocol was developed through the adaptation of various reports including those listed in the references and from a personal communication with Si Lok.

References

Belyavsky A, Vinogradova T, Rajewsky K (1989) PCR-based cDNA library construction: general cDNA libraries at the level of a few cells. Nucleic Acids Res 17:2919–2932

Brady G, Iscove N (1993) Construction of cDNA libraries from single cells. Methods Enzymol 225:611–623

Brady G, Barbara M, Iscove NN (1990) Representative cDNA amplification from individual hemopoietic cells and colonies. Methods Mol Cell Biol 2:17

Broglie R, Bellemare G, Bartlett S, Chua N-H, Cashmore AR (1981) Cloned DNA sequences complementary to mRNAs encoding precursors to the small subunit of ribulose-1,5-bisphosphate carboxylase and a chlorophyll a/b binding polypeptide. Proc Natl Acad Sci USA 78:7304–7308

Duguid JR, Dinauer MC (1989) Library subtraction of in vitro cDNA libraries to identify differentially expressed genes in scrapie infection. Nucleic. Acids Res 18:2789–2792

Elrich HA (1989) PCR Technology. Stockton Press, New York

Feinberg AP, Vogelstein B (1983) A technique for radiolabelling restriction endonuclease fragments to high specific activity. Anal Biochem 132:6–13

Feinberg AP, Vogelstein B (1984) Addendum: A technique for radiolabelling restriction endonuclease fragments to high specific activity. Anal Biochem 137:266–267

Foote HCC, Ride JP, Franklin-Tong VE, Walker EA, Lawrence MJ, Franklin FCH (1994) The cloning and expression of a distinctive class of self-incompatibility (S) gene from *Papaver rhoeas* L. Proc Natl Acad Sci USA:91 2265–2269

Hames BD, Higgins SG (1988) Nucleic acid hybridization: practical approach. IRL Press, Oxford

Hara E, Kato T, Nakada S, Sekiya S, Oda K (1991) Subtractive cDNA cloning using oligo$(dT)_{30}$-latex and PCR: isolation of clones specific to undifferentiated human embryonal carcinoma cells. Nucleic Acids Res 19:7097–7104

Hodge R, Paul Y, Draper J, Scott R (1992) Cold-plaque screening: a simple technique for the isolation of low abundance, differentially expressed transcripts from conventional cDNA libraries. Plant J 2:257–260

Huyn TV, Young RA, Davis RW (1985) Construction screening of cDNA libraries in λgt ID and λgt 11. In: Glover DM (ed) DNA cloning vol 1. A practical approach. IRL Press, Oxford

Lebeau M-C, Alvarez-Bolado G, Wahli W, Catsicas S (1991) PCR driven DNA-DNA competitive hybridization: a new method for sensitive differential cloning. Nucleic Acids Res 19:4778

Lo Y-MD, Mehal WZ, Kleming KH (1988) Rapid production of free biotinylated probes using the polymerase chain reaction. Nucleic Acids Res 16:8719

McConlogue L, Brown MAD, Innis MA (1988) Structure-independent DNA amplification by PCR using 7-DEAZA-2′-deoxyguanosine. Nucleic Acids Res 16:9869

Sambrook T, Fritsch EF, Maniatis T (ed) (1989) Molecular cloning, a laboratory manual. Cold Spring Harbor Laboratory Press, Cold Spring Harbor

Sive HL, St John T (1988) A simple subtractive hybridisation technique employing photoactivatable biotin and phenol extraction. Nucleic Acids Res 16:10937

Timberlake WE (1980) Developmental gene regulation in *Aspergillus nidulans*. Dev Biol 78:497

Timblin C, Battey J, Kuehl WM (1990) Application for PCR technology to subtractive cDNA cloning: identification of genes expressed specifically in plasmacytoma cells. Nucleic Acids Res 18:1587–1593

Chapter 4

Characterization of Clones

4.1 DNA Sequencing

J.R. CLARK

4.1.1 Introduction

The purpose of this chapter is to introduce the plant researcher with little or no experience of DNA sequencing to a technology central to many molecular biology projects. DNA sequencing technology continues to develop at a rapid rate, and refinements to methods are reported almost daily, largely in response to the demands of ambitious projects aiming to sequence the complete genome of organisms as diverse as man, yeast (*Saccharomyces cerevisiae*), plants (*Arabidopsis*) and nematodes (*Caenorhabditis elegans*). It is beyond the scope of this chapter to provide a comprehensive review of every new development, but by using the protocols provided, good-quality sequence should be obtained routinely. An increasing number of laboratories have access to automated sequencing facilities. However, for many small laboratories, and those in which molecular biology is a new venture, manual methods of DNA sequencing will continue to be required. It is at plant molecular biologists working in these laboratories that this chapter is aimed.

As in so many areas of molecular biology, it is now possible to buy kits containing all the components necessary for DNA sequencing reactions, and there is a bewildering array of products available. It is certainly convenient to buy sequencing reagents in kit form and consistency of results should be assured, but convenience often needs to be weighed against cost. Sequencing reagents are simple to prepare, but it is vital to buy high-quality, sequencing grade chemicals and enzymes.

The fundamental principle behind DNA sequencing is the resolution on a denaturing polyacrylamide gel of radiolabelled or fluorescently labelled single-stranded DNA fragments, each differing in size by a single base. Two quite different approaches have been used to generate these fragments. In the method developed by Maxam and Gilbert (1977), fragments are produced by base-specific chemical cleavage of DNA end-labelled with ^{32}p. In the chain termination method of Sanger et al. (1977), DNA polymerase (Klenow fragment of DNA polymerase I in the original method) is used to synthesise a new strand of DNA from an oligonucleotide primer annealed to one strand of the template. Synthesis is carried out in the presence of small amounts of dideoxynucleotide (ddNTP) analogues of natural nucleotides (dNTPs). When a ddNTP is incorporated into a growing strand, synthesis is terminated at that base because the chemical nature of the analogue prevents addition of a further nucleotide. By carefully controlling the proportion of ddATP, ddCTP, ddGTP and ddTTP during synthesis, it is possible to generate a nest of fragments terminating at every base in a given sequence.

The protocols provided in this chapter use the chain termination method, which is more widely used than chemical cleavage. It has the advantage of not requiring noxious chemicals, and is more amenable to high-throughput sequencing. Maxam-Gilbert sequencing is still preferred in some laboratories, and does have particular uses, but it is not dealt with further in this chapter. Many refinements have been made to the original chain termination method, enabling more and better quality sequence to be obtained from fewer sets of reactions. Improvements have arisen from the discovery of more efficient DNA polymerases, the availability of radiolabels and fluorescent labels which allow greater resolution, and development of gel systems which allow more efficient separation of DNA fragments. Klenow fragment of DNA polymerase I has now been virtually superseded by the more processive T7 DNA polymerase (such as the recombinant enzyme Sequenase version 2). Methods have also been developed using thermostable enzymes (Innis et al. 1988), which enable sequencing reactions to be carried out at high temperatures, an advantage when dealing with DNA templates with a high degree of secondary structure. Sequencing with a thermostable polymerase also allows cycles of sequencing (Murray 1989) to be undertaken, with consequent amplification of the product. This can be very useful when only small amounts of DNA template can be obtained.

^{35}S is now more commonly used than ^{32}P for radiolabelling in chain termination sequencing because of the improvements in resolution it offers than autoradiography. ^{32}S produces lower energy β-particles than ^{32}P, resulting in less scatter and, consequently, clearer bands. This is particularly noticeable on the upper regions of the gel, and allows more sequence to be read. ^{32}S is also less hazardous to handle than ^{32} P because of its lower energy emission. Protective screens are not required, and the same precautions as observed for handling ^{14}C are sufficient. Another advantage to using ^{35}S is that it is a more stable isotope than ^{32}P. It has a half-life of approximately 87 days compared to approximately 14 days for ^{32}P, allowing the labelled dNTP to be stored for longer. The reaction products can be stored for at least 1 week compared to no more than overnight for sequencing reactions labelled with ^{32}P. The only slight disadvantages to using ^{35}S are that a longer exposure time is needed than for ^{32}P (24–36 h compared to 12 h), and, because the β particles emitted by ^{35}S are less penerating and more easily quenched, gels need to be fixed and dried to a thin layer prior to autoradiography. The first disadvantage can be overcome by using more sensitive film or by simply loading more of each reaction onto the gel. Although gel fixing and drying does take time, it can, in fact, be useful to have the gel in a dried form, for example if a further exposure is required.

Recently, a second isotope of phosphorus, ^{33}P, has been introduced for DNA sequencing (Zagursky et al. 1991). It emits β-particles with a lower energy than ^{32}P and is consequently safer to handle, but the particles have more energy than those emitted by ^{35}S and so are less easily quenched. Unfortunately, ^{33}P is more expensive than ^{32}P or ^{35}S. For most routine sequencing in which radiolabel is incorporated into the synthesised strand of DNA, ^{35}S remains the radiolabel of choice. However, for certain protocols, such as cycle-sequencing using Tag polymerase, in which the primer itself is end-labelled (and for which ^{35}S cannot be used), ^{33}P is a safer alternative to ^{32}P and also offers improved resolution.

An alternative to radiolabelling is end-labelling of primers with fluorescent dyes. This is the method used in automated sequencing, which requires specialist equipment for detection and is outside the scope of this chapter. If you do hold reservations about using radioisotopes in molecular biology, then there are non-radioactive alternatives available for manual sequencing. These methods , however, are not without technical difficulties, which is perhaps why they are not more widely used. One method is to biotinylate the primer (Creasey et al. 1991). Biotin is bound with high affinity by streptavidin, and so biotinylated DNA can be detected by means of a conjugate of streptavidin and an enzyme such as horseradish peroxidase or alkaline phosphatase. DNA fragments on a sequencing gel are blotted onto a nylon membrane, and the membrane is then incubated with the streptavidin conjugate. Bound conjugate can be detected by using an appropriate colorimetric or chemiluminescent substrate.

Before starting to sequence a region of DNA, it is important to consider the aim of the project, as this will dictate overall strategy. For example, the approach needed to determine the sequence of a novel cDNA or genomic clone will require more steps than if simply confirming the sequence of a clone before and after site-directed mutagenesis, or that an insert is in the correct reading frame after subcloning into an expression vector. Indeed, one of the most routine, but important uses for sequencing is to verify constructs after manipulation. Much wasted effort can be avoided by the simple expedient of sequencing vector-insert junctions of new constructs.

The limits of resolution on DNA sequencing gels mean that it is unlikely that the entire sequence of a cDNA or genomic clone can be obtained from a single sequencing run. Usually, 300–500 bp of readable sequence can be obtained from each set of sequencing reactions. Therefore, for most cDNAs (1.5–30 kb) and genomic clones (often several kb), multiple sequencing runs will be required. It is also necessary to sequence both strands of DNA, both to confirm the sequence and to highlight artefacts. You will find that most of the sequence can be obtained quite easily with standard protocols. You can then concentrate on problem regions and decide on the best method to resolve them. Some laboratories claim that up to 1 kb of readable sequence can be obtained from a single set of reactions. This may well be so, but to obtain this much sequence at sufficiently good resolution requires a great deal of practice and optimisation of conditions. In practice, it is far better to run more gels with shorter stretches of good-quality sequence than to be tempted to read too far.

Several methods have been devised to allow sequencing of an entire clone on both strands. In the earliest protocols, DNA was fragmented by sonication or with DNAseI. The fragments were then cloned into vectors engineered from the *E.coli* filamentous bacteriophage M13 (Messing and Vieira 1982). However, there was no control over how large the fragments were that were cloned. The sequence was gradually pieced together by identifying overlapping regions, a task that could only be achieved efficiently through the development of computer programmes. Since then, computers have become an essential part of managing and analysing sequence data, and programmes for sequence assembly have become more user-friendly.

One way in which the generation of new data can be made more efficient is to adopt a more systematic approach to creating overlapping DNA fragments. This can be achieved by generating sets of unidirectional deletion subclones with a common starting point and extending for varying distances into the target sequence (Henikoff 1984, 1987). Such "nested" sets of deletions can be generated by using exonuclease III (Exo III), an enzyme which digests double-stranded DNA from 5′ overhangs and blunt ends, but not from 3′ overhangs of at least four bases. This makes it possible to create deletions in the target sequence (but not in the vector DNA) by digesting the clone with a pair of appropriate restriction endonucleases (that is ones which do not cut within the insert, have unique sites in the vector and leave the required overhangs). An example of this strategy is illustrated in Fig. 4.1.1.

In this case, it has been determined by restriction mapping that the cloned DNA does not contain either EcoRI or PstI sites. The vector contains a single site for each enzyme. PstI produces a 3′ overhang resistant to digestion with ExoIII and EcoRI produces a 5′ overhang vulnerable to digestion. ExoIII is added to the EcoRI/PstI-digested DNA. Samples are stopped at set time points to allow different lengths of DNA to be digested. The ends of the digested DNA are then blunt-ended with a mung bean nuclease and religated. The priming site in the vector remains constant, thereby allowing the same primer to be used in all reactions. The effect of the deletion strategy is to bring all regions of the gene within a suitable distance of the universal priming site. By cloning the cDNA or genomic insert into both versions of a vector which differ only in orientation of the polylinker region, it is possible to prepare nested sets of deletions from both ends of the clone. Although they do take time to prepare, nested sets of deletions are extremely useful. They simplify data collection and management by allowing sequencing to be undertaken in a systematic manner. In addition, they provide a bank of deletion mutants which can be used in further studies, for example, identification of motifs required for tissue-specific transcriptional regulation, or investigation of the effect of truncating the encoded protein.

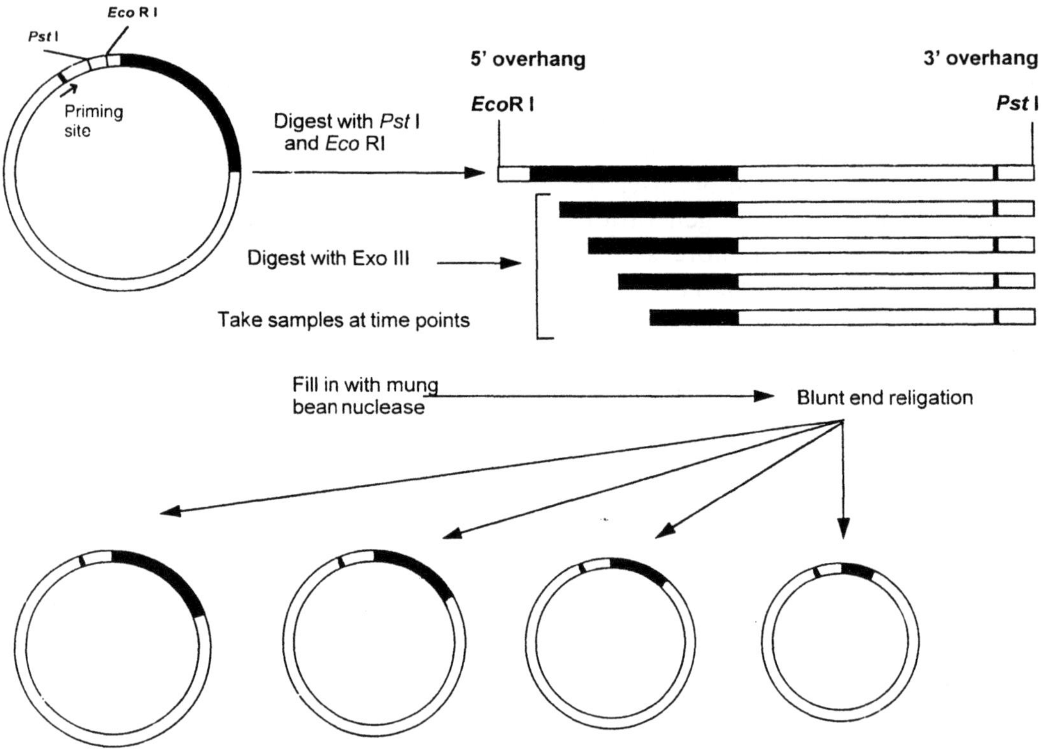

Fig. 4.1.1. Production of nested deletions using ExoIII

An alternative approach to obtaining overlapping sequence is to use oligonucleotides to move the priming site along the clone in a stepwise fashion, using the sequence obtained from one set of reactions to design a new primer for the next stretch of DNA. At first sight, this would appear to be a rapid alternative to generating sets of nested deletion, and it can indeed be very effective. It can also prove expensive, particularly as it may take practice to design good primers. It is vital to ensure that there are no sites which might cause mispriming. This is a particular hazard if your sequence contains repeated elements, which can be common in genes encoding structural proteins. Repeats are also frequently encountered in some plant storage protein genes. This approach may also prove problematic when sequencing the untranslated regions of genomic clones where repeated motifs may occur.

Now that the polymerase chain reaction (PCR) is used routinely by molecular biologists either directly for cloning, or to generate probes for cDNA or genomic library screening, there is often a need to sequence PCR products directly. This may be to ensure that the correct product has been obtained prior to cloning, or to confirm the identity of a fragment to be used as a probe. The protocols for sequencing PCR products are essentially the same as for sequencing cloned genes. The most important factor is to remove the components of the PCR reactions, which may interfere with the sequencing reactions, in particular unused primers, excess nucleotides and spurious amplification products. There are a number of kits available commercially specifically for this purpose. If this is not an option, then it is worth scanning journals such as Nucleic Acids Research and BioTechniques for the most up-to-date and effective means of obtaining PCR products of sufficient purity for sequencing.

If you develop an interest in more advanced sequencing methodologies, the aforementioned journals are excellent sources of information. In addition, many molecular biology suppliers publish their own bulletins on the latest developments. These are usually available free on request. In conclusion, do not look for difficulties before you start. Keep your sequencing methods as simple as possible, but be prepared to adopt more sophisticated approaches as needed.

4.1.2
Generation of Nested Unidirectional Deletions

4.1.2.1
Selection of Cloning Vector for ExoIII Deletion

The requirements for a suitable plasmid, phagemid or M13 phage vector are that it should possess a wide range of unique restriction sites in the polylinker and that it is available with the polylinker in both orientations. Phagemid vectors such as pBluescript II (Stratagene) and the pGEM series (Promega) have been designed with the ExoIII deletion strategy in mind and they offer a range of restriction sites yielding 5', blunt-end and 3' overhangs after digestion. Vectors of this type also contain a number of widely used priming sites. If your cDNA clone was isolated from a lambda ZAPII library, you will be able to rescue it directly as a pBluescript II construct. Phagemid vectors such as pBluescript II and pGEM offer the option of recovering either ds or ssDNA as the sequencing template. A selection of unique restriction sites present in commonly available vectors is listed below, classified according to the type of overhang they produce.

- Restriction enzymes production 3′ overhangs of at least four bases: ApaI, BstxI, KpnI, PstI, SacI, SphI
- Restriction enzymes producing blunt ends: EcoRV, SmaI, StuI
- Restriction enzymes producing 5′ overhangs: AccI, BamHI, EagI, EcoRI, HindIII, NotI, SalI, SpeI, XbaI, XhoI

Note which restriction enzyme cleavage sites in the polylinker of your chosen vector will produce 3′ overhangs of at least four bases, and which will yield blunt ends or 5′ overhangs. The next step is to determine experimentally which enzymes do not cut within the cloned insert. This is not an exhaustive list and it is possible that you will have to consider alternative sites and vectors. If a suitable 3′ overhang restriction site cannot be found, it is still possible to construct a set of nested deletions by protecting a 5′ overhang from Exo III digestion by in-filling with deoxythionucleotides, which are resistant to digestion (Guo and Wu 1983; Jasin et al. 1983).

If you are working with a large cDNA or genomic insert from a lambda library, it may be necessary to subclone restriction fragments of the insert into a suitable vector for ExoIII deletion. A series of nested deletions from overlapping fragments can be constructed to ensure full coverage of the insert.

The approximate rates at which ExoIII digests DNA are as follows:

Temperature (°C)	Bases per minute
22	100
25	150
30	200
35	400
37	500
39	600

These rates can be used as a guide to the conditions most suitable for your clone. If you find too much, or too little is being deleted, then the experiment can be repeated at a more appropriate temperature.

The overall aim is to generate a nest of deletion subclones, each differing in length by 150–300 bp. This should afford good overlap for sequencing. you will probably find that as you get more practised at sequencing, you dispense with the more closely spaced subclones.

4.1.2.2
Preparation of Closed Circular DNA

The successful generation of ordered sets of deletions requires a uniform rate of digestion. Unfortunately, Exo III will digest from nicks in double-stranded DNA molecules (Barnes et al. 1983). The presence of nicked molecules in starting material leads to the generation of random deletion mutants. The quality of the input DNA is, therefore, of utmost importance. Closed circular DNA can be prepared on CsCl gradients (Sambrook et al. 1989) or by commercially available plasmid DNA purification kits. Such a kit may be a worthwhile investment to ensure that DNA of the requisite quality is obtained. An effective alternative is to exploit the tendency of nicked and linearised plasmid DNA to selectively partition into phenol at low pH. Although the reason for the partitioning is not fully understood, accurate adjustment of pH is important. At pH 4.2, nicked and linear DNA will remain in the aqueous phase, but at pH 4, it will partition into the phenol. Phenol preequilibrated to the correct pH can be purchased; however, the following protocol is very straightforward.

Caution: Observe recommended safety precautions at all times when handling phenol.

Acid-Phenol Preparation

Reagents

- Phenol
- 50 mM sodium acetate pH 4
- chloroform/isoamyl alcohol (24:1)

Equipment

- pH meter
- Magnetic stirrer
- Vortex
- Vacuum desiccator
- Glass pipettes (10- and 25-ml) plus pipette filler
- 1-l glass beaker
- 500 ml glass or polypropylene bottle

Procedure

1. In a fume hood add 250 ml 50 mM sodium acetate pH 4 to 250 g phenol in a glass beaker. Stir at room temperature to dissolve fully. This will take 2–3 h.
2. Switch off the stirrer and allow the phases to separate. Carefully pipette off the upper aqueous phase.
3. Add a further 250 ml of sodium acetate pH 4 and stir to mix. An emulsion will form. Leave to stand overnight to allow the phase to separate fully.
4. Take a small sample of the aqueous phase and determine the pH using a pH meter. Pipette off the remainder of the aqueous phase if the pH is above 4.1
5. Repeat steps 3 and 4 until the pH of the aqueous phase is less than 4.1. If the pH is 4.2 or above, nicked DNA will not be removed from the preparation.
6. When the correct pH has been achieved, remove all but 50 ml of the aqueous phase. Transfer the phenol and remaining buffer to a suitable glass or polypropylene bottle. Store in the dark at 4 °C. Ensure that the phases have separated before use.

Acid Phenol Extraction of Plasmid DNA

Reagents

- Phenol equilibrated to pH 4
- 2 M sodium acetate pH 4
- 3 M sodium acetate pH 5
- 2 M NaCl
- 1 M Tris.HCl pH 8.6
- TE buffer (10 mM Tris.HCl pH 8/1 mM EDTA)
- Isopropanol
- 70% ethanol
- SDW

Equipment

- 1.5-ml microcentrifuge tubes
- Micropipettes (range 2–200 μl) and disposable tips
- Microcentrifuge

Procedure

1. Prepare plasmid DNA from 500 ml of culture by the alkaline lysis method detailed in Chapter 2.2, ensuring that it is free of RNA

and protein. Determine the concentration and purity of the DNA spectrophotometrically.

2. Transfer an aliquot containing 200 μg DNA to a 1.5-ml Micro centrifuge tube

3. Add 0.1 vol 3 M sodium acetate pH 5 and precipitate the DNA with 0.6 vol room temperature isopropanol. Collect the pellet by centrifugation at 12 000 *g* for 10 min.

4. Wash the pellet with 5 vol 70% ethanol by centrifuging at 12 000 *g* for 5 min.

5. Remove the ethanol and dry the pellet under vacuum.

6. Dissolve the pellet in 200 μl SDW. Add 5 μl 2 M sodium acetate pH 4 and 7.5 μl 2 M NaCl. Mix by vortexing briefly.

7. Add an equal volume of phenol pH 4 and mix well by vortexing for 1 min.

8. Separate the phases by centrifugation at 12 000 *g* for 5 min. Transfer the aqueous phase to a clean 1.5-ml tube. Do not disturb the interface. The organic phase may either remain opaque or be clear with a thick white interface.

9. Repeat steps **7** and **8** on the transferred aqueous phase until the organic phase and interface are clear.

10. Add 0.05 vol 1 M Tris.HCl pH 8.6 to the aqueous phase and an equal vol chloroform/isoamyl alcohol. Vortex to mix and separate the phases as before. Repeat.

11. Transfer the aqueous phase to a clean 1.5-ml tube. Precipitate, wash and dry the DNA as in steps **3–5**. Dissolve the DNA in 50 μl TE pH 8 and store at 4 °C.

4.1.2.3
Restriction Endonuclease Digestion

Because Exo III can initiate digestion from nicks in the DNA, it is important to use highly supercoiled DNA and to use high-quality restriction enzymes to prevent nicking by contaminating non-specific nucleases. It is essential to ensure that the first restriction endonuclease digest is complete before starting the second. This can be checked by analysing a sample on an agarose gel. For the first digest, use the enzyme which will generate a blunt end or 5′ overhang. If any molecules then escape digestion by the second enzyme, they will be digested from both ends by ExoIII, and it is unlikely that they will generate viable clones.

Reagents

- Restriction enzymes and appropriate buffers
- Agarose
- TAE or TBE electrophoresis buffer
- Sample loading buffer
- Ethidium bromide for gel staining
- 1:1 phenol/chloroform mix (not phenol pH 4)
- 24:1 chloroform/isoamyl alcohol
- 3 M sodium acetate pH 5
- Isopropanol
- 70% ethanol

Equipment

- Water bath or heating block set to appropriate temperature for chosen enzymes
- Microcentrifuge
- 1.5-ml microcentrifuge tubes
- Micropipettes (range 2–200 µl) and disposable tips
- Agarose gel electrophoresis apparatus
- UV transilluminator

Procedure

1. Digest 50 µg DNA in a 500-ml reaction volume using 1 unit of enzyme per µg. Ensure that all components of the digest are thoroughly mixed.
2. After 2 h analyse a 5 µl sample on a 1% agarose gel to check that the digest has gone to completion. Run some of the undigested plasmid for comparison. If digestion is not complete, add more enzyme and continue incubation for a further hour. Analyse another 5 µl sample.
3. When you are satisfied that the digest is complete, commence digestion with the second enzyme. If your chosen enzymes have different salt requirement, digest in with the enzyme requiring lower-salt buffer first and adjust the salt concentration for the second enzyme. Alternatively, desalt the first digest on a Sephacryl-400 spun column (Pharmacia) before proceeding with the second digest.

4. Purify the digested DNA by extraction with phenol/chloroform. Precipitate the DNA with isopropanol, wash the pellet with 70% ethanol and dry under vacuum until all traces of ethanol have gone.

4.1.2.4
ExoIII Digestion

Reagents

- Exonuclease III
- Mung bean nuclease
- 50 mM 2-mercaptoethanol
- 5x ExoIII buffer
  250 mM Tris.HCl pH 8
  25 mM $MgCl_2$
- 10x mung bean nuclease buffer
  300 mM sodium acetate pH 5
  500 mM NaCl
  10 mM $ZnCl_2$
  50% (v/v) glycerol
- 1x mung bean nuclease dilution buffer
  10 mM sodium acetate pH 5
  1 mM zinc acetate
  1 mM cysteine
  0.1% Triton x-100
  50% (v/v) glycerol
- 8 M LiCl
- 10% SDS
- 1 M Tris.HCl pH 9.5
- 3 M sodium acetate pH 7.0
- 1:1 phenol/chloroform
- 24:1 chloroform/isoamyl alcohol
- TE buffer pH 8.0
- SDW
- Molecular biology-grade glycogen suspension

Equipment

- Water bath or heating block set to desired digestion temperature (22–39 °C)
- Water bath or heating block set to 30 °C
- Water bath or heating block set to 68 °C
- Microcentrifuge

- 1.5 ml microcentrifuge tubes
- Micropipettes (2–200-μl range) and disposable pipette tips

Procedure

1. Dispense 175 μl of 1x mung bean nuclease dilution buffer into five marked microcentrifuge tubes on ice. The buffer will stop the activity of Exo III.

2. Calculate the number of picomoles of unprotected ends of DNA in 5 μg (this will vary with the length of the clone).

3. Prepare sufficient reaction mix in one 1.5-ml tube for the number of timepoints required. For five timepoints, prepare 6 vol of reaction mix (25 μl per timepoint) to allow for carry-over on pipette tips. Each 25 μl will contain:
   5.0 μg double-digested DNA
   5.0 μl 5x ExoIII buffer
   5.0 μl 50 mM 2-mercaptoethanol
   x ml Exo III (20 U/pmol susceptible ends)
   SDW to 25 ml
   Add ExoIII last, mix well and place tube in water bath or heating block.

4. Remove 25 μl aliquots at 1-min intervals and dispense into the tubes of stop buffer on ice.

5. When all the samples have been taken, heat the tubes at 68 ° C for 5 min and then place back on ice.

6. Dilute mung bean nuclease to 15 U/μl in 1x buffer and add 1 μl to each tube. Mix well and incubate at 30 °C for 30 min.

7. To each tube add:
   10 μl 1 M Tris.HCl pH 9.5
   20 μl 8 M LiCl
   8 μl 10% SDS 250 μl phenol/chloroform mix
   Vortex to mix and centrifuge at 12 000 *g* for 5 min to separate the phases.

8. Extract the aqueous layer with chloroform/isoamyl alcohol. Repeat.

9. Add 25 μl 3 M sodium acetate pH 7 and 1 μl glycogen suspension to act as a carrier for precipitation. Add 0.6 vol isopropanol to

precipitate the DNA. Collect pellets by centrifugation, wash in 70% ethanol and dry as previously described.

10. Redissolve the DNA in 10 ml TE buffer pH 8. Take a 1 µl sample for ligation and store the remainder at 4 °C.

4.1.2.5
Religation and Transformation of E. coli

Reagents

- T4 DNA ligase
- 5x T4 DNA ligase buffer
  250 mM Tris.HCl pH 7.5
  35 mM $MgCl_2$
  5 mM DTT
- 10 mM ATP
- SDW

Equipment

- Refrigerated water bath set to 16 °C (optional)

Procedure

1. To each 1-µl sample of ExoIII-deleted DNA add:
  - 4.0 µl 5x ligase buffer
  - 1 ml 10 mM ATP
  - 1 U of T4 DNA ligase
  - SDW to 10 ml

2. Incubate at room temperature for 2 h, 16 °C for 4 h or 4 °C overnight.

3. Use 1 µl aliquots to transform competent cells of the same strain of *E.coli* used to maintain the full-length clone. Plate out using the appropriate antibiotic selection for the strain and vector (for transformation protocol see Chap. 2.2)

4.1.2.6
Screening Transformants

Transformants can be screened in a number of ways to determine whether the desired range of deletions has been achieved. The most rapid is to directly amplify the inserts from a few colonies selected from each plate, using primers designed to anneal to the vector on either side of the insert and to compare the lengths of the amplification products with the full-length insert. Alternatively, prepare plasmid DNA from 5 ml cultures. Linearise the plasmid by digestion with an enzyme having a single recognition site in the vector but no sites within the insert. Also linearise the intact plasmid. Analyse by agarose gel electrophoresis.

When you are satisfied that the required nest of deletions is represented in your library of transformants, toothpick colonies onto fresh plates in a gridded array, or into 96- well plates to make it easier to keep track of which clones have been sequenced. For long-term storage, add sterile glycerol to 15% to each well and mix well. Seal the plate with Parafilm and freeze on dry ice. Store the plate at –70 °C.

4.1.3
Plasmid DNA Sequencing

In the original chain-termination sequencing method, ssDNA isolated from M13 bacteriophage was used as the template. Since then, however, methods have been devised to allow sequencing to be performed directly from plasmid DNA (Wallace et al 1981; Chen and Seeburg 1985). For some applications (such as automated sequencing), ssDNA is still the template of choice. To avoid problems when using plasmid DNA, it is essential to ensure that the DNA is of high quality and free of contaminating RNA and protein. Plasmid DNA of the required quality can be prepared using kits, CsCl gradient purification or from standard alkaline lysis mini-prep methods followed by spun-column gel filtration. After alkaline denaturation, the DNA is passed through a second spun-column to remove the alkali. This second round of gel filtration serves as a further purification step.

The advantage of sequencing from plasmid DNA is that from a single preparation both strands of DNA can be sequenced. This is achieved by setting up two sets of reactions, primed by oligonucleotides which anneal to opposite strands (forward and reverse primers). Before the sequencing reactions can be performed, the two strands of DNA must be separated to allow the primer to anneal. When using a thermostable polymerase for sequencing, this is achieved by melting the DNA at 95 °C. For sequencing with T7, polymerase alkaline denaturation (Chen and Seeburg 1985) has been found to be a reliable method. Sequencing reactions can be carried out in microcentrifuge tubes (using different colours for A, C, G and T is helpful). An alternative, if processing a large number of templates, is to use microassay plates (which float on water). These have V-shaped wells which take 10–12 µl of sample.

The stability of many enzymes is enhanced by the presence of glycerol in the storage buffer, as is the case for Sequenase. However, it has been found that high concentrations of glycerol in sequencing reactions can produce distortions in the spacing and shape of bands on a stranded TBE-buffered sequencing gel, typically in the region 300–500 bases from the primer (Tabor and Richardson 1987). This artefact is caused by the reaction of glycerol with borate ions to form an anionic complex which co-migrates with DNA fragments 300–500 bases in length (Pisa-Williamson and Fuller 1992). This problem can be overcome by buffering the gel with taurine, which does not react with glycerol. Use of taurine-buffered gels makes it possible to store Sequenase in a diluted form in glycerol, avoiding the need to dilute an aliquot every time (usually more than is needed). This is an important consideration because the enzyme is very expensive. By pre-diluting it, only the exact amount needed is taken each time. Taurine is slightly more costly than boric acid, but the saving on the enzyme outweighs this. In addition, the presence of glycerol at high concentration confers greater tolerance to extended incubation times and higher temperatures (up to 60 °C). This can be of benefit when sequencing templates containing a high degree of secondary structure (Pisa-Williamson and Fuller 1992).

4.1.3.1
Plasmid DNA Preparation

Reagents

- Sephacryl-S200HR (or ready-made Sephacryl S200HR microspin columns)
- TE buffer pH 8
- SDW

Equipment

- 1.5 ml microcentrifuge tubes
- 0.5-ml microcentrifuge tubes or MobiCol microspin columns
- 18-gauge syringe needle
- Very fine glass beads (ballotti beads available from Jencons) autoclaved in water (1 g per 20 ml)
- Micropipettes (2–200-µl range) and disposable tips.
- Microcentrifuge

Procedure

1. Prepare plasmid DNA from 10 ml of culture using the alkaline lysis method detailed in Chapter 2.2. Treat with RNAse and extract with phenol/chloroform and precipitate. Resuspend the plasmid in 50 µl TE pH 8. Determine the concentration of the DNA. Adjust concentration to 500 ng/µl.

2. Thoroughly mix the Sephacryl-S200HR to achieve an even suspension. Place a microspin column inside a 1.5 ml microcentrifuge tube and fill with Sephacryl-S200HR. If you cannot obtain these columns, you can use a 0.5-ml microcentrifuge tube instead. Pierce the bottom of the tube with the very tip of the hypodermic needle. Place inside a 0.5 ml tube. Resuspend the glass beads and pipette 20 µl into the bottom of the pierced tube. Fill the tube with Sephacryl-S200HR.

3. Spin the columns for 1 min at 3000 *g*. Remove the column from the 1.5-ml tube and discard the flow-through. Replace the column in the 1.5 ml tube and add 200 µl of TE to the column. Spin as before, discard the flow-through and repeat.

4. Place the column in a clean 1.5-ml tube. Pipette 40 µl of plasmid DNA onto the column. Spin as before. The plasmid DNA will be in the flow-through and should be free of RNA fragments and other contaminants.

4.1.3.2
Template Denaturation and Primer Annealing

Reagents

- Sephacryl-S200HR (or ready-made Sephacryl S200HR microspin columns)
- SDW
- Freshly prepared denaturing solution (0.2 M NaOH, 0.2 mM EDTA)
- 5x annealing/reaction buffer (200 mM Tris.HCl pH 7.5, 100 mM $MgCl_2$, 250 mM NaCl)
- Sequencing primers (forward and reverse) at 0.5 pmol/µl

Equipment

- 1.5-ml microcentrifuge tubes
- 0.5-ml microcentrifuge tubes or Mobicol microspin columns
- 18-gauge syringe needle
- Very fine glass beads (ballotti) autoclaved in water (1 g per 20 ml).
- Micropipettes (1–200-µl range) and disposable tips.

- Microcentrifuge
- Water bath or heating block set to 37 °C

Procedure

1. Divide the flow-through from 4.1.3.1 into two. To each aliquot add 0.1 vol denaturing solution. Mix well and incubate at 37 °C for 20 min. while the DNA is incubating, prepare two Sephacryl-S200HR spin columns as in steps **2-5** of Section 4.1.3.2, but using SDW to wash the column rather than TE.
2. For each template, prepare two microcentrifuge tubes containing 2 µl of 5x annealing/reaction buffer. To one, add 1 µl of forward primer and to the other.
3. Pipette the denatured DNA onto the Sephacryl. Centrifuge as before.
4. Immediately pipette 7 µl of the flow-through into the tubes containing primer and annealing buffer. Incubate at 37 °C for 20 min. After annealing, place on ice.

4.1.3.3
Sequencing with T7 DNA Polymerase

DNA synthesis for chain-termination sequencing is carried out in two steps. In the first, the primed strand of DNA is extended and at the same time labelled by the incorporation of ^{35}S nucleotide. In the second step, dideoxynucleotides are added to the population of labelled DNA molecules (ranging in length from a few to many hundreds of nucleotides) and synthesis continues until a ddNTP is incorporated, thus terminating the chains.

Reagents

- [^{35}S] dATp
- Labelling mix (1.5 µM dCTP, 1.5 µM dGTP, 1 µM dTTP)
- Termination mixes (all contain 50 mM NaCl)
  A (80 µM each dATP, dCTP, dGTP and dTTP, 8 µM ddATP)
  C (80 µM each dATP, dCTP, dGTP and dTTP, 8 µM ddCTP)
  G (80 µM each dATP, dCTP, dGTP and dTTP, 8 µM ddGTP)
  T (80 µM each dATP, dCTP, dGTP and dTTP, 8 µM ddTTP)
- 0.1 M dithiothreitol (DTT)
- T7 DNA polymerase (Sequenase Version 2)

- Ice-cold enzyme dilution buffer (10 mM Tris.HCl pH 7.5, 5 mM DTT, 0.5 mg/ml BSA) or glycerol enzyme dilution buffer (20 mM Tris.HCl; pH 7.5, 2 mM DTT, 0.1 mM EDTA, 50% glycerol)
- Stop solution (95% deionised formamide, 20 mM EDTA, 0.05% bromophenol blue, 0.05% xylene cyanol)
- SDW

Equipment

- Water baths or heating blocks set at 37 and 75 °C
- Micropipettes (1–20-µl range) plus disposable tips
- Microcentrifuge tubes or microassay plates
- Timer

Procedure

1. Label tubes or microassay plate wells A, C, G and T plus the code for your primer and template. Dispense 2.5 µl of each termination mix into the appropriate tube or well. Keep at room temperature.

2. Dilute the 5 labelling mix five fold in SDW. For each set of sequencing reactions you will need 2 µl of the diluted labelling mix.

3. Dilute an aliquot of Sequenase eight fold in dilution buffer and place on ice. you will require 2 µl of diluted enzyme per set of reactions. Unused diluted enzyme should be discarded. Alternatively, predilute the entire vial of enzyme eight fold with glycerol dilution buffer. After dispensing the required amount, return the diluted enzyme to storage at –20 °C.

4. On ice, add the following to each annealing reaction:
   1 µl 0.1 DTT
   2 µl diluted labelling mix
   1 µl [^{35}S]dATP
   2 µl diluted Sequenase.
   Mix well and incubate at room temperature for 5 min.

5. While the labelling reaction is proceeding, place the tubes or microassay plate containing the termination mixes at 37 °C.

6. After 5 min, dispense 3.5 µl of the labelling reaction into each of the termination mixes (A, C, G and T). Mix well and continue incubating at 37 °C for 5 min.

7. After 5 min add 4 µl of stop solution to each reaction. The stopped reactions can be stored at –20 °C for up to 1 week.

4.1.3.4 *Sequencing Gel Casting*

Sequencing gels are cast between two glass plates (usually 40–50 cm in length and 25–50 cm in width) held apart by plastic spacers sandwiched at the long edges. The usual thickness for spacers is 0.4 mm. Gels are run vertically. Spacers which are 1.2 or 0.8 mm thick at the bottom and tapering to 0.4 mm at the top are available commercially. In gels cast with wedge-shaped spacers, a voltage gradient is established which slows the migration of the DNA fragments in the lower, thicker part of the gel. Smaller fragments are retained on the gel rather than running off the bottom, bands become more evenly spaced, and so sequence can read better from one run than on gels of even thickness. A similar effect can be achieved by casting gels with an ionic strength gradient; however, wedged gels are easier to cast and results are more consistent.

The inner glass plate forms a seal with the upper buffer tank via a silicone rubber gasket. The type of plate used will depend on the electrophoresis apparatus (Fig. 4.1.2).

In one, the inner plate is shorter than the outer and so, to allow a seal to be made with the buffer tank, foam rubber blocks which press

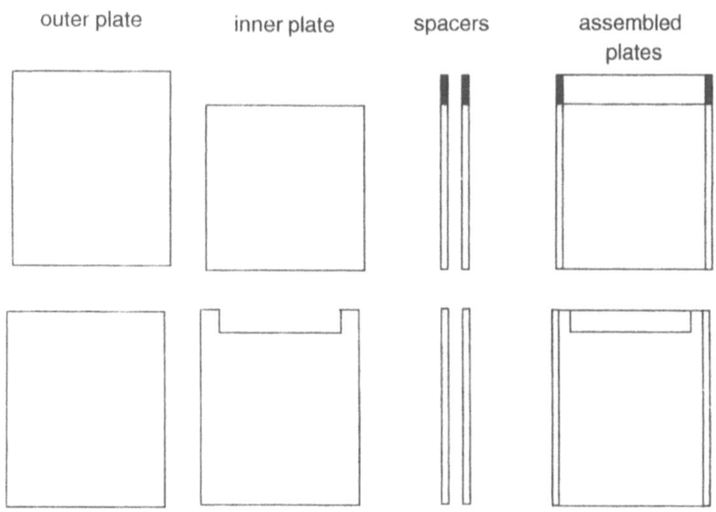

Fig. 4.1.2. Types of glass plate used in sequencing

tightly against the top edge of the plate are attached to the tops of the spacers. Other sequencing apparatus is designed to be used with a notched, or 'bunny-eared' inner plate. In this case, the glass itself forms a seal with the gasket. There is little to choose between the two systems in terms of ease of use. If you are going to invest in some new apparatus, however, it is worth noting that notched plates are difficult to make and, consequently, are more expensive than plain plates. In addition, the 'bunny ears' are rather vulnerable and tend to snap off if not handled carefully. The form blocks used with plain plates need to be replaced regularly to ensure a good seal.

Another consideration when selecting a gel rig is whether or not it has an integral metal plate to ensure even heating of the gel. Gels run without a heat diffuser to maintain a constant temperature in all parts of the gel tend to "smile". Those fragments at the middle (where it is hotter) migrate faster than those at the cooler edges. This can hinder accurate reading of the gel. If an apparatus does not have an integral heat diffuser, then an aluminium or copper plate clipped to the outer plate works very well; but be warned, the metal can get very hot.

Reagents

- 40% acrylamide stock solution (38 g acrylamide, 2 g N, N′-methylene bisacrylamide per 100 ml)
- Sequencing grade urea
- Ammonium persulphate (AMPS). Freshly prepared 10% solution in deionised water
- TEMED (N, N, N, N′, N′-tetramethylenediamine)
- 10x TBE buffer (0.9 M Tris/0.9 M boric acid/0.025 M EDTA) or 20x glycerol-tolerant gel buffer (1.78 M Tris/0.57 M taurine/ 0.01 M EDTA)
- Deionised water
- Water-repellent siliconising solution (Repelcote)

Caution: NB Acrylamide monomer is a potent neurotoxin. Wear suitable gloves when handling it and avoid inhaling the solid. An approved dust mask should be worn when weighing the solid. If at all possible, purchase ready-made acrylamide solutions. If this is not an option, try to buy acrylamide/bisacrylamide premixed in the correct proportions and dissolve the solid in the bottle with slightly less water than required for the final concentration. When fully dissolved, pour into a measuring cylinder and adjust the volume. In this way, you can avoid exposure to acrylamide dust.

Equipment

- Sequencing gel electrophoresis apparatus
- Pair of glass plates (one long, one short or one plain, one notched, depending on apparatus)
- Plastic spacers (0.4 mm thick or wedged)
- Foam blocks (if required)
- 0.4-mm-thick sharkstooth comb (24-or 48-well)
- Roll of gel sealing tape (40 mm wide)
- Hot plate stirrer (set to 37 °C) and magnetic stirring bar
- Glass beaker and measuring cylinder
- 100 ml side-arm flask (or beaker, 50 ml syringe or small squeezy bottle with nozzle)
- A number of large (50-mm) fold-back clips
- Disposable gloves
- Photographic tray sufficiently large to take the plates with at least 2–3 cm clearance on each side.

Procedure

1. In a fume hood, and wearing protective gloves, pour a small amount of siliconising reagent onto the surface of one of the glass plates (preferably the notched or shorter plate). Spread the solution over the entire surface with lint-free tissue. Allow to dry. The treated side should be on the inside when the plates are assembled.

2. Calculate the volume needed to fill the space between the plates, depending on the thickness of the spacers. Add an extra 10 or 20 ml to allow for spillages or leakage. Prepare the appropriate volume of gel mix using slightly less water than required to achieve the final concentration. A standard sequencing gel contains 6% acrylamide, 7 M urea and 1x running buffer (TBE or taurine buffer). For 100 ml of gel mix combine the following in a beaker:
   15 ml 40% acrylamide stock solution
   42 g urea, 10 ml 10x TBE or 5 ml 20x glycerol-tolerant buffer
   30 ml deionised water

   You will need to heat the solution to around 37 °C while stirring to allow the urea to dissolve. If necessary, the solution can be heated to up to 45 °C. Do not be tempted to heat it to more than 45 °C because the urea may start to decompose (the pungent smell of ammonia may be evident). Prepare sufficient running buffer (1x TBE or 1x glycerol-tolerant buffer) for your electrophoresis apparatus.

3. When the urea has fully dissolved, allow the solution to cool to room temperature. Pour into a measuring cylinder and adjust the final volume with deionised water. Transfer the gel mix to the side-arm flask (or other vessel) for pouring. The gel solution does not need to be degassed.

4. While the urea is dissolving, assemble the plates. Place the longer or unnotched plate inside on the bench with the long edge parallel to the edge of the bench. Align the spaces carefully at the edges, ensuring that they do not protrude over the side edge and that they are level with the bottom edge.

5. Holding the shorter, or notched plate by the edges, carefully lower onto the spacers. Make sure that all edges are aligned and that the spacers have not been disturbed. If using spacers with attached foam blocks, ensure that the blocks fit tightly against the top of the short plate.

6. Carefully move the glass plate sandwich so that it protrudes slightly over the edge of the bench. Seal the protruding long side with a piece of tape 3-5 cm longer than the plates, sticking it to the top side of the sandwich first, then folding over the aligned edges, and finally, folding it under to stick to the lower edge. Avoid introducing wrinkles. Press the tape down well to remove air bubbles and fold the extra length of tape along the bottom edge of the plates. Carefully turn the plates round and repeat for the other long edge. Repeat for the bottom edge, sticking the excess length along the long edges. Ensure that where the tape overlaps at the corners it forms a tight seal.

7. Place the assembled plates in the photographic tray to catch spills when pouring.

8. When you are ready to pour the gel, add 0.5 ml freshly prepared 10% ammonium persulphate and 100 μl of TEMED. Swirl gently to mix.

9. Hold the plates at an angle of 45° resting on the bottom corner nearest to you. Pour the gel solution through the side arm of the flask between the plates in a steady stream so that it flows along the spacer. When the solution reaches the bottom, tilt the assembly to let it flow across the bottom and at the same time begin to lower it, so that by the time the gel solution reaches the top, the plates are at

an angle of approximately 5°. If a bubble should form while pouring (more of a problem with wedge-shaped gels), gently raise the plates to the vertical and carefully tilt the solution away from the bubble. Lower the plates again and continue pouring. If the bubble cannot be dislodged in this way, insert a thin strip of plastic about 0.2 mm thick and gently push the bubble to the surface. Purpose-made bubble removers can be obtained from Promega. When the gel has been poured, rest the plates on a object at an angle of 5°. Insert the sharkstooth comb and clip the edges of the plate as shown in Fig. 4.1.3A. Allow the gel to polymerise fully before removing the clips and comb.

10. Rinse the top of the gel with running buffer to remove crystallised urea and fragments of gel. Wash the comb thoroughly in deionised water.

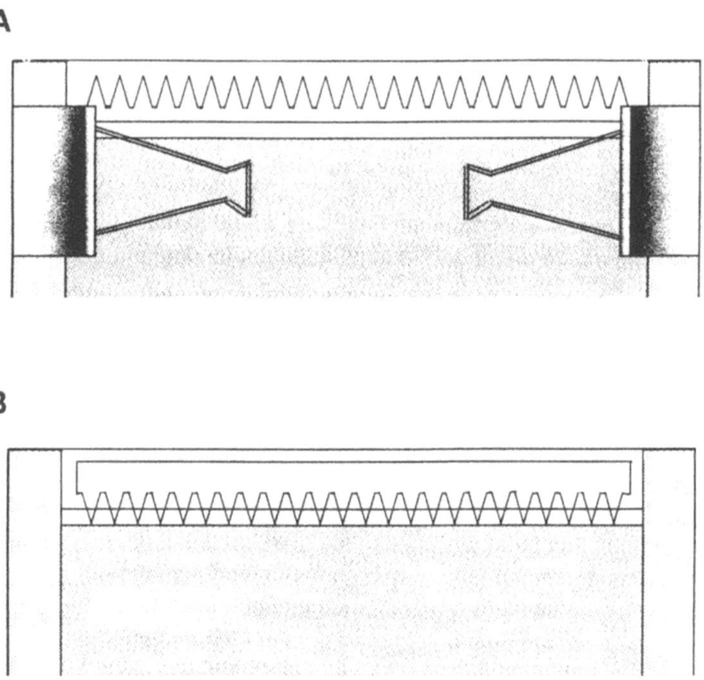

Fig. 4.1.3. Positioning of the sharkstooth comb. **A.** immediately after the sequencing gel has been poured. **B.** prior to loading the samples: the tips of the comb slightly indent the top of the gel

11. Insert the comb as shown in Fig. 4.1.3B. Peel the sealing tape off the plates and clamp them onto the apparatus with the shorter or notched plate on the inside.

12. Add running buffer to the top and bottom tanks to cover the edges of the plates.

4.1.3.5
Sample Preparation and Electrophoresis

Equipment

- Sequencing gel electrophoresis apparatus
- Power pack with constant power setting capable of supplying 2000 V
- Water bath or heating block set to 75 °C.
- Micropipettes (2–20-µl range) and gel loading tips (round-end or duck-billed)
- 20-ml syringe fitted with a 21-gauge needle

Procedure

1. Preelectrophorese the gel on constant power at 35 W (1500–1800 V) for 30 min to warm the gel to 50 °C. Self-adhesive liquid crystal thermometers (obtainable from IBI Ltd.) can be used to check the temperature of the glass plates. Running gels at too high a temperature can cause a range of problems from cracked plates, decomposing urea and fuzzy bands.

2. When you are ready to load the gel, heat the sequencing reactions to 75 °C for 5 min.

3. Switch off the power supply and unplug the gel. Using the syringe, flush each well with running buffer to remove urea before loading. Load 4 µl of each sample into wells 1–12 (2 µl into wells 1–24 for 48 well-combs) in the order A, C, G, T as shown in Fig. 4.1.4. Take care not to introduce bubbles into the wells. Store the remaining samples at –20 °C.

4. Attach the power cords to the apparatus and power supply and electrophorese the samples at 35 W until the bromophenol blue marker dye is 1–2 cm from the bottom. Switch off the gel.

5. Heat the remaining samples to denature as before. Rinse the wells and load the samples in the same order as previously. Electrophorese as before until the bromophenol blue dye is 1–2 cm from the bottom. By loading the samples twice, a long and short run is

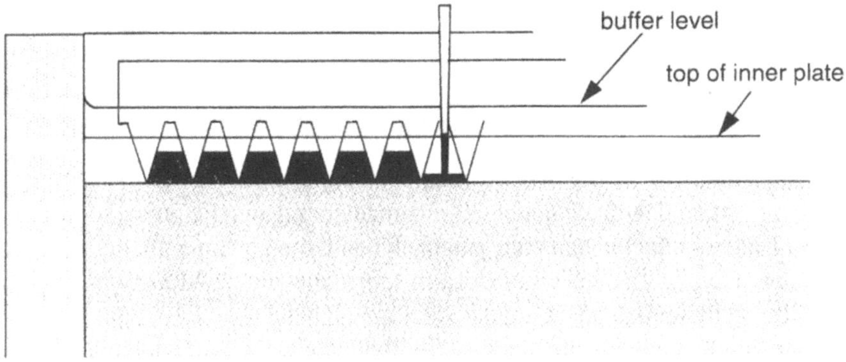

Fig. 4.1.4. Loading the samples on the sequencing gel. The samples are loaded between the sharkstooth comb. The positions of the top of the inner plate and buffer level are indicated

obtained. This allows more sequences to be obtained from the samples on one gel. The long run will resolve the sequence 200 plus bases from the primer, and the short run will resolve the sequence 30–250 bases from the primer.

4.1.3.6
Gel Fixing, Drying and Autoradiography

Reagents

- 2 l fixing solution (10% acetic acid, 10% methanol)

Equipment

- Heated gel drier and vacuum source
- Photographic tray sufficiently large to take the plates with at least 2–3-cm clearance on each side.
- Thin plastic spatula
- Rubber tubing and large plastic beaker
- Two pieces of Whatman 3 M chromatography paper cut 2–3 cm wider and longer than the larger gel plate
- Saran wrap or cling film
- Scissors
- X-ray cassette
- Autoradiography film (such as Kodak X-Omat or Amersham Hyperfilm-βmax)

Procedure

1. Drain the buffer from the upper tank. Remove the plates and place in photography tray with the shorter or notched plate uppermost.

2. Insert the spatula (avoid metal, as this can chip the glass) between the plates at a bottom corner. Gently press down on the spatula to prise the plates apart. This will happen quite suddenly. Carefully lift off the upper plate, starting at the bottom.

3. Gently pour fixing solution onto the gel to cover it completely. Use the entire 2 l. Leave for 20–30 min to allow the urea to wash out. The tray can be agitated gently from time to time, but do not shake the tray, or the gel may detach from the plate. Meanwhile, preheat the geldrier to 65 °C.

4. Siphon the fixing solution into the beaker. Do not tip the solution off. If you try to lift the plate directly out of the solution, the gel may be dislodged.

5. When the surface of the gel is no longer submerged, carefully lift the plate out and rest it on several layers of tissue. Allow the excess solution to drain away for about 5 min. Mop up remaining buffer around the gel with tissues, taking care not to touch the gel. The surface of the gel should not be wet.

6. Take one of the pieces of Whatman paper. Hold it at either end and bend it slightly in the middle. Lower the paper onto the gel (placing the middle section down first) and apply gentle pressure to smooth the paper. Check for adhesion by lifting the paper at one corner. The gel should be firmly attached to the paper. If not, smooth down more firmly.

7. Carefully peel the paper away from the plate in one movement and lay it gel side up on the second piece of paper.

8. Cut a length of Saran wrap to cover the gel. Stretch it taut and smooth it over the gel. Trim the edges.

9. Lay the gel on the drier and lower the rubber gasket. Apply the vacuum and leave to dry (the length of time needed will depend on the vacuum source). Wedged gels will take much longer to dry than standard gels. The surface of the gel should be smooth to the touch with no trace of tackiness. If the gel is fully dry, the Saran wrap should peel away with no resistance.

10. When the gel is dry, allow it to cool to room temperature. Remove the cling film (β-particles emitted by ^{35}S will be quenched by it) and place the gel in the X-ray cassette. Under safe-light illumination,

place a sheet of film on the gel. Cut a corner from the sheet of film to aid orientation. If using film coated with emulsion only on one side, ensure that the coated side is in contact with the gel.

11. Develop the film after 24-h exposure at room temperature. If necessary, repeat the exposure for 36–48 h.

12. Orientate the autoradiograph and label with the date plus template and primer codes. Mark each set of reactions.

4.1.3.7 *Reading and Recording the Sequence*

You will require a large light box. The sequence is read 5′–3′ from the bottom of the gel. You may find it helpful to mark the set of reactions being read with a piece of card and to move it up the autoradiograph as you read. This will allow you to keep your place if distracted. It is also helpful to write the sequence on squared paper in even blocks. This makes it easier when typing the sequence into a computer and to detect typing errors. Many laboratories now have light boxes equipped with a stylus which allows the sequence to be scanned directly into the computer. Reading sequencing gels is largely a matter of practice, and you will quickly become proficient. It is a good idea to have another person read the sequence independently. In this way, mistakes in reading and artefacts in the sequence itself can be highlighted. You will find that most of the sequence can be read quite easily but there will always be sections that are unclear. When you encounter such regions, do not attempt to read through them. You may find that the problem section is resolved quite easily on a different subclone or on the opposite strand. The troubleshooting section in this chapter offers advice on how to resolve a variety of artefacts.

There are a number of sequence assembly and analysis packages available for use on stand-alone PCs or Macs. Most of these are user-friendly and easy to use. Many organisations provide sequencing software, such as the widely used UWGCG package, as part of their central computing facilities. The programmes may vary in detail and ease of use, but all work by searching for regions of overlapping sequence (on both strands) and subsequently melding the matched fragments into contiguous stretches of sequences (called contigs in some packages and melds in others). This process is continued until all gaps have been filled and a single, contiguous sequence is obtained.

Sequence analysis packages enable you to search for restriction sites, other features such as open reading frames, and can quickly derive the amino acid sequence of the encoded protein. You can also scan your sequence for similarity to known sequences in databases such as Genbank and EMBL.

Troubleshooting

The following is a guide to solving common problems and artefacts that may be encountered during a sequencing project.

- Gel does not polymerise
  - Component missing from gel mix. Use a check list when preparing gels.
  - AMPS solution too old. Prepare a fresh solution each time.
  - Acrylamide solution has deteriorated. Acrylic acid is a breakdown product of acrylamide which can inhibit polymerisation. Deionise acrylamide solutions with mixed bed resin after preparation and store at 4 °C. Do not prepare more than can be used within 3 months. Alternatively, purchase ready-made acrylamide solutions which contain a stabilising agent. Buying acrylamide in this form also eliminates exposure to acrylamide dust.

- Gel polymerises while being poured
  - Solution not cooled to room temperature after dissolving urea.
  - Too much AMPS added. Check concentration of stock solution.

- Air bubbles in gel
  - Plates not thoroughly clean. Ensure that the surfaces are free of dust, grease and dried-on gel. Rinse plates in mild detergent, followed by deionised water. Wipe with ethanol just before use. If you cannot remove the bubbles before the gel polymerises, it is best to discard the gel and pour another.

- Samples leak between wells when loading
  - Comb not held tightly against glass plates during gel polymerisation. Use fold-back clips to clamp the comb in place as described in Section 4.1.3.3. You may need to use an additional clamp at the middle of the comb.
  - Keep combs and spacers in matched sets, as they might vary in thickness.
  - Points of teeth damaged. Do not bend the teeth when removing the comb. Store with care.

- Samples do not enter gel.
  - Samples not properly denatured. Heat samples to at least 75 °C using a direct heat source such as a water bath or heating block for 2 min before loading. An oven is not recommended because it is difficult to ensure that all samples reach the required temperature.

- Gel sticks to plates after electrophoresis.
  - Ensure that the inner surface of one plate is treated with water-repellent siliconising solution. Before assembling plates, check that the chosen plate is sufficiently siliconised by placing a drop of water on the surface. This should form a discrete bead with no trailing when tilted. If trails do form, treat the plate again.

- Gel detaches from plate during fixing.
  - All is not lost. A detached gel can be gently floated back onto the plate but first make sure that the edges are free first, otherwise the gel may tear. Make sure you work under good illumination. Siphoning off some of the fixing solution first may make it easier to manoeuvre the gel.
  - Ensure that only one plate is siliconised before pouring the gel.

- Film blank or nearly blank
  - If using single-sided film, the emulsion side must be placed facing the dried gel.
  - Cling film not removed from dried gel, preventing low-energy β-particles emitted by ^{35}S from reaching the emulsion.
  - Labelled nucleotide too old.
  - Incorrectly labelled nucleotide used.
  - Component of the sequencing reaction missing: prepare a check list when setting up the reactions to ensure that all components are added.
  - Enzyme may, have lost activity. Always store according to the manufacturer's instructions and dilute using the recommended buffer.

- Complete sets of blank lanes on autoradiograph
  - Incorrect primer used.
  - Primer contains regions of self-complementarity. Design primers to avoid this. Keep primers as short as possible (17–25 bp in length).

- Primer synthesis incorrect. Purchase the best quality primers possible. Ensure that someone else double-checks the sequence to be synthesised. If synthesising primers yourself, make sure that reagents are stored and handled correctly. Reagents should be used within the recommended time after opening and should be kept anhydrous at all times. Perform trityl assays to check for complete synthesis. Ensure that the correct synthesis column is selected and that the sequence is entered in the correct orientation. Double-check the entered sequence before commencing synthesis.
- Template DNA may be degraded: make a new preparation using fresh solutions if nuclease contamination is suspected. Autoclave solutions where possible and wear disposable gloves when preparing and handling DNA.
- The priming site may have been deleted from the template DNA. This can occur during the preparation of nested sets of deletions using ExoIII nuclease if the plasmid DNA contains nicks. Use only high-quality supercoiled plasmid when preparing nested deletions.

• Bands appear in all four lanes
  - DNA template insufficiently pure. Repurify to remove traces of RNA and oligonucleotides resulting from degraded chromosomal DNA which can act as non-specific primers.
  - Template and/or primer concentration too high. Check concentration by spectrophotometry.
  - Components of reaction not mixed thoroughly. Avoid introducing bubbles when mixing. Ensure that droplets do remain on the sides of the reaction tubes. If using microcentrifuge tubes, centrifuge briefly.
  - Annealing carried out at too high a temperature and/or for too long.
  - The template may contain regions prone to strong secondary structure formation causing the enzyme to pause, especially during the labelling/primer extension step. Reducing the concentration of nucleotides during this step can help, as can the addition of more polymerase. Sequencing from a priming site closer to the problem region is often effective. Sequencing a different clone from a set of nested deletions may be all that is required. Alternatively, design an oligonucleotide from the

available sequence which will prime 25–50 bp closer to the problem region. If secondary structure remains a problem, consider using a thermostable polymerase to allow the sequencing reactions to be performed at a higher temperature.
- If bands are seen in all four tracks near the bottom of the gel, the primer-extension labelling step may have taken place above 20 °C and/or for longer than 5 min. If labelling is to be performed at temperatures greater than 20 °C, the enzyme must first be diluted in buffer containing glycerol.
- Bands across all four lanes on the upper part of the gel can be the result of the chain termination step being performed at room temperature. Always perform this step at at least 37 °C. Better results can be obtained for some templates by incubation 50 °C in the presence of glycerol.

• Bands in two or three lanes
  - Insufficient mixing of reaction components.
  - Regions of secondary structure (usually a stem-loop) which do not remain fully denatured during electrophoresis migrate anomalously and produce an artefact in which bands become too compressed to be resolved. Bands in adjacent lanes migrate to the same position. Do not be tempted to read through a compression. An apparently single band may, infact, be compression, the sequence obtained will undoubtedly be incorrect. In many cases, the correct sequence can be from the complementary strand. Methods by which compressions can be resolved are described below.

• Irregular band spacing.
  - A compression may not always be apparent at first sight. In some cases, the only indication that a compression is present may be the irregular spacing of bands. A fragment of DNA containing a potential stem-loop structure will migrate more quickly through the gel matrix when folded than when fully denatured, resulting in a region of closely spaced bands below a region of abnormally widely spaced bands.
  - Bands within a compression may be abnormally intense. Spotting compressions is largely a matter of practice. If in doubt, seek another opinion of the sequence. Attempting to read through such a region will undoubtedly produce errors in the final sequence. Sometimes, discrepancies caused by compressions only come to light when the opposite strand is sequenced, or when an

unexpected frameshift or deletion is discovered in the sequence detected. The nucleotide sequence of regions prone to compressions (typically G-C-rich) can be resolved by incorporating analogues of dGTP (such as dITP or 7-deaza-dGTP) into the newly synthesised strand, which give rise to less stable secondary structures. An alternative is to use a more strongly denaturing gel containing 40% formamide in addition to 7 M urea. Observe appropriate safe-handling procedures when using formamide, which is a teratogen.

- Double bands or additional "ghost" sequence visible
  - A second template is present. Always start cultures for plasmid preparation from a single, well-isolated colony. This is good microbiological practice.
  - The primer is binding to more than one site. This may happen if primers are being designed from previously sequenced sections of the clone. Check sequence for the presence of repeated sequences before designing primers. Repeats are common in some genes such as those encoding plant storage proteins.
  - Heterogeneity in the length of the primer. Purify full-length primer away from truncated products.

- Bands smeared
  - Template DNA may be contaminated with RNA and DNA. Repurify or make a new preparation.
  - Sequencing gel may be defective. Ensure that fresh gel mix is prepared each time.
  - Gel temperature too high. Glass surface temperature should be about 50 °C. Run gel at constant power (35 W).
  - Gel dried at too high a temperature. Avoid drying gels at greater than 70 °C.

- Sequence fades early in one lane (usually T)
  - If the template possesses a poly (A) tail, dTTP and ddTTP may become depleted, causing the T lane to fade before the others. Try increasing the concentration of dTTP in the labelling reaction (0.5 ml of a 1-mM solution).
  - If the sequence fades early in a lane other than T, the template may have a biased nucleotide composition. Try increasing the concentration of the appropriate dNTP in the labelling reaction as above.

- Pipetting may be inaccurate. Ensure that an equal amount of labelling mix is dispensed into each termination mix.

- One blank track in set of reactions
  - Probably due to pipetting error. Use a check list to ensure that all components are added to each reaction.

- Sequence faint near the primer
  - If the template concentration is too low, it becomes limiting, and long DNA molecules will be synthesised during the labelling step. To obtain sequence close to the primer, use 0.5 pmol template DNA (3–4 mg of double-stranded template or 1 mg single-stranded template).
  - Add manganese buffer to the sequencing reaction.
  - Primer concentration too low. A minimum of 0.5 pmol should be used. For some primers, a slightly higher concentration may be needed. The molar ratio of primer to template should be between 1:1 and 5:1.

- Lanes distorted
  - Glycerol in enzyme interacting with borate ions in TBE. If diluting enzyme with glycerol-based buffer, use taurine in place of boric acid in gel mix and running buffer.

- Distorted or wavy bands
  - Urea not removed from wells prior to sample loading.

- Distortions across the autoradiogram
  - Wrinkles in gel. Ensure that excess fixing solution is drained from the gel before applying chromatography paper.

- Spots on autoradiogram
  - Urea incompletely dissolved prior to casting gel.
  - Gel still contains urea when dried. Ensure that gel is left in fixing solution for at least 20 min prior to drying.

References

Barnes WM, Bevan M, Son PH (1983) Kilo-sequencing: creation of an ordered nest of asymmetric deletions across a large target sequence carried on phage M13. Methods Enzymol 101:98–122

Chen EJ, Seeburg PH (1985) Supercoil sequencing: a fast and simple method for sequencing plasmid DNA. DNA 4:165–170

Creasey A, D'Angio L, Dunne TS, Kissinger C, O'Keeffe T, Perry-O'Keeffe H, Moran LS, Roskey M, Schildkraut I, Sears LE, Slatkoh B (1991) Application of a novel chemiluminescence-based DNA detection method to single-vector and multiplex DNA sequencing. BioTechniques 11:102–109

Guo L-H, Wu R (1983) Exonuclease III: Use for DNA sequence analysis and in specific deletions of nucleotides. Methods Enzymol 100:60–96

Henikoff S (1984) Unidirectional digestion with exonuclease III creates targeted breakpoints for DNA sequencing. Gene 28:351–359

Henikoff S (1987) Unidirectional digestion with exonuclease III in DNA sequence analysis. Methods Enzymol 155:156–165

Innis MA, Myambo KB, Gelfand DH, Brow MAD (1988) DNA sequencing with Thermus aquaticus DNA polymerase and direct sequencing of polymerase chain reaction-amplified DNA. Proc Natl Acad Sci (USA) 85:9436–9440

Jasin M, Regan L, Schimmel PR (1983) Modular arrangement of functional domains along the sequence of an aminoacyl tRNA synthetase. Nature 306:441–447

Kusukawa N, Vemori T, Asada K, Kato I (1990) Rapid and reliable protocol for direct sequencing of material amplified by the polymerase chain reaction. BioTechniques 9:66

Maxam AM, Gilbert W (1977) A new method for sequencing DNA. Proc Natl Acad Sci (USA) 74:560–564

Messing J, Vieira (1982) A new pair of M13 vectors for selecting either strand of double-digest restriction fragments. Gene 19:269–276

Murray V (1989) Improved double-stranded DNA sequencing using the linear polymerase chain reaction. Nucl Acids Res 17:8889

Pisa-Williamson D, Fuller CW (1992) Glycerol-tolerant DNA-sequencing gels. Editorial Comments 17:29–36. United States Biochemical Corp. Cleveland, Ohio

Sambrook J, Fritsch EF, Maniatis T (1989) Molecular cloning: laboratory manual. Cold Spring Harbor Laboratory Press, Sanger F, Nicklen S, Coulson Ar (1977) DNA sequencing with chain terminating inhibitors. Proc Natl Acad Sci (USA) 74:5463–5467

Tabor S, Richardson CC (1987) DNA sequence analysis with a modified bacteriophage T7 DNA Polymerase. Proc Natl Acad Sci (USA) 84:4767–4771

Wallace RB, Johnson MJ, Suggs SV, Miyoshi K, Bhatt, Itakurah J (1981) A set of synthetic oligonucleotide primers for DNA sequencing in the plasmid vector pBR322. Gene 16:21–26

Zagursky RJ, Conway PS, Kashdan MA (1991) The use of ^{33}P for Sanger DNA sequencing. BioTechniques 11:36

4.2 Expression of Cloned Genes

T. PEHU

4.2.1 Introduction

Production of biologically active proteins coded for by cloned DNA molecules is a prerequisite for studying the structure and function of the cloned genes as well as for immunological studies to detect and localize the proteins in vivo.

The cloned DNA fragments can be expressed either in vitro in combined transcription and translation systems (rabbit reticulosyte and wheat germ lysates) or in vivo in heterologous systems. One of the advantages of the first method is its sensitivity because it is more suitable for radioactive labelling than in vivo techniques are. However, the yields are relatively low, ranging from 1 ng up to 1 ug. For most of the methods used for characterization of the protein products of the cloned DNA fragments, including immunological studies, a large scale production of the purified protein (mg quantities) is needed. This can be achieved only in vivo in suitable host organisms. A variety of single cell production systems for recombinant proteins employing both prokaryotic (bacteria) and eukaryotic (e.g. yeast and insect cells) have been developed. However, since the first demonstration of heterologous expression of a bacterial gene in 1974 by Chang and Cohen and a eukaryotic gene (Struhl et al. 1976), *Escherichia coli* has remained the best understood and most widely used host for this purpose.

Several expression systems are described in the following.

Promoters

Most of the expression vectors for *E. coli* used today are derivatives of the widely used cloning vectors (e.g. pBR322 and pUC series). Foreign proteins can be expressed basically in two forms, as fusion proteins between a protein from *E. coli* and the foreign protein, or as intact native proteins. For efficient production of the heterologous protein, whichever strategy is used, the isolated gene has to be placed downstream of a strong promoter. Among a large variety of *E. coli* promoters are the natural *lac* (Goeddel et al. 1979), *trp* (Seeburg et al. 1983), and *PL* (Hwang et al. 1990) (Studier and Moffat 1986) and synthetic *tac* and *trc* (Amann et al. 1983; Amann and Brosius 1985) promoters. Depending on the vector system used, heterologous proteins in the cytoplasm of *E. coli* characteristically accumulate to levels ranging from 1 up to 40% of the total protein (Beernink and Tolan 1992).

Fusion Proteins

A heterologous protein gene can be fused to the N-terminal coding region of a bacterial gene (e.g. to a portion of the *lacZ* gene of *E. coli*) to synthesize a fusion protein. Transcription and translation initiation of a fusion protein is directed by native *E. coli* sequences. The expression level is often improved compared to the expression level of a native heterologous protein, which is often recognized as foreign and degraded by endogenous proteases (Marston 1986). The main application of fusion proteins is their use as antigens in antibody production, where large-scale production of the protein is desirable. A typical vector system is the pUR series of vectors (Rüther and Müller-Hill 1983), which target the expression of *E. coli lacZ* gene fused to virtually any cloned gene, which can be cloned to an appropriate cloning site in all three reading frames. If a proper cleavage site has been placed between C-terminus of the prokaryotic sequence and the N-terminus of the eukaryotic coding sequence, it can be cleaved enzymatically or chemically (Marston 1986). Fusion strategy can also be used for purification of the produced proteins (see below).

Native Proteins

When studying the function of the gene product (e.g. enzymes), it has to be expressed in its native form. An "ATG vector" is then a good choice. In these vectors, an ATG initiation codon is included in the vector itself right after the ribosome-binding site. A coding sequence cut with an appropriate restriction enzyme can be ligated to a corresponding restriction site immediately preceding the ATG codon, resulting in in-frame expression of the native protein. For efficient expression of eukaryotic genes, in addition to a strong promoter, an efficient ribosome-binding site (e.g. lacZ RBS) at the correct distance from the start codon must also be included in the vector (Amann and

Brosius 1985). An example of an "ATG vector" is pKK233-2, which has been used to express a large variety of proteins in *E. coli* (Amann and Brosius 1985). A good description of most commonly used vector systems for expressing both fusion and native proteins is given in and Ausubel et al. (1987) and Sambrook et al. (1989). A more comprehensive list of expression vectors can be found in Pouwels et al. (1985).

Secreted Proteins

The expression systems described above accumulate the produced proteins in the intracellular space. However, *E. coli* also secretes proteins either to the periplasmic space between the cytoplasmic membrane and outer membrane or across the outer membrane to the extracellular space (Michaelis and Beckwith 1982). The secreted proteins are synthesized as precursors with an N-terminal signal sequence, which may be cleaved during secretion. Secretion of proteins has some advantages: some proteins that are degraded in the cell are stable when secreted to the periplasm (Talmadge and Gilbert 1982), correct folding of the proteins have been shown to happen simultaneously with secretion (Pollitt and Zalkin 1983) and proteins do not have the amino-terminal methionine, if the signal sequence is correctly processed (Marston 1986). Prokaryotic signal sequences from β-lactamase (Villa-Komaroff et al. 1978), OmpA protein (Ghrayeb et al. 1984) and alkaline phosphatase (phoA, e.g. in expression vector pTA1529) (Oka et al. 1985) have been used to secrete eukaryotic polypeptides into the periplasm of *E. coli*. However, if a large-scale production of a heterologous protein by secretion is required, *Bacillus subtilis* provides a more efficient system (Palva et al. 1982; Nicaud et al. 1986).

Eukaryotic Hosts

To be biologically active it is often necessary to express the eukaryotic genes in eukaryotic environment, e.g in yeast or insect cells. For instance, for correct post-translational processing of the proteins (e.g. signal cleavage, glycosylation, proteolysis and phosphorylation) the baculovirus/insect cell system has proven to be functional (Miller 1988). A good description of the vector systems used in these hosts is given in Old and Primrose (1989) and Miller (1988).

4.2.2
Vector Construction and Expression of Protein

In the following example, the construction of an expression vector pET3a with a cloned foreign gene inserted for production of a native protein is described. pET3a (Novagen) can be used for efficient production of prokaryotic and eukaryotic proteins both as fusion proteins and as native forms (Studier et al. 1990; Studier and Moffat 1986). pET3a carries the bacteriophage T7 gene 10 promoter, translation start and transcription terminator. In addition, it has an NdeI restriction site immediately preceding the ATG initiation codon. The promoter is regulatable, i.e. the host bacteria can be first grown to log phase, before the production of the protein is induced (by IPTG), diminishing the possible toxic effect of the protein on plasmid stability and growth of the cells. Vector construction and transformation are performed using conventional cloning techniques and are not explained here in detail. Expression of the gene is a modification of the methods of Citovsky et al. (1990) and Sambrook et al. (1989).

Procedure

1. Preparing the gene fragment. Add the NdeI restriction site in the correct reading frame upstream of the coding sequence and a BamHI site downstream of the coding sequence using for instance PCR or synthetic linkers. When using PCR, the restriction sites can be included in the oligonucleotide primers.

2. Digest the vector and the insert with NdeI and BamHI. Note that NdeI is sensitive to impurities. Use phenol-chloroform-purified DNA preparations. Digest first with NdeI.

3. Purify the digested insert and vector in an agarose gel.

4. Ligate the purified vector and insert. Check the result in a gel. The insert should now be placed immediately downstream from the ATG codon of the vector in the correct reading frame. This can be checked by sequencing.

5. Transform the vector in *E. coli* BL21(DE3) pLysE cells (Novagen), which contain the bacteriophage T7 RNA polymerase recognizing only bacteriophage T7 promoters. pLysE cells contain also a plasmid that provides a small amount of T7 lysozyme, a natural inhibitor of T7 RNA polymerase, which helps to stabilize the target gene (Moffatt and Studier 1987).

6. Spread the transformation mix on LB plates, containing chloramphenicol 34 μg/ml (pLysE selection) and 60 μg/ml ampicillin or carbenicillin (pET3a selection).
7. Screen the O/N grown colonies for inserts, e.g. by plasmid DNA extraction and restriction or by PCR.
8. Screen a few positive colonies for expression by inoculating NZCYM or LB medium (containing chloramphenicol 34 μg/ml, ampicillin 100 μg/ml) with one colony/tube and incubate O/N at 37 °C to obtain a saturated culture.
9. Inoculate 5 ml of NZCYM or LB medium (containing the antibiotics) with 50 μl of saturated culture and allow the bacterial cell culture to grow them at 37 °C shaking at 225–250 rpm until the cell density is 0.6–0.8 (A600).
10. Induce expression by adding IPTG to a final concentration of 1 mM and incubate the cultures for a further 2.5 h.
11. Remove 1 ml aliquots of the cultures and centrifuge 12000 *g* for 1 min. Remove the supernatant.
12. Resuspend each pellet in 100 μl of 1x SDS gel-loading buffer, heat to 100 °C for 3 min.
13. Load 10 μl of each sample in an SDS-polyacrylamide gel of appropriate concentration. Use as a control a sample of an uninduced bacterial population and another induced and uninduced sample transformed with the vector alone. The method for running an SDS-PAGE gel is described in Section 4.2.3.2.
14. After the run, stain the gel with Coomassie Brilliant Blue or carry out a Western blot, if the antibody is already available. A positive result in a stained gel is a "fat" band that is usually clearly distinguishable from the negative controls and from the native *E. coli* proteins.

Troubleshooting

Heterologous protein production is a sensitive process, which can be negatively affected by several different factors, which all give the same negative end result in the gel.

- *Plasmid instability.* Toxicity of the protein product to the host cell often causes plasmid instability. If the plasmid is to be maintained in a significant fraction of the cells, the culture must not be allowed to grow in the absence of selection of the plasmid. Also raising the ampicillin concentration up to 200 μg/ml or changing it to carbenicillin often has a stabilizing effect on the plasmid. Because of the potential loss of the plasmid, it is recommended to determine the composition of the cells in the culture by plating them on LB and LB + amp agar just before induction. Some genes give the best expression right after transformation.

- *Degradation.* Toxicity of the protein product to the host cell can also lead to rapid degradation of the foreign protein. This can be at least partly avoided by using protease inhibitors, such as PMSF (1 mM), Pepstatin A (20 mg/ml), Leupeptin (20 mg/ml) or Aprotinin (1% (v/v)) and working in a cold room.

- *Insolubility.* In the majority of cases, the expressed proteins are present in a protease protected form as insoluble inclusion bodies. In this case, proteins may move in the gel as only partly denatured aggregates giving rise to bands of unexpected size. Solubilization of the inclusion bodies is described in Section 4.2.3.

- *Poor staining.* In some cases, the proteins cannot be stained with Coomassie Brilliant Blue, preventing their detection in the gel. In these cases, alternative staining methods, such as 0.2 M KCl and 4 M ammonium acetate precipitation of the proteins in the gel or other commercially available staining systems can be used (see below). Also Western blotting can be used to visualize proteins which cannot be detected with Coomassie Blue.

4.2.3 Protein Purification

Inclusion Bodies

Several different purification methods can be used depending mainly on whether the protein is produced in a fusion or native form. In both cases, the majority of the expressed proteins appear as insoluble inclusion bodies as mentioned before. Although, as explained earlier, the proteins have to be solubilized, the formation of the aggregates is useful for the purposes of purification in many cases. Inclusion bodies are dense and thus can be separated and pelleted from the bulk of the cell debris. The degree of purification varies between 30 and 90% (Marston 1986). For immunological studies, purified inclusion bodies can often be used as antigens without further solubilization. Methods for solubilization depend on the nature of the expressed protein and it has to be experimentally determined for each protein (Marston, 1986). Soluble and solubilized proteins can be further purified by ion-exchange chromatography, gel filtration (George et al. 1985; Gill et al. 1985) and preparative gel electrophoresis (Kleid et al. 1981).

Fusion Proteins

Purification of fusion proteins from prokaryotic sources has been facilitated by the introduction of techniques which allow rapid, single-step affinity purification. In these cases, recombinant proteins are synthesized as fusions with polypeptides such as maltose, polyhistidine and glutathione-S-transferase that can be purified using an appropriate affinity matrix, maltose (Guan et al. 1987), Ni^{2+} metal affinity resin (Hochuli et al. 1988) and glutathione affinity agarose (Smith and Johnson 1988), respectively. The native protein can be further cleaved enzymatically or chemically.

Re-Folding

If the protein is purified under denaturing conditions, removal of the denaturing agent may allow refolding of the protein with the same disulphide bonds as the original protein and reconstitution of activity. Several methods can be used including dialysis and dilution in a proper buffer (Marston 1986; Hoess et al. 1988; Darby 1990; Kohno et al. 1990).

In this example two protein purification methods are described. In the first method, inclusion bodies of heterologous protein from *E. coli* cells are purified and solubilized (a modification of a method by Citovsky et al. 1990). The second method consists of the use of SDS-polyacrylamide gel electrophoresis for the detection and purification (preparative gel electrophoresis) of the soluble or solubilized protein (a modification of a method by Kleid et al. 1981). The protein is first run in an SDS-polyacrylamide gel (SDS-PAGE). The gel slice corresponding to the correct band is then cut out of the gel. The cut slice containing the protein in denatured form can be either fragmented and used for immunization without further processing or it can be electroeluted out of the gel for dialysis to obtain the protein in a refolded form. A combination of these methods can also be used to purify the protein.

4.2.3.1
Purification of Inclusion Bodies

Reagents

- Buffer L (10 mM Tris-Cl, pH 80, 200 mM NaCl, 1 mM EDTA, 10% glycerol, 1 mM dithiotreitol, 1 mM PMSF)
- Buffer L with 1 M NaCl
- Buffer L with 1 M Nacl and 6 M urea

Equipment

- 37 °C shaker incubator
- Refrigerated centrifuge capable of at least 15000 *g*
- French cell press
- Waterbath at 70 and 56 °C

Procedure

1. Grow a 20-ml culture of BL21(DE3) pLysE cells transformed by the recombinant pET3a vector and express the gene of interest as described in Section 4.2.2.
2. Centrifuge the culture at 7000 *g* for 10 min at 4 °C.
3. Re-suspend the pellet in 2 ml of buffer L.
4. Lyse the cells in a French press cell at 20 000 psi.
5. Centrifuge the lysate at 12 000 *g* for 5 min at 4 °C.
6. Re-suspend the pellet containing most of the overproduced protein in 1 ml of buffer L with 1 M NaCl.
7. Centrifuge the suspension as above.
8. Re-suspend the pellet in 0.5 ml of buffer L with 1 M NaCl and 6 M urea and incubate for 10 min at 70 °C.
9. Centrifuge the suspension as above.
10. Re-suspend the pellet in 1.5 ml of buffer L with 1 M NaCl and 6 M urea and incubate 20 min at 56 °C.
11. Separate the solubilized protein from insoluble protein aggregates by centrifuging at 15 000 *g* for 10 min at 4 °C.

12. Dialyze the protein in a dialysis bag of the appropriate pore size in 3 × 2 l of buffer L. Aliquot the protein and store at –20 °C for further use, e.g as antigen or for functional studies.

This method can yield between 0.5 and 1 mg of purified protein.

4.2.3.2
Gel Electrophoresis

Electrophoresis is the separation of charged molecules by differential migration through a porous solid matrix (polyacrylamide, starch, paper etc.) when exposed to an electric field. In general, fractionation by gel electrophoresis is based on size, shape and net charge of the macromolecules. Among different types of electrophoresis, SDS-polyacrylamide gel electrophoresis (SDS-PAGE) is the most widely used technique to analyze protein mixtures. In this method, proteins are denatured by treatment with an anionic detergent sodium dodecylsulphate, reducing agent and heat. In the gel, denatured molecules bind to SDS and become negatively charged. While migrating towards the anode they are separated according to their molecular weight (Laemmli 1970). Polyacrylamide gels are formed by polymerization of acrylamide monomers and by cross-linking with N′-methylene bisacrylamide. The proteins are first concentrated in highly porous stacking gel and then separated in the resolving gel.

The effective range of separation of the gel depends on the absolute concentrations of acrylamide and bisacrylamide used to cast the gel usually varying from 5 to 15% (separation range of the proteins 212–12 kDa). The volume of the gel mixtures depends on the measures of the electrophoresis apparatus. In the following example, a 15% resolving gel (separation range 12–43 kD), in a volume of 10 ml and a stacking gel in a volume of 2 ml, are enough for two gels with the size of 8 cm × 7.3 cm and thickness 0.75 mm. This gel size can be mainly used for analytical studies. For preparative purposes, a gel size 16 mm × 14 cm with a thickness from 0.75 to 1.5 gives a better separation of protein bands. Mix the components in the given order. Note that polymerization of the gel starts after addition of TEMED. Cast the gel right after adding TEMED.

Reagents

- 29% (w/v) acrylamide; 1% (w/v) N,N′ -methylenebisacrylamide stock solution. Note that they are potent neurotoxins

- 10% sodium dodecyl sulphate stock solution (electrophoresis grade)
- 10% ammonium persulphate (make a fresh solution daily)
- TEMED (N,N,N',N',-tetramethylenediamine (electrophoresis grade)
- 1.5 M Tris (base), pH 8.8
- 1.0 M Tris (base), pH 6.8
- 1x Tris-glycine electrophoresis buffer (25 mM Tris, 250 mM glycine (electrophoresis grade), pH 8.3, 0.1% SDS). Prepare a 5×stock solution
- 1x SDS gel-loading buffer (50 mM Tris-Cl pH 6.8, 100 mM dithiothreitol, 2% SDS, 0.1% bromophenol blue, 10% glycerol)
- Resolving gel

H_2O	2.3 ml
30% acrylamide mix	5.0 ml
1.5 M Tris (pH 8.8)	2.5 ml
10% SDS	0.1 ml
10% ammonium persulphate	0.1 ml
TEMED	0.004 ml

- stacking gel:

H_2O	1.4 ml
30% acrylamide mix	0.33 ml
10 M Tris (pH 6.8)	0.25 ml
10% SDS	0.02 ml
10% ammonium persulphate	0.02 ml
TEMED	0.002 ml

Sample Preparation

The amount of protein to be applied in each well depends on the nature of the sample, and the optimal concentration should be determined in preliminary experiments. Generally, 1–5 µg of protein (e.g. cell extracts in protein extraction buffer) per band is adequate for good separation and visualization of the bands. Before electrophoresis the samples are boiled for 3–5 min in 1xSDS gel loading buffer (pellet) or 2xSDS gel loading buffer (solution).

Procedure

1. Thoroughly clean the glass plates with alcohol and water (wear gloves).
2. Assemble the electrophoresis apparatus according to the manufacturer's instructions.

3. Pour the resolving acrylamide mixture between the plates. Leave sufficient space for the stacking gel (the length of the teeth of the comb plus 1 cm). Carefully overlay the solution with isobutanol or water (prevents oxygen inhibiting polymerization).

4. After the gel has set (20–30 min), pour off the overlay and dry carefully with paper.

5. Pour the stacking gel solution directly onto the polymerized resolving gel and put the appropriate comb into the solution, being careful not to trap air bubbles.

6. After polymerization, carefully remove the comb. Add running buffer to the top reservoir.

7. Load samples (volume depends on the teeth size of the comb and thickness of the gel) into the bottom of the wells with a microlit syringe or a micropipette fitted with a long narrow tip. As a protein size marker it is advisable to use pre-stained markers (Sigma Co.), because their migration on the gel can be followed during the run, allowing the optimization of the time length of the run.

8. Attach the apparatus to the power supply and apply a voltage of 8 V/cm in the stacking gel and 15 V/cm for resolving gel (follow the dye front).

9. When the dye front reaches the bottom of the gel, turn off the power pack and disassemble the apparatus.

The gel is now ready for staining and/or blotting.

Analytical Studies

Visualization of the gel

If the gel is used for analytical purposes, staining with Coomassie Brilliant Blue (detection level 0.1 μg of protein in a single band) is most commonly used. A simple staining procedure is described below.

Staining solution. Mix 2.5 ml 1% Coomassie Brilliant Blue R 250 in ethanol with 50 ml 6% trichloroacetic acid. Scale the volume up or down according to the size of the gel.

Incubate the gel in the staining solution from 4 h to O/N. Note that prolonged incubation may stain the background violet. De-stain the gel in distilled water if necessary (Fig. 4.2.1).

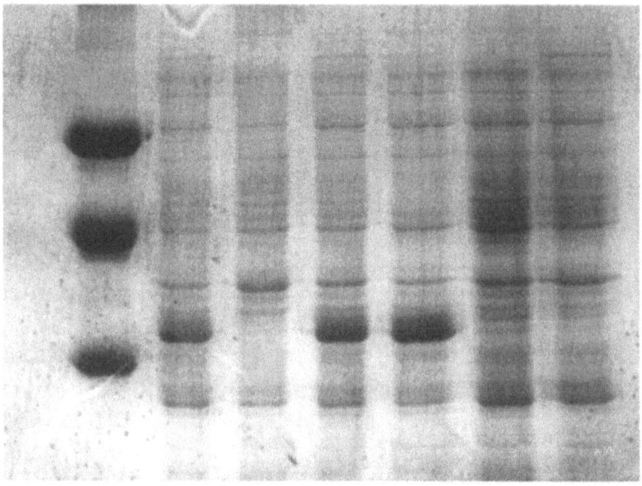

Fig. 4.2.1. Polyacrylamide gel showing: lane 1 molecular weight marker, 66, 45 and 29kDa; lane 2 pET3a with P1 gene sequence of potato virus Y, induced; lane 3 pET3a with P1 gene sequence of potato virus Y, uninduced; lane3 4–5 as *2* except that the amount of protein loaded onto the gel was 2x and 3x the amount loaded in *2;* lane 6 "empty" pET3a plasmid, induced; lane 7 "empty" pET3a plasmid, uninduced. The *E.coli* host strain was pLysE. Expression of the protein was induced with 1 mM IPTG (final concentration)

For a more sensitive detection, silver staining (detection level 0.1–1.0 ng of protein in a single band) is recommended (Merril et al. 1984).

Preparative Studies

Visualize the protein bands incubating the gel in 0.2 M KCl at 4 °C for 3 min. The protein bands can be seen as opaque zones against a clear background. Alternatively, the gel can be incubated in 4 M sodium acetate for 60 min. In this case, the protein bands appear as clear zones against a clouded background. Cut out the band of the correct size. At this point, the protein can either be electroeluted out of the gel piece (see below) or the gel piece can be fractionated, e.g. by freezing and thawing, and mixed with an equal volume of adjuvant and injected into an animal.

Electroelution

Electrophoretic separation of proteins in various types of polyacrylamide gels is used from analytical to a preparative scale. After separation, it is often necessary to extract or elute a specific protein from the gel for further studies such as amino acid composition, sequence analysis, peptide mapping or to raise specific antisera. Several devices can be used for efficient electroelution. A simple way to carry out the elution is to use an electrophoresis apparatus for agarose gels.

1. Place the gel piece in a dialysis bag with SDS-PAGE running buffer and place the bag in a transverse position in an electrophoresis tank. Fill the tank with the SDS-PAGE running buffer so that the dialysis bag is completely immersed in the buffer.
2. Attach the electrophoresis tank to the power supply and apply 80 V voltage for the gel pieces for 1–2 h. In the end of the elution reverse the current for 5 min to remove the protein from the inner walls of the dialysis bag.
3. Discard the gel and dialyze the buffer with the protein in 3 × 2 l of an appropriate buffer such as buffer L or PBS. Aliquot the protein and store in –20 °C for further use.

Troubleshooting

- *Insolubility.* If the method described above does not work for solubilization of the inclusion bodies, several other agents can be tried. These include 5–8 M guanidinium chloride, detergents such as NP-40 and sodium dodecyl sulphate (SDS) and under different pH conditions. Also the incubation times and temperatures can be changed. However, the conditions should be kept as mild as possible, because these stringent denaturing conditions can chemically alter the protein in an irreversible manner (Marston 1986).
- *Precipitation of the protein.* When the denaturant is removed during dialysis, the protein can precipitate back to an insoluble form. Often, this can be avoided by gradually decreasing the concentration of the denaturant using low-protein concentrations during the dialysis. However, in many cases, for antibody production the insoluble protein precipitate can be successfully used. To improve the immunizing effect, inclusion bodies can be broken into smaller pieces by sonication.

- *Poor visualization.* The accuracy of the bands visualized by 4 M ammonium acetate and 0.2 M KCl is sometimes insufficient. However, they are easy to use and the gels do not have to be destained before recovering the protein. Coomassie Brilliant Blue can also be used, but quite stringent staining and destaining conditions, which can also alter the chemical structure of the protein, have to be used. There are commercially available staining systems (Promega) that are sensitive, give sharp bands and are easy to use. Also destaining of the gels is easy, allowing good recovery of the protein.

4.2.4 Western Blotting

Blotting, defined as the process of transferring macromolecules from gels to an immobilizing matrix (Gershoni and Palade 1983), was first performed by Southern in 1975 with the transfer of DNA from agarose gels to nitrocellulose membranes (Southern blotting). Since then, blotting has been applied also to RNA (northern blotting; Alwine et al. 1979) and proteins (Western blotting; Towbin et al. 1979; Bowen et al. 1980) in both agarose and polyacrylamide gels. In Western blotting, protein mixtures, e.g. from cell extracts, separated in a polyacrylamide gel, are electrotransferred to a solid membrane like nitrocellulose or polyvinylidene difluoride (PVDF), where they are accessible for immunodetection with specific antibodies. The proteins are incubated with the antibodies, which in turn are coupled to enzymatic labels (e.g. alkaline phosphatase and horseradish peroxidase) or radioactive labels to visualize the antigen-antibody complexes allowing detection and identification of the proteins. In addition to identification of specific peptides western blotting can be used also for purification and concentration of antigen (Olmsted 1981). A major limitation of this method is that the proteins bound to the membrane are denatured and may not contain the same conformational and structural antigenic determinants present in the native protein. Therefore, only antibodies which recognize determinants consisting of a specific amino acid sequence will be useful in this technique. In practice, this means that only polyclonal antibodies or mixtures of monoclonal antibodies can be successfully used.

4.2.4.1
Electroblotting

Reagents

- Transfer buffer (39 mM glycine, 48 mM Tris base, 0.037% SDS (electrophoresis grade), 20% methanol)

Equipment

- Two types of electrophoresis apparatus are available for electroblotting. In the older model the polyacrylamide gel and the membrane between Whatman 3-MM paper are sandwiched together in a buffer tank and an electric field is applied across the sandwich. This procedure requires large volumes of buffer, an extended period of transfer (1 h to overnight) and only one or two gels can be processed at the same time. Recently, a semi-dry electroblotting technique has been developed in which six or more gels can be simultaneously processed in about 30 min using a reduced volume of buffer. The latter technique is described below.

Procedure

1. Prepare protein samples and run an SDS-PAGE gel as described in Section 4.2.3.2.
2. Rinse electrode plates of the semi-dry apparatus with distilled water.
3. Wearing gloves, cut six pieces of Whatman 3-MM paper and one piece of nitrocellulose, which are exactly the size of the gel.
4. Float the nitrocellulose filter on the surface of deionized water or transfer buffer and allow it to wet from beneath by capillary action, then submerge the filter for several minutes. Wet also the 3-MM paper in transfer buffer.
5. Place on the bottom electrode (the anode) three pieces of wet 3-MM paper so that they are exactly aligned.
6. Place the nitrocellulose filter on the stack of 3-MM paper, exactly aligned. Remember to mark the membranes with soft pencil.
7. Place the polyacrylamide gel, which is rinsed briefly in deionized water, on the nitrocellulose, exactly aligned.
8. Finally place three pieces of wet 3-MM paper on the top of the gel, again exactly aligned.

9. Remove gently the air bubbles either using a gloved hand or by rolling a pipe over the sandwich.

10. Place the upper electrode and connect the electrical leads. Apply a current of 0.65–0.8 mA/cm^2 of gel for 1.5–2.0 h.

11. After the transfer, the nitrocellulose membranc can be now stained and/or used for immunological visualization of proteins.

4.2.4.2
Immunological Visualization of Proteins

There is a large variety of Western blotting procedures, and in many cases. the choice is just a matter of personal preference. It only has to be remembered that the sensitivity in Western blotting is primarily a function of the specific antibody titre of the immune serum being utilized (Burnette 1981). Especially antibodies with low affinity are influenced by buffer conditions, incubation times and relative concentrations. Because polyclonal antibodies are undefined mixtures of individual immunoglobulins, the specificity, affinity and concentration of which are usually unknown, it is important to include proper controls, i.e. irrelevant antibodies or pre-immune serum and control preparations with known amounts of target antigen or lack it altogether, to the design of the experiment.

Reagents

- Specific antiserum or purified antibodies
- Anti-IgG AP conjugate
- PBS, pH 7.4
- 150 mM NaCl, 50 mM Tris-Cl, pH 7.5
- Ponceau S
- Blocking solution (7.5% or 1% (w/v) non-fat dried milk, 0.01% antifoam A, 0.02% sodium azidc in PBS, pH 7.4)
- Phosphate-free, azide-free blocking solution (1% (w/v) nonfat dried milk, 150 mM NaCl, 50 mM Tris-Cl, pH 7.5)
- NBT (0.5 mg/10 ml of 70% dimethylformamide)
- BCIP (0.5 g/10 ml of 100% dimethylformamide)
- Alkaline phosphatase buffer (100 mM NaCl, 5 mM $MgCl_2$, 100 mM Tris-Cl, pH 9.5)

Procedure

In the following example, carry out all the incubations with gentle shaking at room temperature in shallow vessels, the size of which is only slightly larger than the size of the membranes. For the antibody incubations, use the minimum amount of solution to cover the membranes. Do not let the membranes dry out at any point during the procedure.

1. After blotting, stain the membrane with Ponceau S according to manufacturer's instructions to check that blotting has succeeded. Mark the stained molecular markers and protein lanes with pencil. Ponceau S will stain only transiently and is washed away during the incubations without interfering in the detection of antigens.
2. Block the membranes in a blocking solution for 1 h.
3. Incubate with specific antibodies in 1% blocking solution for 2 h. Recommended dilutions for polyclonal antibodies are 1:100 to 1:5000.
4. Wash the membranes 3 × 5–10 min in PBS and once in 150 mM NaCl, 50 mM Tris, pH 7.5.
5. Incubate with the anti-IgG AP (dilution according to the manufacturer's instructions) for in phosphate-free, azide-free blocking solution for 1 h.
6. Wash the membranes 4 × 5–10 min in 150 mM NaCl, 50 mM Tris-Cl, pH 7.4.
7. Mix 66 μl of NBT stock with every 10 ml of alkaline phosphatase buffer. Add 33 μl of BCIP stock for every 10 ml of alkaline phosphatase buffer. Use this mixture in 30 min.
8. Incubate the membranes in the solution, monitoring the progress of the colour reaction. When the bands are of desired intensity, stop the reaction rinsing the membranes in distilled water.
9. Store the membranes at room temperature protected from light.

Troubleshooting

- *Poor transfer from the gels.* Especially when using high-percentage gels and when the transferred protein molecules are large, the proteins may remain in the gel matrix. This can be easily checked by staining the gel with Coomassie Brilliant Blue after blotting. Reducing the gel percentage and, in some cases, in situ proteolysis of the proteins (Gibson 1981) may be tried. Also too short transfer times and too low voltages can impair the transfer. Poor contact between the membrane and the gel (air bubbles, excess moisture) can lead to a diffuse transfer and swirls.

- *Poor binding to the membrane.* Proteins < 15 kDa may show diminished binding to 0.45 micron nitrocellulose. Usage of 0.2 micron nitrocellulose is recommended. Also SDS and NP-40 used in immune assay detections may remove proteins from the nitrocellulose. They can be often replaced with Tween-20 detergents.

- *High background.* Several different factors can contribute to high background staining: improper blocker protein (e.g. BSA) or its concentration and blocking time; the primary or the secondary antibody are in too high a concentration or post antibody reaction washes are not effective enough.

- *Poor detection.* Enzyme conjugate or substrate is inactivated or the antibodies are inactive. Also antibody reaction times can be insufficient or too little sample is loaded on the gel.

- *Nonspecific detection.* Main reason usually is the contamination of the primary antibody with nonspecific or species cross-reactive IgG. In these cases, antibodies can be immunoaffinity purified (Harlow and Lane 1988; Sambrook et al. 1989) or by including in the antibody preparation a suspension of competitor proteins, prepared by acetone precipitation, that do not contain the antigen of interest (Harlow and Lane 1988). Also too high concentration of the primary or secondary antibody can lead to non-specific detection.

References

Alwine JC, Kemp DJ, Parker BA, Reiser J, Renart J, Stark GP, Wahl GW (1979) Detection of specific RNAs or specific fragments of DNA by fractionation in gels and transfer to diazobenzyloxymethyl paper. Methods Enzymol 68:220–242

Amann E, Brosius J (1985) "ATG vectors" for regulated high level expression of cloned genes in *Escherichia coli*. Gene 40:183–190

Amann E, Brosius J, Ptashne M (1983) Vectors bearing a hybrid *trp-lac* promoter useful for regulated expression of cloned genes in *Escherichia coli*. Gene 25:167–178

Ausubel FM, Brent R, Kingston RE, Moore DD, Seidman JG, Smith JA, Struhl K (eds) (1987) (Suppl 1988 and 1989) Current protocols in molecular biology. Greene Publishing Associates and Wiley -Interscience, John Wiley & Sons, New York

Beernink PT, Tolan DR (1992) Construction of a high-copy "ATG vector" for expression in *Escherichia coli*. Prot Exp Purif 3:332–336
Bowen P, Steinberg J, Laemmli UK, Weintraub H (1980) The detection of DNA-binding proteins by blotting. Nucleic Acids Res 8:1–20
Burnette WN (1981) "Western blotting": electrophoretic transfer of proteins from sodium dodecyl sulfate-polyacrylamide gels to unmodified nitrocellulose and radiographic detection with antibody and radioiodinated protein A. Anal Biochem 112:195–203
Chang ACY, Cohen SN (1974) Genome construction between bacterial species in vitro: replication and expression of *Staphylococcus* plasmid genes in *Escherichia coli*. Proc Natl Acad Sci USA 71:1030–1034
Citovsky V, Knorr D, Schuster G, Zambryski P (1990) The P30 movement protein of tobacco mosaic virus is a single-strand nucleic acid-binding protein. Cell 60:637–647
Darby NJ, Creighton TE (1990) Folding proteins. Nature 344:715–716
George HJ, L'Italien JJ, Pilancinski WP, Glassman DL, Krzyzek RA (1985) High-level expression in *Escherichia coli* of biologically active bovine growth hormone. DNA 4:273–281
Gersoni JM, Palade GE (1983) Protein blotting: principles and applications. Anal Biochem 131:1–15
Ghrayeb J, Kimura H, Takashara M, Hsiong H, Masui Y, Inovye M (1984) Secretion cloning vectors in *Escherichia coli*. EMBO J 3:2437–2442
Gibson W (1981) Protease-facilitated transfer of high-molecular-weight proteins during electrotransfer to nitrocellulose. Anal Biochem 118:1–3
Gill JA, Sur JP, Donaldson EM, Dye HM, Souza L, Berg T, Wypch J, Langley K (1985) Recombinant chicken and bovine growth hormones accelerate growth in aquacultured juvenile Pacific salmon *Oncorhynchus kisutch*. Bio/Technology 3:643–646
Guan C, Li P, Riggs PD, Inouye H (1987) Vectors that facilitate the expression and purification of foreign peptides in *Escherichia coli*. Gene 67:21–30
Goeddel DV, Heyneker HL, Hozumi T, Arentzen R, Itakura K, Yansura DG, Ross MJ, Miozzari G, Crea R, Seeburg PH (1979) Direct expression in *Escherichia coli* of a DNA sequence coding for human growth hormone. Nature 281:544–548
Harlow E, Lane D (1988) Antibodies: laboratory manual. Cold Spring Harbor Laboratory Press, Cold Spring Harbor
Hochuli E, Bannwarth W, Dobeli H, Gentz R, Stuber D (1988) Genetic approach to facilitate purification of recombinant proteins with a novel metal chelate absorbant. Bio/Technology 6:1321–1325
Hoess A, Arthur AK, Wanner G, Fanning E (1988) Recovery of soluble biologically active recombinant protein from total bacterial lysates using ion exchange resin. Bio/Technology 6:1214–1217
Hwang L-H, Tsi H-F, Liu S-T (1990) High level expression of porcine growth hormone in *Escherichia coli* from an expression vector containing bacteriophage PL and N gene untranslated region. Biochem Biophys Res Commun 173:711–717
Kleid PG, Yanasura D, Small B, Dowbenko D, Moore D, Grubman M, McKercher P, Morgan DO, Robertson BH, Bachroch HL (1981) Cloned viral

protein vaccine for foot-and-mouth disease: responses in cattle and swine. Science 214:1125–1129
Kohno T, Carmichael DF, Sommer A, Thompson RC (1990) Refolding of recombinant protein. Methods Enzymol 185:187–195
Laemmli UK (1970) Cleavage of structural proteins during the assembly of the head of bacteriophage T4. Nature 227:680–685
Marston AO (1986) The purification of eukaryotic polypeptides synthesized in *Escherichia coli*. Biochem J 240:1–12
Merril CR, Goldman D, van Keuren ML (1984) Gel protein stains: silver stain. Methods Enzymol 104:441–447
Miller LK (1988) Baculoviruses as gene expression vectors. Annu Rev Microbiol 42:177–199
Michaelis S, Beckwith J (1982) Mechanism of incorporation of cell envelope protein in *Escherichia coli*. Annu Rev Microbiol 36:435–465
Moffat BA, Studier FW (1987) T7 lysozyme inhibits transcription by T7 RNA polymerase. Cell 49:221–227
Nicaud J-M, Mackman N, Holland IB (1986) Current status of secretion of foreign proteins by micro-organisms. J Biotechnol 3:255–270
Oka T, Sakamoto S, Miyoshi K-I, Fuwa T, Yoda K, Yamasaki M, Tamura G, Miyake T (1985) Synthesis and secretion of human epidermal growth factor by *Escherichia coli*. Proc Natl Acad Sci USA 82:7212–7216
Old RW, Primrose SB (1994) Principles of gene manipulation: an introduction to genetic engineering, 5th EDN. Blackwell, Oxford
Olmsted JB (1981) Affinity purification of antibodies from diazotized paper blots of heterogenous protein samples. J Biol Chem 256:11955–11957
Palva I, Sarvas M, Lehtovaara P, Sibakov M, Kääriäinen L (1982) Secretion of *Escherichia coli* β-lactamase from *Bacillus subtilis* by the aid of α-amylase signal sequence. Proc Natl Acad Sci USA 79:5582–5586
Pollitt S, Zalkin H (1983) Role of primary structure and disulfide bond formation in lactamase secretion. J Bacteriol 153:27–32
Pouwels PH, Enger-Valk BE, Brammar WJ (1985, suppl. 1988) Cloning vectors: a laboratory manual. Elsevier, Amsterdam
Rüther U, Müller-Hill B (1983) Easy identification of cDNA clones. EMBO J 2:1791–1794
Sambrook J, Fritch EF, Maniatis T (1989) Molecular cloning: a laboratory manual, 2nd edn. Cold Spring Harbor Laboratory Press, Cold Spring Harbor
Seeburg PH, Sias S, Adelman J, de Boer HA, Hayflick J, Jhurani P, Goeddel DV, Heyneker HL (1983) Efficient bacterial expression of bovine and porcine growth hormones. DNA 2:37–45
Smith DB, Johnson KS (1988) Single step purification of polypeptides expressed in *Escherichia coli* as fusions with glutathione-s-transferase. Gene 67:31–40
Southern EM (1975) Detection of specific sequences among DNA fragments separated by gel electrophoresis. J Mol Biol 98:503–517
Struhl K, Cameron JR, Davis RW (1976) Functional genetic expression of eukaryotic DNA in *Escherichia coli*. Proc Natl Acad Sci USA 73:1471–1475
Studier FW, Moffatt BA (1986) Use of bacteriophage T7 RNA polymerase to direct selective high-level expression of cloned genes. J Mol Biol 189:113–130

Studier FW, Rosenberg AH, Dunn JJ, Dubendorff JW (1990) Use of T7 RNA polymerase to direct expression of cloned genes. Methods Enzymol 185:60–88
Talmadge K, Gilbert W (1982) Cellular location affects protein stability in *Escherichia coli.* Proc Natl Acad Sci USA 79:1830–1833
Towbin H, Staehelin T, Gordon J (1979) Electrophoretic transfer of proteins from polyacrylamide gels to nitrocellulose sheets: procedure and some applications. Proc Natl Acad Sci USA 76:4350–4354

Part II

Characterization of Plant DNA

Chapter 5

Organelle DNA Isolation

M. LANDGREN and K. GLIMELIUS

5.1 Introduction

Green plants and photosynthetic organisms are unique among eukaryotes since they possess three genomes: the nuclear, the mitochondrial and the plastid. The first indication that genes were present in the cytoplasm was discovered by the inheritance of certain traits that did not follow Mendel's rules. Even though these observations were made as early as the beginning of the 20th century, it was not until the 1960s that convincing physical evidence was obtained regarding presence of unique DNA in chloroplasts (Chun et al. 1963; Sager and Ishida 1963) and mitochondria (Luck and Reich 1964). Since then, the rapid development in molecular techniques has contributed enormously to our knowledge about genome organization, gene content and gene structure. For details of the genetics and molecular biology of the organelles in higher plants, the readers are referred to review articles in the field (Lonsdale 1989; Gray 1990; Bogorad and Vasil 1991; Herrmann 1992).

Size, Base Composition and Buoyant Density

The organelle genomes vary in size depending on species. The size of chloroplast DNA (cpDNA) for higher plants varies between 120 and 160 kb, but can be as large as 217 kb, which has been found in, e.g. *Geranium* (Palmer 1991). Mitochondrial DNA (mtDNA) varies between 200 kb in *Brassica* species to 2500 kb in Cucurbitaceae (Lonsdale 1989). Both the chloroplast and mitochondrial genomes are present as circular molecules and, in the case of mtDNA, in a multicircular organization (Palmer 1992). The buoyant density for cpDNA in higher plants ranges from 1.696–1.706 g cm^{-3} (Kirk and Tilney-Bassett 1978), and is equivalent to 37–40% G+C. For most higher plants, mtDNA has a remarkably uniform buoyant density from 1.705–1.707 g cm^{-3}, with ca. 40–47% G+C (Leaver and Gray 1982; Gray 1989).

Genome Structure

Repeated sequences present in the genomes of both cpDNA and mtDNA play an important role in the physical organization of such genomes. The chloroplast genomes have very few repeated sequences; however, most cpDNAs have a large inverted duplication. In the case of mtDNA, most genomes contain several repeats of different sizes. Most of these repeats are directly oriented duplications and are of importance for the intragenomic recombination processes resulting in re-organizations of the master circle into sub-genomic circles (Palmer 1991). The number of generated re-arrangements depends on the number of repeats involved in the recombinations, which varies between species. Besides these major recombinational events, recombination between short, repeated mtDNA sequences, also occurs at low frequency. Such rare recombinations and the sudden amplification of a newly created arrangement present in low concentration (sub-limon) have been reported by Small et al. (1987).

The mtDNA of several plant species differs from chloroplasts in that they contain extra-chromosomal elements. Small linear or circular DNA and RNA molecules have been localized to the mitochondria (Lonsdale and Grienenberger 1992). Some of these molecules have no detectable homology to high-molecular-weight mtDNA and can be lost without any apparent phenotypic effect. In other species, e.g. from the genus *Zea*, linear plasmids are present which are characterized by a polypeptide attached to the 5'-termini and a terminal inverted repeat. These episomes are under nuclear control (Escote-Carlson et al. 1990), and share homology with the master chromosome. They can integrate with the high-molecular-weight DNA via recombinations involving the inverted repeat (Schardl et al. 1984; Lonsdale et al. 1988).

Promiscuous DNA

Another interesting difference between cpDNA and mtDNA is that DNA from both the nucleus and the plastids have been integrated into mtDNA, whereas integration of foreign DNA into the cpDNA has not been found (Stern and Lonsdale 1982; Schuster and Brennicke 1988; Lonsdale and Grienenberg 1992). Some of the chloroplast genes found in the mitochondrial genome act as functional genes, while others remain inactive. In contrast, the cpDNA contains neither episomes nor promiscuous DNA, which might be due to some barrier preventing uptake of foreign DNA, or to the fact that the cp-genome is so densely packed that any random DNA insertion would lead to lethal mutations (Lonsdale and Grienenberger 1992).

Genes and Coding Capacity

Even though higher plant cpDNA, in general, is smaller than mtDNA, the chloroplast genome contains a larger number of genes than the mitochondrial genome. CpDNA has been completely sequenced in the species *Marchantia* (Ohyama et al. 1986), tobacco (Shinozaki et al. 1986) and rice (Hiratsuka et al. 1989). Higher plant cpDNA contains 120 different genes, of which 80 are protein genes (Whittier and Sugira 1992). The genes are active in two processes, gene regulation and photosynthesis. Furthermore, about 30 conserved open reading frames (ORFs) of unknown function have been reported (Palmer 1992). Gene content is relatively stable in spite of differences in size, and the same linear order of genes has been found among most vascular species investigated (Dowine and Palmer 1992). The cpDNA is compact and the genes are densely packed with short intergenic regions. Many genes are parts of operons and expressed in a poly-cistronic fashion (Whittier and Sugiura 1992).

The mtDNA largely differs from cpDNA with regard to genome organization and expression. In the case of the mitochondrial genome, a complete sequence map has not been published yet; however, sequencing of mtDNA in *Arabidopsis thaliana* is in progress (Unseld et al. 1993). According to investigations using hybridization of heterologous probes, analysis of cDNA libraries and isolation of cytoplasmic male sterility associated sequences, it has been demonstrated that mtDNA in most vascular plants contains about 50 genes, of which at least 20 are protein genes (Levings and Brown 1989; Lonsdale 1989; Lonsdale and Grienenberger 1992).

In the absence of a complete sequence of mtDNA, clear evidence for differences in gene content among different species is not available. However, DNA-DNA hybridization experiments made between species in the Cucurbitaceae having large differences in size of the mtDNA genome revealed that only a small portion, the "core DNA" was conserved and coded for genes (Stern and Newton 1985). The genes characterized are genes involved in the translation, 20 tRNA, 3 rRNA, (26S, 18S and 5S), and 18 protein genes, the latter coding for subunits of the electron transport chain complexes and the F_0 - F_1 ATP synthase (Lonsdale and Grienenberger 1992). Furthermore, a growing number of unidentified ORFs have been found that may code for essential gene products. The amount of the mt-genome needed to code for the identified genes amounts to only approximately 20 kb. Thus, compared to the densely packed cpDNA, the mtDNA contains a large portion that is non-coding, the reason for which is still unknown. Furthermore, the gene order is not conserved among different species. Intergenic regions are quite large and most of the genes are transcribed singley and not in a poly-cistronic fashion as in chloroplasts (Palmer 1992). It is apparent, though, that the large and variable size of the mtDNA is due to gene duplications, unidentified ORFs, chimeric genes and promiscous DNA. The presence of introns in the genes does not contribute much to the genome size variation. In the angiosperm mtDNA eight to ten introns have been described (Palmer 1992).

Mutants

Both chloroplasts and mitochondria are involved in the basic energy metabolism of the cell, photosynthesis and respiration. Mutations in any of the cpDNA or mtDNA genes would be likely to result in drastic effects leading to lethality. The lack of mutations has unfortunately restricted and complicated the identification and characterization of organellar genes. However, the unicellular green algae found in the genus *Chlamydomonas* have been suitable models for investigation of chloroplast genes. Since photosynthetic function can be replaced by culture of *Chlamydomonas* in a medium containing a reduced carbon source as, e.g. acetate, it has been possible to isolate mutants deficient in the photosynthetic activity. Furthermore, the rapid cell division, short sexual cycle and the fact that the algae contains only one single chloroplast makes this material excellent for analysis of the chloroplast genome. Plastid mutations have been found related to deficiences in the light and dark reactions of photosynthesis, changes in colour, and herbicide and antibiotic resistance (Hagemann 1992). Detailed analysis of these mutants has made it possible to identify genes involved in the chloroplast transcription and translation system, the electron transport complexes of the thylakoids and the CO_2-fixing enzyme Rubisco (Boynton et al. 1992; Rochaix 1992). Regarding the basic gene constitution in mtDNA, yeast has been the most extensively studied organism (Dujon 1981). However, there are also a few mtDNA mutants known affecting development of higher plants. The mutations found can result in changes in flower morphology (Beillard et al. 1979; Kofer et al. 1991), male fertility or cytoplasmic male sterility (Hanson and Conde 1986; Kaul 1988), leaf morphology (Newton and Coe 1986; Bonnett et al. 1993), overall plant development (Newton and Coe 1986; Newton et al. 1989) and sensitivity to pathogens (Dewey et al. 1987; Rottman et al. 1987; Wise et al. 1987).

Leaf Morphology Mutants

Several mutants affecting leaf morphology which cause light green or yellow stripes in the leaves have been found in *Zea mays* (Coe 1983; Newton and Coe 1986; Newton et al. 1990; Newton 1993). They are called non-chromosomal stripe (NCS) mutants, since they are maternally transmitted. Although the pigmentation phenotypes are associated with the chloroplasts (Roussell et al. 1991), the mutants are of mitochondrial nature (Newton and Coe 1986). The mutants analyzed so far have all been shown to be associated with small deletions in the mitochondrial genomes, resulting from a recombination followed by differential amplification and loss of one of the recombination products (Newton 1993). The genes affected are *nad4* in the NCS2 mutant (Newton 1993), two ribosomal proteins (*rps3* and *rpl16*) in NCS3 (Hunt and Newton 1991) and NCS4 (Newton 1993) and the *coxII* gene in NCS5 and NCS6 (Newton et al. 1990; Newton 1993).

A maternally transmitted mutant similar to the NCS mutants has recently been discovered in *Nicotiana* (Bonnett et al. 1993). A tobacco cybrid, which was obtained after a fusion between *N. tabacum* and irradiated *Petunia hybrida* (Glimelius and Bonnett 1986), exhibited variegated leaves with light and dark green sections (Bonnett et al. 1993). The light green sections result from the substitution of palisade parenchyma for spongy parenchyma. Although the original cybrid contained *Petunia* chloroplasts and mitochondria together with a tobacco nucleus, the correlation between an altered mitochondrial genome and the mutant phenotype could be shown after replacing the

Petunia chloroplasts with tobacco chloroplasts in a second round of fusions. These cybrids retained the variegated phenotype, in spite of regaining tobacco chloroplasts.

Cytoplasmic Male Sterility

Cytoplasmic male sterility (CMS) is a maternally inherited trait resulting in pollen abortion while maintaining female fertility and normal vegetative devlopment. CMS has been associated with the mitochondrial genome, either as an effect of mutations in the mtDNA or by alloplasmic combinations of nuclear and mitochondrial genomes resulting in nuclear-cytoplasmic incompatibility (Hanson and Conde 1986; Kaul 1988; Hanson 1991). In spite of detailed investigations of several different CMS systems, the molecular basis for CMS is still poorly understood. In most of the CMS systems investigated, re-arrangements of the mitochondrial genome in the male fertile progenitors, as a result of intra- or inter-molecular recombination events, have been detected. These re-arrangements or mutations are "phenotypically silent" in the original nuclear background. However, when the re-arranged mt-genome is transferred to a different nuclear background lacking restorer genes, the male-sterile phenotype is obtained. Investigations of mitochondrial genomes associated with CMS in maize (Dewey et al. 1986), petunia (Young and Hanson 1987), sunflower (Köhler et al. 1991; Laver et al. 1991), bean (Johns et al. 1992), rapeseed (Singh and Brown 1991; Bonhomme et al. 1992) and tobacco (Bergman et al. 1995) have demonstrated that the rearrangements have generated novel open reading frames (ORFs). Northern analyses have also shown that these CMS-specific ORFs are transcribed. Furthermore, protein studies revealed that several of the ORFs were translated and expressed CMS-associated polypeptides with hydrophobic characteristics (Dewey et al. 1987; Nivison and Hanson 1989; Grelon et al. 1994; Moneger et al. 1994; Abad et al. 1995). Based on these findings it has been suggested that novel CMS-specific polypeptides might be targeted to the inner mitochondrial membrane and thereby disturb the function of the electron transport chain. However, other explanations have also been suggested. In some cases, promoter sequences seem to be shared by the novel ORF and unaffected ordinary genes. A competition for limiting factors controlling expression of the novel ORF and a normal gene might lead to incorrect expression of the normal gene which could result in impairments of the mitochondrial biogenesis or function. Furthermore, alterations in expression of normal mitochondrial genes by, for instance, altered RNA processing and editing have been suggested to cause male sterility. Hernould et al. (1992) targeted the gene product translated from an unedited *atp9* RNA into the mitochondria of fertile tobacco. This resulted in 50% pollen sterility despite the presence of the endogenous subunit 9 protein. Furthermore, Iwabuchi et al. (1993) also found that the *atp6* transcript in male sterile rice was differently processed when compared to the *atp6* transcript in the fertile line.

Genetic Modifications of Organellar Genomes

In the majority of green plants, inheritance of mitochondria and chloroplasts is maternal, which means that mixing of respective organelle type after fertilization is not obtained. Thus the possibilities for recombination events are excluded. However, in a few genera such as *Oenothera* and *Pelargonium*, biparental inheritance of organelles has been found (Metzlaff et al. 1981; Chiu et al. 1988). Furthermore, paternal inheritance of plastids has been recorded in *Medicago* and *Daucus* (Schumann and Hancock 1989; Boblenz et al. 1990) and in gymnosperms (Neale and Sederoff 1989). Detailed investigations have revealed that, as a result of strong selection, a low frequency of transmission of paternal plastids has been reported in *Nicotiana* (Medgyesy et al. 1986; Avni and Edelman 1991). Paternal inheritance has also been registered as a rare event in *B. napus* (Erickson and Kemble 1990) and in *Sequoia sempervirens* (Neale et al. 1989).

The major way to obtain cytoplasmic genetic diversity, or to induce heteroplasmy, is thus to utilize somatic hybridization (see chapter 10). In all but a few cases reported (Izhar et al. 1983; Kao et al. 1992) the heteroplasmic condition usually persists only for a short period and separation of organelles into a homoplasmic population of one or the other organelle type is usually finished before plant regeneration (Chen et al. 1977; Beillard et al. 1978; Glimelius et al. 1981). Segregation of the organelle genomes in fusion products has been reported as random; however, in several cases a biased segregation has also been found (Bonnett and Glimelius 1983; Kumar and Cocking 1987; Sundberg and Glimelius 1991).

Due to the heteroplasmy in somatic hybrids, opportunities exist for recombination of organellar genomes. Chloroplasts, however, recombine rarely and recombinants have only been recovered after efficient screening by direct selection for a new combination of chloroplast markers (Medgyesy et al. 1985; Thanh and Medgyesy 1989). In contrast, intergenomic mtDNA recombination occurs frequently in somatic hybrids (Beillard et al. 1979; Nagy et al. 1981; Boeshore et al. 1983). In some fusion combinations, as much as 90–100% of the hybrids have new recombined mitochondrial genotypes (Nagy et al. 1981, 1983; Galun et al. 1982; Chetrit et al. 1985; Landgren and Glimelius 1994).

Essential to almost all studies of organellar genomes is the ability to isolate highly purified organellar DNA. A large number of protocols have been published regarding isolation of organelle DNA. In this chapter, we will present two methods which we are currently using. With these methods we obtain highly purified organellar DNA which has been used in several different applications.

5.2 Isolation of Chloroplast and Mitochondrial DNA

The basis for isolation of both cpDNA and mtDNA is essentially the same. The procedure starts with isolation of intact organelles. Organelle DNA is then extracted from the organelles. To obtain highly purified organelle DNA, it is important to separate the organellar fraction from subcellular contaminants. This is achiveved by differential centrifugation and/or gradient centrifugation. The intact organelle fraction is then lysed and the DNA is isolated by CsCl centrifugation or by phenol/chloroform extractions. Often, deoxyribonuclease (DNase) is used to degrade DNA external to the intact organelles before lysis.

The chloroplast isolation method presented here is a modification of the procedure reported by Palmer (1986). The mtDNA isolation is a modified procedure according to Bland et al. (1985). The methods have been used on green leaves. Both DNA isolation methods have been used for characterization of organellar DNA in several different *Nicotiana* species (Archer et al. 1987; Håkansson et al. 1988). Analysis of organelle content has also been performed in a large number of somatic hybrids within *Nicotiana* and Brassicaceae. (Sundberg et al. 1987; Fahleson et al. 1988; Landgren and Glimelius 1990; Kofer et al. 1991; Bonnett et al. 1993; Landgren and Glimelius 1994). The isolated cpDNA and mtDNA have been visualized on agarose gels after re-

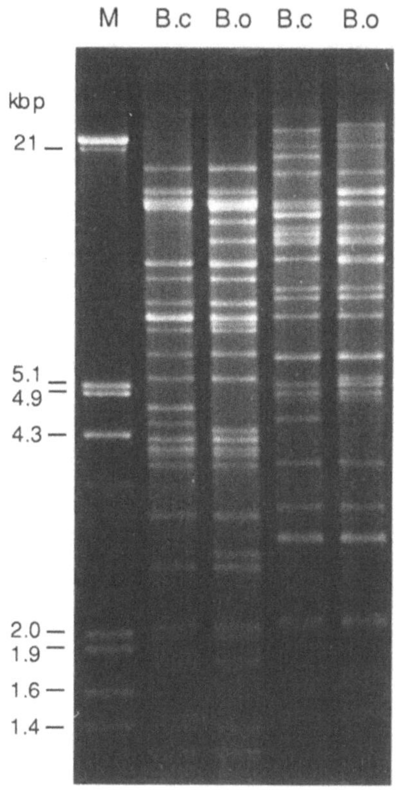

Fig. 5.1. Restriction pattern of mitochondrial DNA isolated from *Brassica campestris* (*lanes B.c*) and *B. oleracea* (*lanes B.o*) digested with BamHI (*left B.c* and *B.o lanes*) or PstI (*right B.c* and *B.o lanes*). Lambda DNA digested with EcoRI/HindIII was used as a molecular weight standard (*lane M*)

striction enzyme digestion and the DNA has also been used for filter hybridizations. Furthermore, mtDNA isolated by this procedure has been used for cloning of several different mtDNA regions in *Nicotiana* as well as in Brassicaceae.

For comprehensive reviews about cpDNA and mtDNA isolation the reader is referred to Palmer (1986) and Hanson et al. (1986).

5.2.1
CpDNA Isolation

Reagents

- Buffer A (extraction buffer): 0.3 M sorbitol, 50 mM Tris-HCl, 20 mM EDTA, 0.1% BSA is added just before use
- Buffer B (lysis buffer): 0.5% SDS, 50 mM Tris-HCl, 0.4 M EDTA, 0.1% proteinase K w/v; pH 8. Proteinase K is added just before use. Buffer B will precipitate at room temperature; thus, heat the buffer prior to use.
- TE buffer: 10 mM Tris, 5 mM Na_2EDTA, pH 7.5
- Sucrose gradient: 60, 45 and 30% sucrose dissolved in buffer A
- Nylon mesh 50 and 20 µm
- Phenol (saturated with TE buffer, pH 8.0)
- Chloroform
- 99 and 70% ethanol
- 3 M NaAc

Equipment

- Waring blender
- Refrigerated centrifuge (Sorvall, Beckman etc)
- Large centrifuge bottles 250–500 ml
- Paintbrush (a fine artist's brush)
- Ultracentrifuge with swing-out rotor and appropriate tubes
- Waterbath 50 °C
- Centrifuge tubes 50 ml for phenol extraction
- Corex tubes 30 ml
- Microcentrifuge tubes 1.5 ml
- Microcentrifuge (1.5-ml tubes)
- –20 °C freezer
- Vacuum dryer

Procedure

Day 1

All the subsequent operations are conducted at 4 °C unless otherwise specified. Pre-cool Waring blender, centrifuge bottles and use refrigerated buffers and work with buckets on ice. The *g* value given in brackets for each centrifugation refers to the R_{max} value found in the different rotor manuals.

1. Prepare 750 ml of buffer A. Add BSA just before use to 500 ml of the buffer.

2. Harvest about 25–30 g young healthy leaves. Prior to extraction, plants are kept in the dark for 24–28 h to reduce the starch level. If the plant material is infected or dirty it can be treated with Clorox (5%, 5 min) and rinsed two to three times in tap water.

3. Remove middle vein. Weigh the leaves. The method is designed for about 20 g of plant material. Cut the leaves into small pieces of about 10 cm^2. Place 10 g cut leaves in 200–250 ml buffer A containing BSA (BSA is only used in buffer A during homogenization) and homogenize the tissue in Waring blender for 2–3x 10-s bursts at high speed.

4. Filter the extract through a 50-μm nylon mesh. Repeat the homogenization and filtration with the remaining leaf pieces in 200–250 ml of buffer A with BSA. The homogenate which does not pass through the 50-μm mesh can be pooled and homogenized once more for 2x 10-s bursts followed by filtration.

5. Make a second filtration of the extract through a 20-μm nylon mesh.

6. Centrifuge the extract 3000 rpm (Sorvall GS3 rotor, 1500 *g*) for 10 min.

7. Resuspend the pellet gently in 30 ml of buffer A without BSA by using a soft paintbrush and repeat the centrifugation. Avoid resuspending the starch, which can be seen as a white spot in the bottom of the green pellet. Repeat this washing procedure four to six times, depending on the starch content.

8. Prepare the sucrose gradient during the washing procedure. The step gradient consists of 3.5 ml 60% sucrose overlayered with 3.5 ml 40% sucrose and on top 3.0 ml 20% sucrose. The different layers should be gently mixed to obtain diffuse interphases. This can be done by pulling a long Pasteur pipette gently up and down through the sucrose layers a few times. (The open end of the pipette should be be sealed.)

9. Gently resuspend the pellet in a total volume of 2–6 ml of buffer A. The suspension is loaded onto two to six tubes containing the step gradient (see Notes and troubleshooting). Balance the gradients and place the tubes in a swing-out rotor (SW 41 rotor, Beckman

L2, 65B centrifiuge) and centrifuge at 26 000 rpm (116 000 *g*) for 1 h at 4 °C.

10. The chloroplast bands at the 45–20% interface are removed with a wide-bore pipette into a 50-ml centrifuge tube.

11. Add 3 vol buffer A slowly (to avoid lysis of the chloroplasts). Start by adding drop by drop and gently mix in between (it should take 10–15 min to add the whole volume). Collect the chloroplasts by centrifugation at 5000 rpm (Sorvall SS34 rotor, 3000 *g*) for 5 min.

12. Gently resuspend the pellet in 3 ml buffer A Lyse the chloroplasts by adding 1/5 vol buffer B (Proteinase K is added just before use) and incubate the suspension for 15 min at 50 °C.

13. Extract chloroplast DNA twice at room temperature. Add 1 vol phenol (saturated with TE), mix by inverting the tube several times. Separate phases by centrifugation at 5000 rpm (Sorvall SS34 rotor, 3000 *g*) for 10 min at room temperature. Collect the upper phase and repeat the extraction with 1 vol phenol-chloroform (1:1 v/v).

14. Transfer the DNA solution (water phase) to a 30-ml Corex tub. Precipitate the DNA with 1/10 vol 3 M NaAc and 2.5 vol 99% ethanol at –20 °C over-night.

Day 2

15. Collect DNA by centrifugation at 10 000 rpm (Sorvall SS34 rotor, 12 000 *g*) for 15 min at 4 °C.

16. Decant supernatant and vacuum dry the pellet (do not over-dry the pellet).

17. Dissolve the DNA in 400 µl of TE buffer. This can take several hours. Keep the tube on ice.

18. Transfer the DNA to an Eppendorf tube and precipitate DNA overnight at -20 °C by adding 1/10 vol 3 M NaAc and 2 vol 99% ethanol.

Day 3

19. Collect DNA by centrifugation in a microcentrifuge at 13 000 rpm for 15 min at 4 °C.

20. Decant the supernatant and wash the pellet with 70% ethanol and centrifuge again. Repeat the washing. Centrifuge and vacuum-dry the pellet.

21. Dissolve the pellet in 50–150 µl of TE buffer for storage at −20 or 4 °C. The yield usually obtained is about 10–100 µg DNA. We usually obtain higher yield from *Nicotiana* compared to *Brassica*.

Notes and troubleshooting

- The amount of chloroplasts loaded per sucrose gradient can affect the purity of the cpDNA. According to Palmer (1986), a rule of thumb is to load chloroplasts derived from 10–20 *g* of fresh weight leaf material on one 25-ml sucrose gradient. The volume of the gradient used in the procedure presented here is 10 ml. The chloroplasts are loaded on two to six tubes to avoid overloading of our smaller gradients. The total volume of the gradient has to be tested for each plant material.
- Chłoroplast DNA is often further purified on CsCl gradients, see Kolodner et al. (1976). for a presentation of that procedure. In general, CsCl purified cpDNA is used for cloning. CsCl gradient centrifugation is, however, rather time-consuming. It is worthwhile testing if successful cloning can be achieved with cpDNA isolated by the method presented here since cpDNA of high purity is obtained.

5.2.2 MtDNA Isolation

Reagents

- Buffer A (grinding buffer): 0.3 M mannitol, 50 mM Tris-HCl, 3 mM Na_2EDTA, pH 7.5; add just before use: 10 mM β-mercaptoethanol, 0.1% BSA, 15 mM $Na_2S_2O_5$, 10 mg/ml polyvinylpyrrolidone (PVP, MV 44 000)
- Buffer B (DNAse incubation buffer): 0.3 M mannitol, 50 mM Tris-HCl, 10 mM $MgCl_2$, 0.1% BSA, pH 7.5
- Buffer C (wash buffer): 0.3 M sucrose, 0.1 M Na_2EDTA, pH 8.0
- DNase I: 2 mg/ml dissolved in water
- Buffer D (lysis buffer): 0.1 M Tris, 0.1 M Na_2EDTA, pH 8.0
- Sarkosyl 20%

- Proteinase K: 2 mg/ml dissolved in lysis buffer
- Etidium bromide: 0.7 mg/ml dissolved in lysis buffer
- Isoamylalcohol saturated with 10 mM Na_2EDTA
- TE buffer: 10 mM Tris, 5mM Na_2EDTA pH 7.5
- 99 and 70% ethanol

Equipment

- Mortar (diameter 12 cm) and pestle
- Cheesecloth
- Miracloth (Calbiochem)
- Paintbrush (a fine artist's brush)
- Refrigerated centrifuge (Sorvall, Beckman etc.)
- Large centrifuge bottles (400–500 ml)
- Water bath 60 °C
- Ultracentrifuge with a vertical rotor and quick seal tubes
- UV-lamp
- 2-ml syringes and 0.9 × 40 mm needles
- Corex tubes 30 ml
- Microcentrifuge tubes 1.5 ml
- Microcentrifuge (1.5 ml tubes)
- Vacuum dryer
- –20 °C freezer

Procedure

Day 1

All the subsequent operations are conducted at 4 °C unless otherwise specified. Pre-cool mortar, pestle, centrifuge bottles and use refrigerated buffers and work with buckets on ice. The *g* value given in brackets for each centrifugation refers to the R_{max} value found in the different rotor manuals.

1. Prepare 2 l buffer A (add before use: β-mercaptoethanol, BSA, $Na_2S_2O_5$ and PVP). Thaw buffers B (50 ml) and C (300 ml).

2. Harvest about 100–150 g young healthy leaves. Prior to extraction, plants are kept in the dark for 24–48 h to reduce starch levels. If the plant material is infected or dirty it can be treated with chlorox (5%, 5 min) and rinsed two to three times in tap water.

3. Remove middle vein. Weigh the leaves. The method is designed for about 50–100 g of plant material. Divide the material into four or five portions.

4. Put one portion of leaves in the mortar and tear the leaves into small pieces. Add about 100 ml of buffer A and grind the leaf tissue until the buffer becomes dark green (3–4 min). Filter the homogenized tissue through six layers of cheesecloth. Rinse and return the ground tissue to the mortar. Repeat the grinding procedure three times with each portion.

5. Complete the grinding and filter the extract through two layers of miracloth.

6. Centrifuge the extract at 2500 rpm (Sorvall GS3 rotor, 1000 *g*) for 10 min.

7. Decant the supernatant into fresh bottles and centrifuge at 3500 rpm (Sorvall GS3 rotor, 2000 *g*) for 10 min.

8. Decant the supernatant into fresh bottles and spin down the mitochondria at 9000 rpm (Sorvall GSA rotor, 13 700 *g*).

9. Discard the supernatant. Resuspend the pellet gently in 4–5 ml buffer B by using a soft paintbrush.

10. Pool the suspensions and add buffer B to a final volume of 40 ml. Add 1 ml DNase 1 (stock 2 mg/ml). Incubate at room temperature for 1 h.

11. Stop the DNase activity by washing the mitochondria in buffer C. Dilute the DNase-treated mitochondrial suspension of 40 ml to about 160 ml with buffer C. Divide the solution in four 50-ml centrifiuge tubes and centrifuge at 12 000 rpm (Sorvall SS34 rotor, 17 400 *g*) for 12 min.

12. Resuspend the pellets gently with a paintbrush in a small volume of buffer C. Bring the total volume to 80 ml with buffer C. Divide the suspension into two tubes and repeat the centrifugation.

13. Resuspend the pellets in a small volume of buffer C and combine the suspension into one tube. Bring the total volume to 40 ml and repeat the centrifugation.

14. Resuspend the pellet in 3 ml of buffer C and freeze (–20 °C) the mitochondrial suspension.

Day 2

15. Thaw the mitochondria.

16. Add 360 µl Proteinase K (stock 2 mg/ml) and 300 µl 20% Sarkosyl. Add buffer D to a final volume of 6.8 ml.

17. Lyse the mitochondria at 60 °C for 1 h.

18. Transfer the lysed mitochondria to a tube containing 7.68 g CsCl. Mix gently until the CsCl is dissolved, work at room temperature. Add 1.2 ml ethidium bromide (stock 0.7 mg/ml): mix gently. The final density (ρ) of the solution should be 1.57 (see Notes and troubleshooting).

19. Divide the sample equally (balance) in two, Quick-seal tubes (Hitatchi). Fill the tubes by adding mineral oil and seal tubes. Note that the weight of the Quick-seal tubes is quite variable. Thus, tubes of the same weight should be used.

20. Centrifuge at 40 000 rpm (RPV 65T rotor, LKB ultracentrifuge, 153 000 *g*) for 20 h at 20 °C.

Day 3

21. Collect the DNA band whilst viewing under UV light. Use a syringe and a needle to remove the DNA. Puncture the tube first at the top with a needle. Then puncture the tube just below the band and remove the DNA band into the syringe. Empty the DNA in a sterile Corex tube.

22. Extract ethidium bromide with isopropanol (saturated with 0.01 M EDTA), at room temperature. Add 1 vol isopropanol, mix gently, allow the layers to separate and take off the isopropanol (top layer). Repeat three to four times until the isopropanol phase is clear.

23. Transfer the DNA solution (water phase) to a 30-ml Corex tube. Dilute the DNA with 2 vol cold TE buffer. Add 6 vol 99% ethanol and precipitate the DNA overnight at –20 °C.

Day 4

24. Collect DNA by centrifugation at 10 000 rpm (Sorvall SS34 rotor, 12 000 *g*) for 2 h at 4 °C.

25. Decant the supernatant and vacuum dry the pellet (do not over-dry the pellet).

26. Dissolve the DNA in 400 µl of TE buffer. This can take several hours. Keep the tubes on ice.

27. Transfer the DNA to an Eppendorf tube and precipitate DNA over night at –20 °C by adding 2 vol 99% ethanol.

Day 5

28. Collect DNA by centrifugation in a microcentrifuge at 13 000 rpm for 15 min at 4 °C.

29. Decant the supernatant and wash the pellet with 70% ethanol and centrifuge again. Repeat the washing. Centrifuge and vacuum dry the pellet.

30. Dissolve the pellet in 20–100 µl of TE buffer for storage at –20 or 4 °C. The yield usually obtained is about 5–30 µg DNA. We usually obtain higher yields of mtDNA from *Nicotiana* compared to *Brassica*.

Notes and troubleshooting

- In the protocol by Bland et al. (1985), a Waring blender was used for disrupting the leaf tissue. We found that grinding the leaves in a mortar resulted in higher mtDNA yields when compared to the blender. This procedure, however, is rather laborious, so if a reasonable yield of mtDNA can be obtained using a Waring blender. This would be the preferred method. The length and force of homogenization with a blender have to be tested for each type of plant material.

- The density of the CsCl gradient has to be carefully checked. Weigh 1 ml of the solution prior to centrifugation. The density should be 1.57. We have noticed a slight variation when working with different

Brassica species. To obtain the DNA band in the middle of the Quick-seal tube, we set the density to 1.58 when running CsCl gradients containing *Brassica* mtDNA.

- We have investigated whether the CsCl gradient could be omitted by extraction of mtDNA directly from DNAse-treated mitochondrial fraction with phenol/chloroform according to Kemble (1987). However, we obtained very poor yields and the DNA was not sufficiently pure.

References

Abad AR, Mehrtens BJ, Mackenzie SA (1995) Specific expression in reproductive tissues and fate of a mitochondrial sterility-associated protein in cytoplasmic male-sterile bean. Plant Cell 7:271–285

Archer EK, Håkansson G, Bonnett HT (1987) The phenotype of a virescent chloroplast mutation in tobacco is associated with the absence of a 37.5 kDa thylakoid protein. Plant Physiol 83:926–932

Avni A, Edelman M (1991) Direct selection for paternal inheritance of chloroplasts in sexual progeny of *Nicotiana*. Mol Gen Genet 225:273–277

Belliard G, Pelleitier G (1978) Morphological characteristics and chloroplast DNA distribution in different cytoplasmic parasexual hybrids of *Nicotiana tabacum*. Mol Gen Genet 165:231–237

Belliard G, Vedel F, Pelletier Gh (1979). Mitochondrial recombination in cytoplasmic hybrids of *Nicotiana tabacum* by protoplast fusion. Nature 281:401–403

Bergman P, Kofer W, Håkansson G, Glimelius K (1996) A chimeric and truncated mitochondrial *atpA* gene is transcribed in alloplasmic cytoplasmic malesterile tobacco with *Nicotiana bigelovii* mitochondria. Theor Appl Genet 91:603–610

Bland MM, Matzinger DF, Levings CS III (1985) Comparison of the mitochondrial genome of *Nicotiana tabacum* with its progenitor species. Theor Appl Genet 69:535–541

Boblenz K, Nothnagel T, Metzlaff M (1990) Paternal inheritance of plastids in the genus *Daucus*. Mol Gen Genet 220:489–491

Boeshore ML, Lifshitz I, Hanson MR, Izhar S (1983) Novel composition of mitochondrial genomes in *Petunia* somatic hybrids derived from cytoplasmic male sterile and fertile plants. Mol Gen Genet 190:459–467

Bogorad L, Vasil IK (eds) (1991) The molecular biology of plastids. Cell culture and somatic cell genetics of plants, vol 7A. Academic Press, San Diego, 340 pp

Bonhomme S, Budar F, Dominique L, Small I, Defrance MC, Pelletier G (1992) Sequence and transcript analysis of the *Nco*2.5 Ogura-specific fragment correlated with cytoplasmic male sterility in *Brassica* cybrids. Mol Gen Genet 235:340–348

Bonnett HT, Glimelius K (1983) Somatic hybridization in *Nicotiana*: behavior of organelles after fusion of protoplasts from male-fertile and male-sterile cultivars. Theor Appl Genet 65:213–217

Bonnett HT, Djurberg I, Fajardo M, Glimelius K (1993) A mutation causing variegation and abnormal development in tobacco is associated with an altered mitochondrial DNA. Plant J 3:519–525

Boynton JE, Gillham NW, Newman SM, Harris EH (1992) Organelle Genetics and transformation of *Chlamydomonas*. In: Herrmann RG (ed) Cell organelles Plant gene research Springer, Berlin Heidelberg New York pp, 3–64

Chen K, Wildman SG, Smith HH (1977) Chloroplast DNA distribution in parasexual hybrids as shown by polypeptide composition of fraction I protein. Proc Natl Acad Sci USA 74:5109–5112

Chetrit P, Mathieu F, Vedel F, Pelletier G, Primard C (1985) Mitochondrial DNA polymorphism induced by protoplast fusion in Cruciferae. Theor Appl Genet 69:361–366

Chiu W-L, Stubbe W, Sears BB (1988) Plastid inheritance in *Oenothera*: organelle genome modifies the extent of biparental plastid transmission. Curr Genet 13:181–189

Chun EHL, Vaughan MH, Rich A (1963) The isolation and characterization of DNA associated with chloroplast preparations. J Mol Biol 7:130–141

Coe EH (1983) Maternally inherited abnormal plant types in maize. Maydica 28:151–167

Dewey RE, Levings CS III, Timothy DH (1986) Novel recombination in the maize mitochondrial genome produce a unique transcriptional unit in the Texas male-sterile cytoplasm. Cell 44:439–449

Dewey RE, Hu WWL, Levings CS III (1987) The *urf-13* mitochondrial gene of *cms-T* maize. In: Plant gene systems and their biology. Alan R Liss, New York pp 65–74

Dewey RE, Timothy DH, Levings CS III (1987) A mitochondrial protein associated with cytoplasmic male sterility in the T cytoplasm of maize. Proc Natl Acad Sci USA 84:5374–5378

Downie SR, Palmer JD (1992) Use of chloroplast DNA rearrangements in reconstructing plant phylogeny. In: Soltis P, Soltis D, Doyle DD (eds) Molecular systematics of plants. Chapman and Hall, London pp 14–35

Dujon B (1981) Mitochondrial genetics and functions. In:Strathern JN, Jones EW, Broach JR (eds) The molecular biology of *Saccharomyces*, life cycle and inheritance. Cold Spring Harbor Laboratory Press, Cold Spring Harbor pp 505–635

Erickson L, Kemble R (1990) Paternal inheritance of mitochondria in rapeseed (*Brassica napus*). Mol Gen Genet 222:135–139

Escote-Carlson LJ, Gabay-Laughnan S, Laughnan JR (1990) Nuclear genotype affects mitochondrial genome organization of CMS-S maize. Mol Gen Genet 223:457–464

Fahleson J, Råhlén L, Glimelius K (1988) Analysis of plants regenerated from protoplast fusions between *Brassica napus* and *Eruca sativa*. Theor Appl Genet 76:507–512

Galun E, Arzee-Gonen P, Fluhr R, Edelman M, Aviv D (1982) Cytoplasmic hybridization in *Nicotiana*: mitochondrial DNA analysis in progenies resulting from fusion between protoplasts having different organelle constitutions. Mol Gen Genet 186:50–56

Glimelius K, Bonnett HT (1986) *Nicotiana* cybrids with *Petunia* chloroplasts. Theor Appl Genet 72:794–798

Glimelius K, Chen K, Bonnett HT (1981) Somatic hybridization in *Nicotiana*:-segregation of organellar traits among hybrid and cybrid plants. Planta 153:504–510

Gray MW (1989) Origin and evolution of mitochondrial DNA. Annu Rev Cell Biol 5:25–50

Gray MW (1990) The mitochondrial genome and its expression. In: Dennis DT, Turpin DH (eds) Plant physiology, biochemistry and molecular biology. Longman Scientific and Technical, Harlow pp 147–159

Grelon M, Budar F, Bonhomme S, Pelletier G (1994) Ogura cytoplasmic male-sterility (CMS)-associated *orf138* is translated into a mitochondrial membrane polypeptide in male-sterile *Brassica* cybrids. Mol Gen Genet 243:540–547

Hagemann R (1992) Plastid genetics in higher plants. In: RG Herrmann (ed) Cell organelles. Plant gene research. Springer, Berlin Heidelberg New York, pp 65–96

Hanson MR (1991) Plant mitochondrial mutations and male sterility. Annu Rev Genet 25:461–486

Hanson MR, Conde MF (1986) Function and variation of cytoplasmic genomes: Lessons from cytoplasmic-nuclear interactions affecting male fertility in plants. Int Rev Cytol 94:213–267

Hanson MR, Boeshore ML, McClean PE, O'Connell MA, Nivison HT (1986) The isolation of mitochondria and mitochondrial DNA. Methods Enzymol 118:437–453

Hernould M, Mouras A, Litvak S, Araya A (1992) RNA editing of mitochondrial *atp9* transcript from tobacco. Nucleic Acids Res 20:1809

Herrmann (ed) (1992) Cell organelles. Plant Gene Research. Springer, Berlin Heidelberg New York, 467 pp

Hiratsuka J, Shimada H, Whittier R, Ishibashi T, Sakamoto M, Mori M, Kondo C, Honji Y, Sun C-R, Meng B-Y, Li Y-Q, Kanno A, Nishizawa Y, Hirai A, Shinozaki K, Sugiura M (1989) The complete sequence of the rice (*Oryza sativa*) chloroplast genome:intermolecular recombination between distinct tRNA genes accounts for a major plastid DNA inversion during the evolution of the cereals. Mol Gen Genet 217:185–194

Hunt MD, Newton KJ (1991) The NCS3 mutation:genetic evidence for the expression of ribosomal protein genes in *Zea mays* mitochondria. EMBO J 10:1045–1052

Håkansson G, van der Mark F, Bonnett HT, Glimelius K (1988) Variant mitochondrial protein and DNA patterns associated with cytoplasmic male-sterile lines of *Nicotiana*. Theor Appl Genet 76:431–437

Iwabuchi M, Kyozuka J, Shimamoto K (1993) Processing followed by complete editing of an altered mitochondrial *atp6* RNA restores fertility of cytoplasmic male sterile rice. EMBO J 12:1437–1446

Izhar S, Schlicter M, Swartzberg D (1983) Sorting out of cytoplasmic elements in somatic hybrids of *Petunia* and the prevalence of the heteroplasmon through several meiotic cycles. Mol Gen Genet 190:468–474

Johns C, Meiqing L, Lyznik A, Mackenzie SA (1992) A mitochondrial DNA sequence is associated with abnormal pollen development in cytoplasmic male-sterile bean plants. Plant Cell 4:435–449

Kao HM, Keller WA, Gleddie S, Brown GG (1992) Synthesis of *Brassica oleracea/Brassica napus* somatic hybrid plants with novel organelle DNA compositions. Theor Appl Genet 83:313–320

Kaul MLH (1988) Male sterility in higher plants. In: Frankel R, Grossman M, Linskens HF, Maliga P, Riley R (eds) Monographs on theoretical and applied genetics, vol 10. Springer, Berlin Heidelberg New York, pp 97–192

Kemble RJ (1987) A rapid, single leaf, nucleic acid assay for determining the cytoplasmic organelle complement of rapeseed and related *Brassica* species. Theor Appl Genet 73:364–370

Kirk JTO, Tilney-Bassett RAE (1978) The plastids:their chemistry structure, growth and inheritance Elsevier/North Holland Biomedical Press, Amsterdam, pp 527–559

Kofer W, Glimelius K, Bonnett HT (1991) Modifications of mitochondrial DNA cause changes in floral development in homeotic-like mutants of tobacco. Plant Cell 3:759–769

Kolodner R, Tewari KK, Warner RC (1976) Physical studies on the size and structure of the covalently closed circular chloroplast DNA from higher plants. Biochim Biophys Acta 447:144–155

Köhler RH, Horn R, Lössl A, Zetsche K (1991) Cytoplasmic male sterility in sunflower is correlated with the co-transcription of a new open reading frame with the *aptA* gene. Mol Gen Genet 227:369–376

Kumar A, Cocking EC (1987) Protoplast fusion: a novel approach to organelle genetics in higher plants. Am J Bot 74:1289–1303

Landgren M, Glimelius K (1990) Analysis of chloroplast and mitochondrial segregation in three different combinations of somatic hybrids produced within Brassicaceae. Theor Appl Genet 80:776–784

Landgren M, Glimelius K (1994) A high frequency of intergenomic mitochondrial recombination and an overall biased segregation of *B. campestris* or recombined *B. campestris* mitochondria were found in somatic hybrids made within *Brassicaceae*. Theor Appl Genet 87:854–862

Laver HK, Reynolds SJ, Moneger F, Leaver CJ (1991) Mitochondrial genome organization and expression associated with cytoplasmic male sterility in sunflower (*Helianthus annuus*). Plant J 1:185–193

Leaver CJ, Gray MW (1982) Mitochondrial genome organization and expression in higher plants. Annu Rev Plant Physiol 33:373–402

Levings CS III, Brown GG (1989) Molecular biology of plant mitochondria. Cell 56:171–179

Lonsdale DM (1989) The plant mitochondrial genome. In: Marcus J (ed) The biochemistry of plants, vol 15. Academic Press, New York, pp 229–295

Lonsdale DM, Grienenberger JM (1992) The mitochondrial genome of plants. In: Herrmann RG (ed) Cell organelles. Plant gene research Springer, Berlin Heidelberg New York, pp 183–218

Lonsdale DM, Brears T, Hodge TP, Melville SE, Rottmann WH (1988) The plant mitochondrial genome:homologous recombination as a mechanism for generating heterogeneity. Philos Trans R Soc Lond B 319:149–163

Luck DJL, Reich E (1964) DNA in mitochondria of *Neurospora crassa*. Proc Natl Acad Sci USA 52:931–938
Medgyesy P, Fejes E, Maliga P (1985) Interspecific chloroplast recombination in a *Nicotiana* somatic hybrid. Proc Natl Acad Sci USA 82:6960–6964
Medgyesy P, Páy A, Márton L (1986) Transmission of parental chloroplasts in *Nicotiana*. Mol Gen Genet 204:195–198
Metzlaff M, Börner T, Hagemann R (1981) Variations of chloroplast DNAs in the genus *Pelargonium* and their biparental inheritance. Theor Appl Genet 60:37–41
Monéger F, Smart CJ, Leaver CJ (1994) Nuclear restoration of cytoplasmic male sterility in sunflower is associated with tissue-specific regulation of a novel mitochondrial gene. EMBO J 13:8–17
Nagy F, Török I, Maliga P (1981) Extensive rearrangements in the mitochondrial DNA in somatic hybrids of *Nicotiana tabacum* and *Nicotiana knightiana*. Mol Gen Genet 183:437–439
Nagy I, Lázár G, Menczel L, Maliga P (1983) A heteroplasmic state induced by protoplast fusion is a necessary condition for detecting rearrangements in *Nicotiana* mitochondrial DNA. Theor Appl Genet 66:203–207
Neale DB, Sederoff RR (1989) Paternal inheritance of chloroplast DNA and maternal inheritance of mitochondrial DNA in loblolly pine. Theor Appl Genet 77:212–216
Neale DB, Marshall KA, Sederoff RR (1989) Chloroplast and mitochondrial DNA are paternally inherited in *Sequoia sempervirens* D. Don Endl Proc Natl Acad Sci USA 86:9347–9349
Newton KJ (1993) Nonchromosomal stripe mutants of maize. In: Brennicke A, Kück U (eds) Plant mitochondria: with emphasis on RNA editing and cytoplasmic male sterility. VCH, Weinheim, pp 341–345
Newton KJ, Coe EH Jr (1986) Mitochondrial DNA changes in abnormal growth (nonchromosomal stripe) mutants of maize. Proc Natl Acad Sci USA 83:7363–7366
Newton KJ, Coe EH Jr, Gabay-Laughnan S, Laughnan JR (1989) Abnormal growth phenotypes and mitochondrial mutations in maize. Maydica 34:291–296
Newton KJ, Knudsen C, Gabay-Laughnan S, Laughnan JR (1990) An abnormal growth mutant in maize has a defective mitochondrial cytochrome oxidase gene. Plant Cell 2:107–113
Nivison HT, Hanson MR (1989) Identification of a mitochondrial protein associated with cytoplasmic male sterility in Petunia. Plant Cell 1:1121–1130
Ohyama K, Fukuzawa H, Kohchi T, Shirai H, Sano S, Umesono K, Shiki Y, Takeuchi M, Chang Z, Aota S-I, Inokuchi H, Ozeki H (1986) Chloroplast gene organization deduced from complete sequence of liverwort *Marchantia polymorpha* chloroplast DNA. Nature 322:572–274
Palmer JD (1986) Isolation and structural analysis of chloroplast DNA. Methods Enzymol 118:167–186
Palmer JD (1991) Plastid chromosomes: structure and evolution. In: Bogorad L, Vasil IK (eds) The molecular biology of plastids. Cell culture and somatic cell genetics of plants, vol 7A. Academic Press, San Diego, pp 5–53
Palmer JD (1992) Comparison of chloroplast and mitochondrial genome evolution in plants. In: Herrmann RG (ed) Cell organelles. Plant gene research. Springer, Berlin Heidelberg New York, pp 99–133

Rochaix J-D (1992) Control of plastid gene expression in *Chlamydomonas reinhardtii*. In: RG Herrmann (ed) Cell organelles. Plant gene research. Springer, Berlin Heidelberg New York, pp 249–274
Rottmann WH, Brears T, Hodge TP, Lonsdale DM (1987) A mitochondrial gene is lost via homologous recombination during reversion of CMS T maize of fertility. EMBO J 6:1541–1546
Roussell DL, Thompson DL, Pallardy SG, Miles D, Newton KJ (1991) Chloroplast structure and function is altered in the NCS2 maize mitochondrial mutant. Plant Physiol 96:232–238
Sager R, Ishida MR (1963) Chloroplast DNA in *Chlamydomonas*. Proc Natl Acad Sci USA 50:725–730
Schardl CL, Lonsdale DM, Pring DR, Rose KR (1984) Linearization of maize mitochondrial chromosomes by recombination with linear episomes. Nature 310:292–296
Schumann CM, Hancock JF (1989) Paternal inheritance of plastids in *Medicago sativa*. Theor Appl Genet 78:863–866
Schuster W, Brennicke A (1988) Interorganellar sequence transfer: plant mitochondrial DNA is nuclear is plastid, is mitochondrial. Plant Sci 54:1–10
Shinozaki K, Ohme M, Tanaka M, Wakasugi T, Hayashida N, Matsubayashi T, Zaita N, Chunwongse J, Obokata J, Yamaguchi-Shinozaki K, Ohto C, Torazawa K, Meng BY, Sugita M, Deno H, Kamogashira T, Yamada K, Kusuda J, Takaiwa F, Kato A, Tohdoh N, Shimada H, Sugiura M (1986) The complete nucleotide sequence of the tobacco chloroplast genome: its organization and expression. EMBO J 5:2043–2049
Singh M, Brown GG (1991) Suppression of cytoplasmic male sterility by nuclear genes alters expression of a novel mitochondrial gene region. Plant Cell 3:1349–1362
Small ID, Isaac PG, Leaver CJ (1987) Stoichiometric differences in DNA molecules containing the *atpA* gene suggest mechanisms for the generation of mitochondrial genome diversity in maize. EMBO J 6:865–869
Stern DB, Newton KJ (1985) Mitochondrial gene expression in Cucurbitaceae: conserved and variable features. Curr Genet 9:395–405
Stern DB, Lonsdale DM (1982) Mitochondrial and chloroplast genomes of maize have a 12-kilobase DNA sequence in common. Nature 299:698–702
Sundberg E, Glimelius K (1991) Effects on parental ploidy level and genetic divergence on chromosome elimination and chloroplast segregation in somatic hybrids within Brassicaceae. Theor Appl Genet 83:81–88
Sundberg E, Landgren M, Glimelius K (1987) Fertility and chromosome stability in *Brassica napus* resynthesised by protoplast fusion. Theor Appl Genet 75:96–104
Thanh ND, Medgyesy P (1989) Limited chloroplast gene transfer via recombination overcomes plastome-genome incompatability between *Nicotiana tabacum* and *Solanum tuberosum*. Plant Mol Biol 12:87–93
Unseld M, Brandt P, Heinze B, Eckert-Ossenkopp U, Brennicke A (1993) The mitochondrial genome of *Arabidopsis thaliana*. In: Brennicke A, Kück U (eds) Plant mitochondria: with emphasis on RNA editing and cytoplasmic male sterility. VCH, Weinheim pp 137–143
Whittier RF, Sugiura M (1992) Plastid chromosomes from vascular plants-genes. In: Herrmann RG (ed) Cell organelles. Plant gene research. Springer, Berlin Heidelberg New York, pp 164–182

Wise RP, Fliss AE, Pring DR, Gengenbach BG (1987) *urf13-T* of T cytoplasm maize mitochondria encodes a 13 kD polypeptide. Plant Mol Biol 9:121–126

Young EG, Hanson MR (1987) A fused mitochondrial gene associated with cytoplasmic male sterility is developmentally regulated. Cell 50:41–49

Chapter 6

RAPD Analysis: Use for Genome Characterization, Tagging Traits and Mapping

R. WAUGH

6.1 Introduction

The detection of nucleotide sequence variability and its exploitation in the form of genetic markers has revolutionized many aspects of plant genetics. Initially, variation at the DNA level was detected as Restriction Fragment Length Polymorphisms (RFLPs; Botstein et al. 1980). However, over the last few years, Polymerase Chain Reaction (PCR) technology has had a significant impact in almost all areas of molecular biology (Saiki et al. 1988) and modifications of the basic procedure have allowed the development of numerous assays for detecting variation at the nucleotide level (e.g. Korner and Livak 1989; Newton et al. 1989; Innes et al. 1990; Dockhorn-Dworniczak et al. 1991). Unfortunately, most of these assays require prior DNA sequence information to allow the design and synthesis of oligonucleotide primers which flank the target sequence and allow selective DNA amplification by PCR. In 1990, two groups working independently described a novel approach for detecting nucleotide sequence polymorphisms which was based on PCR but did not require DNA sequence information (Welsh and McClelland 1990; Williams et al. 1990). The randomly amplified polymorphic DNA (RAPD) and arbitrarily primed PCR (AP-PCR) approaches have since been widely adopted by the plant and other research communities (Waugh and Powell 1992). In this chapter, I shall focus my discussions primarily on RAPD markers and refer the reader to the articles emanating primarily from the laboratories of J. Welsh and M. McClelland for a description of the AP-PCR approach (Welsh and McClelland 1990, 1991; Welsh et al. 1991a,b; Al Janabi et al. 1993) and from G. Caetano-Annoles' laboratory for a further arbitrary sequence primer approach termed DNA Amplification Fingerprinting (DAF) (Caetano-Annoles et al. 1991a,b, 1992).

Rationale

The generation of RAPD markers is based on the probability that a DNA sequence, homologous to that of a single, short oligonucleotide primer (usually 10-mers for RAPDs), will occur at different sites on opposite strands of a DNA template within a distance that is amplifiable by PCR. When these criteria are met, the single oligonucleotide will prime exponential DNA amplification in a PCR reaction. Practically, when using 10-mer oligonucleotides, each primer routinely promotes the generation of several discrete DNA products (3–10), and these can usually be considered to originate from different genetic loci. Polymorphisms result from mutations or rearrangements either at or between the primer binding sites and are most frequently detected as the presence or absence of an amplification product. This means that RAPDs generally behave as dominant markers. While this has some obvious disadvantages (e.g. the inability to identify heterozygotes), these are balanced by the numerous advantages of the approach. Ease of assay, accessibility by non-specialists and the potential for easy automation are features which have attracted widespread application. In this chapter, I shall describe some of the approaches and techniques taken with RAPD markers over the last few years for plant genome characterization.

6.2 Genetic Fingerprinting

Traditionally, the genetic characterization of individuals within a species and the relationship between them has been determined by a combination of morphological and agronomic traits or by biochemical tests such as isozyme analysis. However, these approaches are subject to environmental influences and their overall effectiveness in estimating genetic relatedness has been the subject of some debate (Gottlieb 1977; Brown 1979). As a result, it is now widely accepted that information generated from DNA-based polymorphisms provides the best estimate of genetic diversity. The ability to fingerprint individuals genetically through the establishment of diagnostic molecular profiles has important commercial and scientific applications. Unequivocal identification of breeding lines and cultivars is crucial for plant breeders and seed companies, allowing the protection of plant breeders rights, the regulation and marketing of novel germplasm and the quality control of distributed products. Yang and Quiros (1993) applied RAPD markers to the identification and classification of celery cultivars, and 28 primers detected sufficient variation to sub-divide the 21 cultivars into three separate groupings, which agreed with previous classifications. In a similar study, Hu and Quiros (1991) were able to distinguish 14 broccoli and 12 cauliflower cultivars using only four random 10-mers and, as expected, larger differences were found between cultivars from different seed companies than within the same company. Accession or cultivar-specific phenograms have also been generated for a range of other plant species including *Coffea* (Orozco-Castillo et al. 1994) *Theobroma* (Wilde et al. 1992), potato (Demeke et al. 1993), *Allium* (Wilkie et al. 1993), oat (Dweikat et al. 1993), wheat (He et al. 1992; Dweikat et al 1993; Joshi and Ngugen 1993) and apple (Koller et al. 1993; Mulcahy et al. 1993). Naturally, genome characterization at this level has wider implications. A general role in parentage analysis, in determining the homozygosity of breeding lines, the homogeneity of varieties by official regulatory agencies or detection of seed contamination could all be envisaged. RAPDs have been used to examine the uniformity of F_1 hybrid seed to establish the level of inbreeding (Boury et al. 1992) to demonstrate gene introgression from *Solanum phureja* dihaploid inducing clones into *S. tuberosum* (Waugh et al. 1992) and to estimate phenetic relationships between tomato accessions grouped into vintage or modern South American regional cultivars and wild species (Williams and St Clair 1993). However, in addition to a regulatory role, the information obtained through genetic fingerprinting has strategic importance. For example, the identification of suitably diverse parents or the identification of heterotic parental combinations by RAPD profiles could become routine procedures. By screening 20 accessions of wild tetraploid emmer wheats and 10 cultivated tetraploid durum wheats from geographically diverse regions, Joshi and Nguyen (1993) were quickly able to identify highly diverse parents for inclusion in future mapping studies. Such information could likewise be used specifically to modify traits in highly adapted or elite germplasm by crossing individuals which differ at only a few genetic loci, thus preserving the majority of the genetic background.

6.2.1
Methods

The following methodologies are routinely used in my laboratory for RAPD analysis on a wide range of crop species. The procedures have also been used successfully on plant pests such as aphids, whiteflies and nematodes.

6.2.1.1
DNA Isolation

A detailed discussion of total genomic DNA isolation protocols is given in Section 1.1. We routinely use modified versions of the procedures reported by Saghai-Maroof et al. (1984) for maxipreps (used routinely for potatoes, barley, *V. faba*, *Gliricidia*, *Phaseolus*, Wheat and groundnuts) and Edwards et al. (1991) for mini-preps. Other procedures which we have successfully tried and which should be borne in mind if the DNA isolated by the above methods is unsuitable for PCR are described by Dellaporta et al. (1983), Rogers and Benedich (1985), Gawel and Jarret (1991), Kochert et al. (1991), and Langridge et al. (1992), to name a few.

6.2.1.2
Basic RAPD Procedure

Depending on the crop species, we routinely use 10–100 ng DNA per RAPD assay (although this can be reduced or increased in certain cases, see below).

Reagents

- Target DNA (10–100 ng)
- Oligonucleotide primers (20 μm stocks) (Available commercially from Operon Technologies or Perkin Elmer, or can be synthesized in-house)
- Sterile distilled water (SDW)
- Taq polymerase
- Taq polymerase buffer
- $MgCl_2$ (10 mM)
- 2 mM dNTP stocks (x10 dNTPs)
- Light mineral oil or Ampliwax depending on make and model of thermal cycler.
- Molecular biological grade agarose

- 10x TBE buffer (0.89 m Tris pH 8.3, 0.89 m boric acid, 0.1 M EDTA)
- Molecular marker (size 200–4000 bp)
- Gel loading buffer (40% glycerol, 0.5% bromophenol blue, 0.5% xylenecyanol FF)
- Ethidium bromide (10 mg/ml)
- λ DNA

Equipment

- Thermal cycler (e.g. Perkin Elmer 9600, Techne PHC-1, 2 or 3)
- Agarose gel electrophoresis tanks, combs etc. (20 × 25 cm, 60 well format and mini-gel format)
- UV transilluminator
- Darkroom
- Polaroid or other photographic equipment
- UV face shields
- Set of micropipettes (e.g. Gilson P1000, P200, P100, P20, P2)
- Sterile microfuge tubes (2.2, 1.5 and 0.5 ml)
- Sterile tubes/plates for thermal cycler
- Sterile microtitre plates
- Electrophoresis power pack
- Sterile micropipette tips

Procedure

We routinely set up either 50 or 100 PCR reactions at a time varying either the DNA sample or the primer in each tube. We therefore make up a reaction mix containing all but the variable component which can then be aliquoted into individual tubes. For agarose gels, the final reaction volume is 20 μl.

1. On ice, make either: Reaction mix for 100 PCRs with different DNAs (added in the following order):

Water	1475 μl
x10 buffer	200 μl
x10 dNTPs	200 μl
Primer (20 μm)	20 μl
Taq polymerase	5 μl (25 units)
Final volume	1900 μl
DNA (10–100 ng)	1 μl per tube

or Reaction mix for 100 PCRs with different primers:

DNA (optimized conc.)	100 µl
Water	1475 µl
x10 buffer	200 µl
x10 dNTPs	200 µl
Taq polymerase	5 µl (25 units)
Final volume	1980 µl
Primer (20 µm stock)	0.2 µl per tube

2. Aliquot reaction mix into reaction tubes.
3. Add either DNA or primer.
4. Add one drop of mineral oil or Ampliwax pellet to each tube (if required).
5. Seal tubes and place in thermal cycler.
6. Run PCRs. We use a three-component PCR program for RAPDs using 10-mer oligonucleotides (Note 4):

a)	One cycle	94 °C	4 min
		36 °C	1 min
		72 °C	1.5 min followed by
b)	45 cycles	94 °C	1 min
		36 °C	1 min
		72 °C	1.5 min followed by
c)		72 °C	5 min
		4 °C	hold

7. Pour 2% agarose gel in 1x TBE.
8. After PCR add 4 µl gel loading buffer to each of the samples.
9. Load 20 µl per track alongside a suitable molecular weight marker (e.g. *λ Hind* III/*Eco* RI)
10. Run gel until Bromophenol Blue reaches the end of gel (150 V for 3 h for a 20 x 25 cm submersed gel).
11. Stain gel by immersion in 1 µg/1 ml EtBr for 30–60 min.
12. Wash gel in water (10 min).
13. View on UV transilluminator and photograph.

Notes

- *Note 1.* We (and others) have found that varying the DNA concentration can alter the abundance of individual fragments within a PCR reaction. For each crop we therefore standardize the amount of starting DNA in the 10–100-ng range. In a few cases, using less than 10 ng can result in cleaner profiles. We attribute this to the copurification of PCR-inhibiting substances in the DNA isolation procedure. Optimal DNA concentration for a given crop species and DNA isolation procedure should be determined experimentally. We add the primer or DNA after aliquoting out the reaction mix, as it is almost impossible to pipette such a small volume into an empty microfuge tube.

- *Note 2.* Buffer (x10) distributed by the Taq polymerase suppliers usually contains either no Mg^{2+} or 15 mM Mg^{2+} ions. RAPD profiles can vary according to the Mg^{2+} concentration, and a number of groups routinely optimize their RAPD protocol by testing various Mg^{2+} concentrations (usually up to a maximum of about 4 mM final concentration) with a range of primers.

- *Note 3.* The enzyme we use is distributed at a concentration of 5 U/μl. We routinely use 0.1–0.25 units per 20 μl PCR reaction. We have, however, tried many sources of Taq polymerase, and there is enormous variation between them. Both the polymerase and the number of units used per reaction should therefore be determined experimentally.

- *Note 4.* Although this is the programme we use, there are a large number of alternative cycling times and temperatures which are equally effective. Many of the differences arise from variations in the machinery, and microtube wall thickness. It is advisable to vary the times etc. according to your own hardware until a suitable program is obtained. When using longer arbitrary primers in the AP-PCR approach, we follow two initial low-stringency cycles (e.g. annealing at 37 °C) with 30–35 primer-specific high-stringency cycles (e.g. 50–60 °C annealing). These can be performed with or without the addition of a radioactive label and separated accordingly on agarose or polyacrylamide gels (see papers by Welsh and McClelland).

Troubleshooting

Many of the problems which can occur are presented in a troubleshooting format with suggestions as to how they may be solved compiled from our experiences on a variety of crops.

- Erratic or no amplification
  - A component is missing from some or all the tubes. Repeat a few reactions ensuring all PCR components are added
  - Inhibitors of PCR may have co-purified with the DNA. Vary the DNA concentration
  - Include a clean up step in procedure for DNA isolation (e.g. phenol extraction)
  - Make fresh DNA dilutions
  - Primer stock is going off. Make a new stock. Use a different primer or try increasing primer concentration
  - Increase annealing time (segment **2** in PCR protocol) or decrease the ramp rate between the increasing temperature transition steps
  - Increase concentration of Taq polymerase per reaction
  - Make up fresh components and autoclave where necessary

- Poor amplification resulting in bands being fuzzy and/or indistinct
  - Replace Taq polymerase buffer
  - Check the primer (use another primer or check primer by endlabelling and running on a 16% 6 M urea polyacrylamide gel)
  - Check activity of Taq polymerase (compare with a different batch)
  - Vary DNA concentration

- High molecular weight smears >4kb
  - Decrease DNA concentration
  - Decrease Taq polymerase concentration
  - Load less on gel
  - May be the result of excessive cycling (Bell and Demarini 1991). Do fewer cycles
  - Ensure gels are made with the correct buffer (not water!)
  - Conduct electrophoresis at a lower voltage

- Single, monomorphic, very intense band for all samples analyzed and in the control
  - Ensure primer sequence is not palindromic. Product may be a primer artefact. Use a different primer

- Nonreproducible banding patterns
  - Too little or too much DNA. Always keep concentration of genomic DNA in the range 10–100 ng. (Too little DNA has been suggested to result in inefficient priming of "real" target sequences

giving spurious bands as a result of primer amplification (i.e. primer artefact). Too much DNA can promote mis-match (binding of primer to genomic DNA). Run two DNA concentrations of each sample
  - Only score major bands

- Gel has strong background after staining which affects resolution
  - Decrease staining time
  - Destain gel for longer
  - Too much DNA or Taq in PCR reactions

- Inadequate separation of low molecular weight products
  - Separate products on higher concentrations of agarose gel, specialist agaroses or polyacrylamide gels.

Factors Affecting the Reproducibility of RAPD Profiles

A number of groups have reported the sensitivity of RAPD profiles to changes in the assay procedure or conditions (e.g. Devos and Gale 1992; Yu and Pauls 1992). While the general consensus of opinion considers reliability and reproducibility within a single laboratory to be high, some concern remains over transferability between groups. It is, therefore, important that between collaborators, rigorous attempts be made to standardize assay conditions. Several of the factors which have been reported to affect the number, size and intensity of bands have been summarized by Black (1993). These include PCR buffers, dNTP and Mg^{2+} concentration, cycling parameters (annealing temperature, ramp time, thermal cycler), source of Taq polymerase (Schierwater and Ender 1993), condition and concentration of DNA (Black et al. 1992), primer concentration (Ellsworth et al. 1993), co-extraction of contaminating DNA and amplification interference (Wilkinson et al. 1993). In addition, the poor resolution of amplification products in standard agarose gels has been increased by adding cross-linkers to agarose gels (Hunt and Page 1992) using standard 5% polyacrylamide gels (Wilde et al. 1992) or denaturing gradient gel electrophoresis (Dweikat et al. 1993). These sensitivities stress the need for within- and between-group standardization, and accurate reporting of materials and methods in publications resulting from RAPD-based assays.

6.2.2 Analysis of Genetic Relationships

An extension of genome characterization by genetic fingerprinting is the detailed analysis of genetic relationships. In combination with a variety of multivariate statistical procedures (Nei and Li 1979; Kempton and McNicol 1990), RAPD data have been used to construct phenetic trees to document graphically the interrelationships between or within related species, populations or gene pools (e.g. Chalmers et al. 1992). Furthermore, by a comparison of RAPD phenotypic frequencies and ecogeographical variables Dawson et al. (1993) were able to suggest that the variation detected at certain RAPD loci in *Hordeum spontaneum* had adaptive significance.

Data Analysis

As indicated above, it is important for all experiments that the data scored are both reliable and reproducible. It is therefore advised that only major bands which very obviously differ between the individuals in the study be analyzed. An example of the types of gel we would score is shown in Fig. 6.1. For phenetic studies or studies of population genetic structure, RAPD products which are present in more than one individual but not all are most informative. Unique products give little information regarding genetic relationships but are of course valuable as accession-specific markers or indicators of gene introgression. In all cases, a binary matrix of marker allele versus accession number is generated and this can be analyzed by a number of statistical algorithms to reveal either numerically or graphically the level of similarity between the individuals under study. These include single linkage and nearest-neighbor analysis, average cluster analysis using arithmetic average unweighted clustering (UPGMA; available in GENSTAT 5; Sneath and Sokal 1973), the Wagner parsimony method [available on phylogenetic analysis using parsimony (PAUP) developed by D.L. Swofford], or Nei's estimate of similarity (Nei and Li 1979). In addition to the raw binary matrix, estimations for a number of population-genetic parameters such as within- and between-population heterozygosities, degree of individual relatedness, inbreeding or population sub-division, or gene and genotype frequencies can be obtained by examining phenotypic frequencies within sub-populations or groups. A number of statistical approaches have been devised for

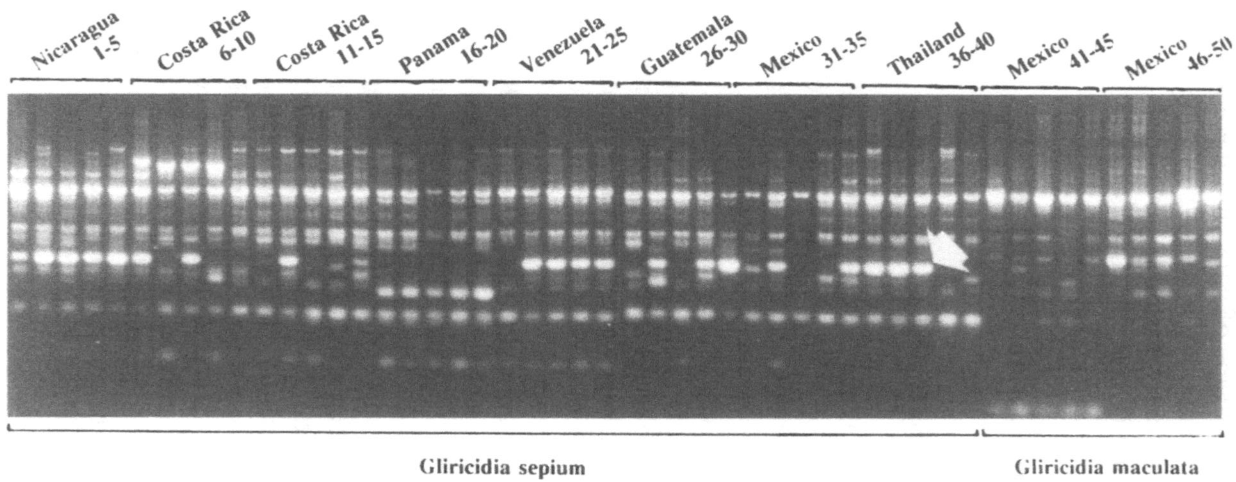

Fig. 6.1. DNA of five individuals from each of ten populations of *Gliricidia sepium* or *G. maculata* from various geographical locations was used for the amplification of RAPD markers. The *arrowed band* represents the type of informative products useful for analysis of genetic relationships

analyzing this type of data, and these include Shannon's Information measure (King and Schaal 1989), gene diversity statistics (Nei 1973), or more recently methods developed specifically for the analysis of RAPD data (Clark and Lanigan 1993; Lynch and Mulligan 1994).

6.3 Development of Genetic Linkage Maps

6.3.1 Global Mapping Strategies

Linkage maps are constructed from the data obtained by following the inheritance of a large number of specific characters in appropriate segregating populations by calculating the degree to which pairs of markers co-segregate. When comparing only two markers, simple statistical tests such as a chi-square analysis will test for linkage. When comparing the data obtained from tens or hundreds of markers, a massive number of statistical tests are required, and several computer software packages are currently available which simplify the process immensely. For a detailed discussion of mapping strategies and statistics, the reader is referred to Chapter 7. However, there are currently a number of good illustrations of how RAPDs have been exploited for the generation of genetic linkage maps. Reiter et al. (1992) placed over 250 RAPD markers on a recombinant inbred population of *Arabidopsis thaliana* in only 4 man-months and a 208-marker map of the polyploid relative of cultivated sugar cane *Saccharum spontaneum* has been constructed using RAPDs which segregate in a 1:1 ratio (single-dose markers; Al-Janabi et al. 1993; Sobral and Honeycutt 1993). More significantly, RAPDs have facilitated the construction of genetic linkage maps from species in which RFLP technology has been difficult to apply. In such species, low levels of resourcing, large genome size, long generation times, lack of suitable mapping populations and recalcitrance to developed DNA extraction techniques have frequently contributed to the comparative lack of progress. However, the small quantity of DNA required and simplicity of the RAPD assay has helped overcome many of these complications. Thus, in conifers, genetic linkage maps of single lobolly pine (Neale and Sederoff 1991) and white spruce trees (Tulsieram et al. 1992) have been constructed using haploid DNA from the megagametophytic tissue. In species such as *Allium* (onion; Wilkie et al. 1993) and *Vicia faba* (field bean; van de Ven et al. 1992), where genome size limits the successful application of RFLPs, and in recalcitrant species like *Theobroma* (Wilde et al. 1992) and *Coffea* (Orozco-Castillo et al. 1994), RAPDs are proving an invaluable tool for genome characterization.

On the negative side, in some crops such as cultivated wheat and barley, the level of polymorphism detected with RAPDs is particularly low (e.g. Devos and Gale 1992; Barua et al. 1993) and much lower than that detected on Southern blots using *PstI* or cDNA clones as RFLP probes. In addition, Ragot and Hoisington (1993) recently compared RFLP and RAPD genotyping costs. They concluded that when genotyping large numbers of individuals (>25), the cost of RFLPs was significantly lower than RAPDs. The basic message is that a careful consideration of all these factors should be made before embarking on a particular strategy. Importantly, where small sample sizes were examined, (<10) RAPDs were a very cost-effective and time-saving approach.

Reagents and Equipment

As indicated in Basic RAPD Procedure

Procedure

The construction of a complete genetic map using RAPD markers generally proceeds in four steps.

1. *Select the parental lines.* To expedite map construction, it is generally advisable to identify parental lines which are genetically divergent. However, genetically diverse lines are not always directly relevant to plant breeding objectives, and a balance between the two should be reached if possible. The speed of map construction will be largely influenced by the level of polymorphism between the parents. If genetic distance is unknown, candidate parental lines can be quickly screened with a selection of RAPD primers to estimate the level of polymorphism. Having identified the parental lines, polymorphic primers should be identified prior to mapping in the segregating population. (This is of course not advisable in allogamous plant species such as potato or when examining intercrossed F_1s).

2. *Produce (or identify) a suitable mapping population.* A number of population types are suitable for a mapping exercise. The most common are F_2, backcross, recombinant inbred, intercrossed F_1 or doubled haploid. A population size of about 50 individuals is sufficient to generate a fairly detailed map using the markers which

segregate in a 1:1 ratio. With F_2 populations informative markers should segregate in a 3:1 ratio and a larger mapping population is more appropriate. Recombinant inbred, doubled haploid or vegetatively propagated (backcross, F_2 or intercrossed F_1) populations are easiest to work with as the mapping population is genetically fixed.

3. *Segregation analysis.* The inheritance of polymorphic products identified by screening the parental lines are scored directly in the amplification products generated from the mapping population. When scoring RAPD data, in our experience, it is important only to score the obviously polymorphic major bands. An example is given in Fig. 6.3.
4. *Linkage analysis.* Linkage analysis is based upon the degree to which RAPDs co-segregate. As soon as two markers segregating in a population have been scored, their linkage can be tested. If linked, they will tend to co-segregate. When testing linkage between many markers, a number of computer software packages (which use different statistical algorithms) have been developed specifically for genetic linkage analysis. These are generally available at a nominal cost e.g. MAPMAKER (Lander et al. 1987), JOINMAP (Stam 1993).

6.3.2 Targeted Approaches

The simplicity of the RAPD assay has been maximally exploited in strategies aimed at forcing markers into specifically chosen genomic segments. In general, these strategies have utilized either the products of targeted plant-breeding efforts, aneuploid genetic stocks or existing segregating populations which have been previously used for the construction of genetic linkage maps. In all of the following cases, the RAPD protocol is the same as described previously. The key to success depends largely on the quality of the genetic material and existing segregation data.

6.3.2.1 *Near-Isogenic Lines (NILs)*

Near-isogenic lines are frequently the product of plant breeding programs where the objective has been the introgression of a desirable character from a donor parent into an otherwise agronomically good cultivar (recurrent parent). Near-isogenic lines are generated by a process of repeated backcrossing with selection for the character at each round of crossing. After backcross 7 or 8, selected genotypes are selfed and homozygotes at the target locus identified. The result is a pair of genotypes (the recurrent parent and the NIL) which are essentially identical at all genetic loci with the exception of the region surrounding the gene under selection. The hypothesis is that any differences detected between the NIL and recurrent parent will probably arise from genetic differences between the donor parent and the recurrent parent at or around the selected locus. Screening for RAPDs between NILs therefore represents a quick and effective way of identifying markers linked to the introgressed segment. The intensity of the linkage should always be checked by segregation analysis on an appropriate segregating population because of the frequent co-introgression of donor DNA segments at unlinked loci (Young and Tanksley 1989).

The use of NILs has proved particularly successful for isolating markers linked to plant disease resistance loci. For example, by screening a pair of tomato NILs with 144 random primers, Martin et al. (1991) were able to identify three products which were tightly linked to a *Pseudomonas syringae* resistance locus (*pto*) on tomato chromosome 5. Carland and Staskawicz (1993) isolated a further seven RAPD markers linked to the *pto* locus using a different pair of NILs. Fortunately, one of the primers used by both groups was identical and this allowed a comparative ordering of the markers flanking *pto* to be deduced. A similar level of success was achieved by Paran et al. (1991), who identified four markers linked closely to the *Dm1* and *Dm3* and seven to the *Dm11* resistance loci in lettuce by screening two pairs of NILs with a total of 212 primers. However, it is important to realize that the efficiency with which NILs can be used to identify closely linked markers depends largely on the level of variation between the donor and recurrent parents at and surrounding the locus under study. Thus, while the success rate in the above examples was relatively high, Ronald et al. (1992) had to examine 985 primers in order to identify two products linked to the rice bacterial blight resistance locus, *Xa21* and only a single marker linked to the oat stem rot gene *Pg3* was obtained by screening 204 primers (Penner et al. 1993). Similarly, when Barua et al. (1993) screened a pair of NILs with 300 primers to identify one marker loosely linked to a *Rhynchosporium secalis* resistance locus in barley, the same 300 primers failed to reveal a single polymorphic product between a pair of nematode-resistant NILs (Waugh, pers comm.). An example of a positive result is shown in Fig. 6.2.

Materials

- DNA from near-isogenic lines (donor parent, recurrent parent and NIL)
- Reagents and equipment for RAPD analysis
- Population segregating for same character introgressed into NILs (preferably derived from same genetic "source" as the resistance locus in the NILs)

Procedure

1. Screen for RAPDs on DNA from donor parent, recurrent parent and NIL.

2. After examining 50 or so RAPD primers, a rough calculation of the level of polymorphism between the donor and recurrent parent will allow a theoretical estimate of the number of RAPD-primers which will need to be screened in order to identify a marker linked (at a given distance) to the target locus (Martin et al. 1991):

Expected minimum distance = $[c/2(nx\text{-}1)]/y$
Distance at 95% confidence level (w) = $[(c/2)(1\text{-}0.05^{1/nx})]/y$,

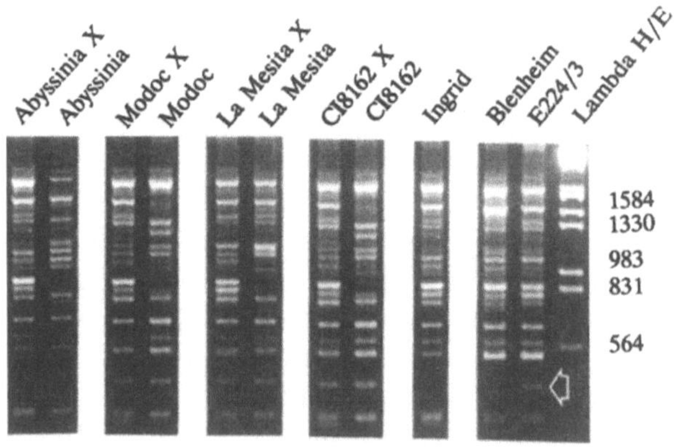

Fig. 6.2. RAPD products from four sets of barley NILs, and their donor parents (Which exhibit resistance to *Rhynchosporium secalis*), the susceptible recurrent parent (Ingrid) and the parents of a population segregating for *Rhy* resistance. The diagnostic RAPD product is indicated

where c = genome size in centimorgans
n = number of primers
x = average number of products/primer
y = frequency of polymorphic fragments
(Assumes a random distribution of PCR products and that the target locus is not at the end of a chromosome).

3. After screening a large number of primers, verify any differences between recurrent parent and NILs by repeating assays. The diagnostic product should be present in the donor line and NIL but not in the recurrent parent (Fig. 6.2).
4. Follow the segregation of polymorphic markers in an appropriate mapping population.
5. Estimate intensity of the linkage by co-segregation analysis.

Notes

- Ideally, in an appropriate mapping population, the locus selected in the NILs should segregate and the polymorphic fragment identified in the NILs should be polymorphic between the parental lines used to construct the population. Alternatively, if the diagnostic products are not polymorphic between the parents of existing populations, the following options should be considered:
  - Clone the polymorphic products (see below) and use them to probe Southern blots of the DNA from the available population. (Procedures described in Chap. 7).
  - Sequence the cloned products and design longer, internal primers for PCR-based assays on the available populations (conversion to sequence-tagged sites).

Conversion of RAPD Markers to Sequence Tagged Sites

The dominant nature of RAPDs, the amplification of multiple loci by a single primer, the sensitivity to reaction conditions and frequent lack of transferability between different crosses places some restrictions on their general applicability. Cloning and sequencing polymorphic RAPD products and designing longer synthetic oligonucleotide primers may overcome many of these problems. For example, Paran and Michelmore (1993) converted eight polymorphic RAPD bands into more reliable PCR-based markers (which they called SCARs). However, only five amplified alternative alleles from their mapping population parental lines while three continued to amplify a dominant v. null allele.

Cloning PCR Products

See Chapter 2.1 for a discussion and protocol on cloning PCR products.

6.3.2.2
Pooling Strategies

One of the drawbacks of NILs is the large number of backcrosses required for their generation. An alternative strategy, initially described by Arnheim et al. (1985) as a means of identifying polymorphic markers at a specific chromosomal location, is to use pooled DNA samples. The approach is based on the linkage disequilibrium which exists at the target locus in the individual's used to construct the pools. The individuals to pool are selected using either phenotypic or genotypic data gathered from an appropriate segregating population and equal quantities of DNA from each of the identified individuals combined to make two pools. The limitation of the approach described by Arnheim et al. (1985) was that it relied on RFLP technology which, as discussed previously, is inefficient when examining small numbers of samples (i.e. two). Michelmore et al. (1991) adapted the technique by using RAPDs in an approach which they termed bulked segregant analysis. They divided an F_2 population segregating for resistance to downy mildew into two groups based on the resistance phenotype, and pooled DNA samples isolated from either group. Screening the bulked DNAs with around 100 RAPD primers allowed three markers linked to the resistance locus to be identified. This approach has been extended by Giovanonni et al. (1991), Reiter et al. (1992) and Williams et al. (1993), who used genotypic information in the form of molecular markers to identify the individuals which should be used to construct the pools. In both cases, a "genetic interval" could be defined based on the marker genotypes of all the individuals in the population. The defined interval could be as small as the distance between two closely linked RFLP markers (or a single locus) up to an entire chromosome (Reiter et al. 1992).

While the majority of the reported applications of pooling strategies based on phenotype have focussed on qualitative characters (e.g. Miklas et al. 1993; Martin et al. 1993; Haley et al. 1993; Barua et al. 1993), by using members of a population which exhibit the extremes of the distribution of a quantitative character (i.e. the majority of positive factors and the majority of negative factors), Chalmers et al. (1993) were able to identify markers linked to genetic factors controlling the milling energy requirement of barley – a polygenically controlled character. Similarly, markers linked to genetic factors controlling the frequency of green plant regeneration in barley anther culture have recently been identified in our laboratory based on a phenotypic pooling strategy (Andersen et al. in prep.). It is commonly considered that DNA pooling approaches are less likely to generate markers which are not linked to the target locus because many segregating individuals can be used to construct each pool. Michelmore et al. (1991) provided a simple formula: $P = 2(1\text{-}[1/4]^n)\ (1/4)^n$ for calculating the probability that polymorphic bands would be unlinked to the target locus in bulks of n individuals constructed from an F_2 population (e.g. if two bulks comprising of six individuals were constructed there is a probability of 5×10^{-4} that a polymorphic band would come from an unlinked locus).

Reagents and Equipment

– As for Basic RAPD Procedure, appropriate segregating population scored for phenotypic or genotypic markers

Procedure

1. *Phenotypic selection.* For practical reasons, the number of individuals which can be bulked will depend on whether the selected character is qualitative or quantitative. For a qualitative character pools of 8–10 individuals are both practical and unlikely to generate false positives. In our experience, RAPD alleles at a frequency of 0.05–0.1 are barely detectable, although this does depend on the amplification potential of individual bands (in some cases weakly amplified bands may be detectable only at a frequency of 0.2 and above). For quantitative characters, it is desirable to bulk the individuals containing the majority of positive alleles and the majority of negative alleles. The restriction on numbers to bulk is frequently the size of the population (which will consequently influence the number of individuals at the extremes of the distribution for the trait) and the number of loci contributing to the overall expression

of the trait. Considering a three component system with equal contributions from each locus, in a cross between two diploid individuals which are heterozygous at each locus (i.e. AaBbCc x AaBbCc) only 1:64 individuals would contain all the positive alleles and 1:64 all the negative alleles. Thus, a population size of approximately 320 would be required to bulk the "best" five (AABBCC) and "worst" five (aabbcc) individuals. Practically, in small populations of 48–60 plants, bulking 5–7 individuals from the tails of the distribution has been effective in our hands for identifying markers linked to loci making the largest contribution towards the expression of the trait (Anderson et al. in prep.).

2. *Genotypic selection.* This approach exploits prior knowledge of the genetic (RFLP or RAPD) composition of all the individuals in a segregating population (Giovannoni et al. 1991; Williams et al. 1993). The idea is to construct two DNA pools which contain opposite alleles within a "genetic window" encompassing the target locus, but which are heterozygous at all loci outside the window. The size of the window can range from a single locus to an entire chromosome (Reiter et al. 1992). The approach involves first dividing the population into halves based on the phenotype or genotype of individuals. Genotypic data at RFLP loci immediately flanking the target locus are then added sequentially until a number of individuals are identified which contain alternative parental alleles within the selected window. Equal quantities of DNA are combined to make the pools and the basic RAPD protocol followed to identify polymorphisms. Polymorphic products must be confirmed as "linked" to the target locus by segregation analysis using the same population (Fig. 6.3). An example of how to select individuals to bulk from the homozygous individuals of an F_2 population segregating for a disease resistance locus is given in Fig. 6.4 and Table 6.1.

6.3.2.3
Genetic Stocks

A number of different plant species offer the opportunity to quickly develop chromosome-specific markers due to the availability of a wide range of aneuploid genetic stocks such as alien addition, nullisomic-

Fig. 6.3. **a** The polymorphic product between two bulked DNA samples and parental DNAs. **b** Same polymorphic product segregating in barley doubled haploid population

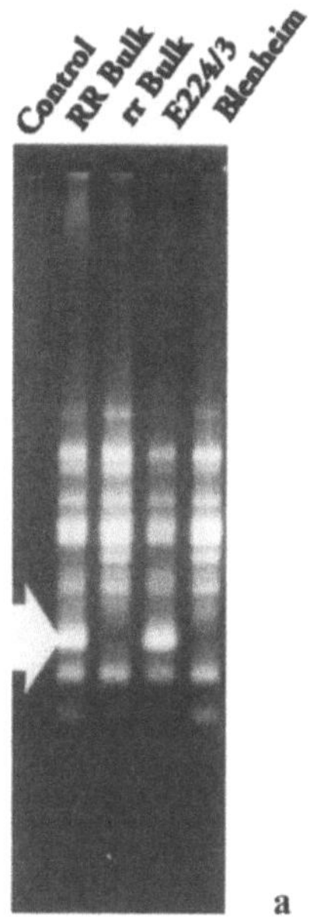

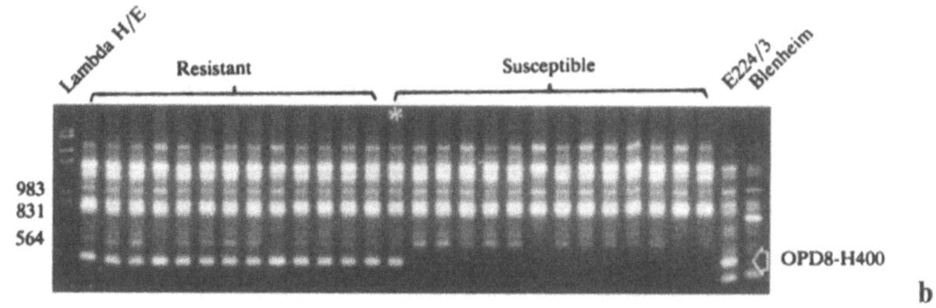

tetrasomic, ditelosomic or single chromosome recombinant lines (Sears 1954; Islam et al. 1981; Janhier et al. 1989; McGrath et al. 1990). Generally, these species are naturally or synthetically derived amphiploids (like wheat and brassicas) which can tolerate the addition or loss of chromosomal segments or entire chromosomes due to the buffering capacity of the homeologous counterparts. Devos and Gale (1992) demonstrated the efficacy of RAPD markers for the analysis of nullisomic-tetrasomic and ditelosomic lines of wheat where they could quickly assign amplification products to both chromosomes and chromosome arms. Further, the same group applied RAPDs for the generation of chromosome-specific markers in barley by using alien addition and substitution lines. Quiros et al. (1991) isolated 16 A

Fig. 6.4. Hypothetical linkage map (see Table 6.1)

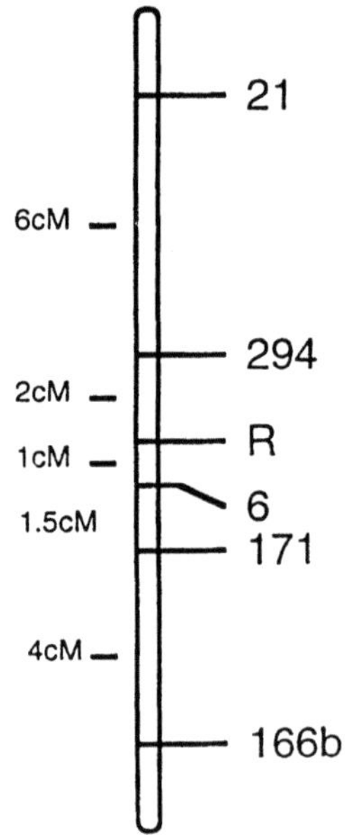

genome, 37 B genome and 12 C genome-specific RAPD markers by utilising six of the eight possible addition lines corresponding to each of the B. nigra chromosomes in B. napus-nigra disomic alien addition lines.

Reagents and equipment

- As for Basic RAPD Procedure
- Genetic stocks e.g.
  Wheat/barley alien addition lines
  Wheat/rye alien addition lines
  Wheat nullisomic-tetrasomic lines
  Brassica napus-nigra disomic addition lines
  Brassica campestris-oleracea addition lines
  Potato disomic addition lines.

Procedure As for Basic RAPD Procedure, using genetic stocks DNA

6.4 Future Applications

RAPD and AP-PCR markers have been successfully applied to DNA isolated from a wide range of organisms and, as I have tried to stress,

Table 6.1. Hypothetical fragment scores at five RFLP loci in two parental lines and a population segregating for a disease resistance locus. Bold numbers represent recombinants at the marker locus. Shaded rows represent individuals to bulk (R lines: bulk nos 1, 3, 11, 33, 50; S lines: bulk 8, 9, 20, 22, 47)

Individual	Locus R/S	6		294		171		21		166b	
P1	R	1	0	0	1	1	0	1	0	0	1
P2	S	0	1	1	0	0	1	0	1	1	0
1	R	1	0	0	1	1	0	1	0	0	1
3	R	1	0	0	1	1	0	1	0	0	1
5	R	**0**	**1**	0	1	**0**	**1**	1	0	**1**	**0**
6	R	1	0	0	1	1	0	**0**	**1**	0	1
11	R	1	0	0	1	1	0	1	0	0	1
14	R	1	0	0	1	1	0	**0**	**1**	0	1
28	R	1	0	0	1	**0**	**1**	1	0	**1**	**0**
29	R	1	0	**1**	**0**	1	0	**0**	**1**	0	1
31	R	1	0	0	1	1	0	1	0	**1**	**0**
33	R	1	0	0	1	**0**	**1**	1	0	0	1
48	R	1	0	0	1	1	0	1	0	**1**	**0**
50	R	1	0	0	1	1	0	1	0	0	1
4	S	0	1	1	0	0	1	**1**	**0**	1	0
8	S	0	1	1	0	0	1	0	1	1	0
9	S	0	1	1	0	0	1	0	1	1	0
18	S	0	1	1	0	0	1	**1**	**0**	1	0
20	S	0	1	1	0	0	1	0	1	1	0
21	S	0	1	1	0	**1**	**0**	0	1	**0**	**1**
22	S	0	1	1	0	0	1	0	1	1	0
39	S	0	1	1	0	0	1	0	1	**0**	**1**
40	S	0	1	1	0	0	1	0	1	**0**	**1**
46	S	0	1	1	0	0	1	**1**	**0**	1	0
47	S	0	1	1	0	0	1	0	1	1	0
49	S	0	1	**0**	**1**	0	1	**1**	**0**	1	0

are particularly applicable to the analysis of small numbers of samples. However, in many species, polymorphic RAPD fragments appear to originate from repetitive DNA sequences, and this can limit their exploitation (although some RAPDs do arise from low copy sequences; Williams et al. 1990). Another feature is illustrated by the low level of RAPD polymorphism compared to that detected by RFLPs in our studies using cultivated barley. This latter finding suggests that much of the highly informative and more easily utilizable variation is going undetected by the RAPD and AP-PCR approach. One potential way of exploiting the variation which undoubtedly occurs in low-copy DNA is to enrich for these sequences in the substrate DNA used for PCR. We are currently using differences detected by arbitrarily primed PCR on cDNA synthesized from pooled RNAs isolated from individuals chosen in the same way as described in the Targeted Approaches outlined in Section 6.3.2. Arbitrarily primed PCR on RNA has been used recently in the approach known as differential display (Liang and Pardee 1992; Welsh et al. 1992, 1993; McClelland et al. 1993a,b; Bauer et al. 1993) as a means of identifying differentially expressed mRNA sequences. However, genotypic differences can also be detected using the RAPD or AP-PCR procedures on RNA or cDNA synthesized from RNA from two different individuals or RNA pools. Furthermore, any polymorphisms detected will reside in low-copy sequences. Like isozymes, RAPD derived from RNA or cDNA will be subject to environmental and developmental changes. Unlike isozymes, RNA- or cDNA-derived RAPD markers should be easily converted to robust and reliable DNA-based PCR markers which are not subject to these influences. An alternative approach has also been described. Lagudah and Appels performed RAPDs on the low-copy fraction of total DNA obtained after the reassociation of denatured total genomic DNA. The next few years will reveal whether the combination of standard and novel approaches can further increase the application of RAPDs to different crops and objectives.

Acknowledgements. The Scottish Crop Research Institute is grant-aided by the Scottish Office Agriculture and Fisheries Department (SOAFD).

References

Al-Janabi SM, Honeycutt RJ, McClelland M, Sobral BWS (1993) A genetic linkage map of *Saccharum spontaneum* L. SES 208. Genetics 134:1249–1260

Arnheim N, Strange C, Erlich H (1985) Use of pooled DNA samples to detect linkage disequilibrium of polymorphic restriction fragments and human disease: studies of the HLA class II loci. Proc Natl Acad Sci USA 82:6970–6974

Barua UM, Chalmers KJ, Hackett CA, Thomas WTB, Powell W, Waugh R (1993) Identification of RAPD markers linked to a *Rhynchosporium secalis* resistance locus in barley using near-isogenic lines and bulked segregant analysis. Heredity 71:177–184

Bauer D, Muller H, Reich J, Riedel H, Ahrenkiel V, Warthoe P, Strauss M (1993) Identification of differentially expressed mRNA species by an improved display technique (DDRT-PCR). Nucleic Acids Res 4272–4280

Bell A, DeMarini (1991) Excessive cycling converts PCR products to random length higher molecular weight fragments. Nucleic Acids Res 19:5079

Black WC (1993) PCR with arbitrary primers: approach with care. Insect Mol Biol 2:1–6

Black WC, DuTeau NM, Puterka GJ, Nectols JR, Pettorini JM (1992) Use of random amplified polymorphic DNA-polymerase chain reaction (RAPD-PCR) to detect DNA polymorphisms in aphids. Bull Entomol Res 82:151–159

Botstein D, White RL, Skolnick M, Davis RW (1980) Construction of a genetic linkage map in man using restriction fragment length polymorphisms. Am J Human Genet 32:314–331

Boury S, Lutz I, Gavalda MC, Guidet F, Schlesser A (1992) Genetic fingerprinting in cauliflower by the RAPD method and determination of the level of inbreeding in a set of F_1-hybrid seeds. Agronomie 12:669–681

Brown AHD (1979) Enzyme polymorphism in plant populations. Theor Appl Genet 15:1–42

Caetano-Anolles G, Bassam BJ, Gresshoff PM (1991a) High resolution DNA amplification fingerprinting using very short arbitrary oligonucleotide primers. Biotechnology 9:553–557

Caetano-Anolles G, Bassam BJ, Gresshoff PM (1991b) DNA amplification fingerprinting: a strategy for genome analysis. Plant Mol Biol Rep 4:294–307

Caetano-Anolles G, Bassam BJ, Gresshoff PM (1992) DNA fingerprinting – mapping out a RAPD redefinition. BioTechnology 10:937

Carland FM, Staskawicz BJ (1993) Genetic characterisation of the *pto* locus of tomato – semi-dominance and co-segregation of resistance to *Pseudomonas Syringae* Pathovar tomato and sensitivity to the insecticide Fenthion. Mol Gen Genet 239:17–27

Chalmers KJ, Waugh R, Sprent JI, Simons AJ, Powell W (1992) Detection of genetic variation between and within populations of *Gliricidia sepium* and *G. maculata* using RAPD markers. Heredity 69:465–472

Chalmers KJ, Barua UM, Hackett CA, Thomas WTB, Waugh R, Powell W (1993) Identification of RAPD markers linked to genetic factors controlling the milling energy requirement of barley. Theor Appl Genet 87:314–320

Clark AG, Lanigan CMS (1993) Prospects for estimating nucleotide divergence with RAPDs. Mol Biol Evol 10:1096–1111

Dawson IK, Chalmers KJ, Waugh R, Powell W (1993) Detection and analysis of genetic variation in *Hordeum spontaneum* populations from Israel using RAPD markers. Mol Ecol 2:151–159

Dellaporta SL, Wood J, Hicks JB (1983) A plant DNA mini-preparation: version II. Plant Mol Biol Rep 1(4):19–21

Demeke T, Kawchuk LM, Lynch DR (1993) Identification of potato cultivars and clonal variants by random amplified polymorphic DNA analysis. Am Potato J 70:561–570

Devos KM, Gale MD (1992) The use of random amplified polymorphic DNA markers in wheat. Theor Appl Genet 84:573–578

Dockhorn-Dworniczak B, Dworniczak B, Bronmelkamp L, Bulles J, Horst J, Bocker WW (1991) Non-isotopic detection of single-strand conformation polymorphism (PCR-SSCP): a rapid and sensitive technique in the diagnosis of phenylketoruria. Nucleic Acids Res 19:2500

Dweikat I, Mackenzie S, Levy M, Ohm H (1993) Pedigree assessment using RAPD-DGGE in cereal crop species. Theor Appl Gene 85:497–505

Eastwood RF, Lagudah ES and Appels R (1994) A directed starch for DNA-sequences tightly linked to ceretal cyst-nematodet resistant genes in Triticum tauschii Genome 37:311–319

Edwards K, Johnstone C, Thomspon C (1991) A simple method for the preparation of plant genomic DNA for PCR analysis. Nucleic Acids Res 19:1349

Ellesworth DL, Ritterhouse KD, Honeycutt RL (1993) Artificial variation in randomly amplified polymorphic DNA banding patterns. Biotechniques 14:214–217

Gawel NJ, Jarret DL (1991) A modified CTAB DNA extraction procedure for *Musa* and *Ipomoea*. Plant Mol Biol Rep 9:262–266

Giovannoni JJ, Wing RA, Ganal MW, Tanksley SD (1991) Isolation of molecular markers from specific chromosomal intervals using DNA pools from existing mapping populations. Nucleic Acids Res 19:6553–6558

Gottlieb LD (1977) Electrophoretic evidence and plant systematics. Ann Mo Bot Gard 64:161–180

Haley SD, Miklas PN, Stavely JR, Byrum J, Kelly JD (1993) Identification of RAPD markers linked to a major rust resistance gene block in common bean. Theor Appl Gen 86:505–512

He S, Ohm H, Mackenzie S (1992) Detection of DNA sequence polymorphisms among wheat varieties. Theor Appl Genet 84:573–578

Hu J, Quiros CF (1991) Identification of broccoli and cauliflower cultivars with RAPD markers. Plant Cell Rep 10:505–511

Hunt GJ, Page RE (1992) Patterns of inheritance with RAPD molecular markers reveal novel types of polymorphisms in the honey bee. Theor Appl Gene 85:15–20

Innes MA, Gelfand DH, Sninsky JJ, White TJ (1990) PCR protocols, 1st edn. Academic Press, San Diego, p 482

Islam AKMR, Shepherd KW, Sparrow DHB (1981) Isolation and addition of euplasmic wheat-barley addition lines. Heredity 46:161–174

Janhier J, Chevre AM, Tanguy AM, Eber F (1989) Extraction of disomic addition lines of *Brassica napus-B. nigra*. Genome 32:374–383

Joshi CP, Nguyen HT (1993) Application of the random amplified polymorphic DNA technique for the detection of polymorphism among wild and cultivated tetraploid wheats. Genome 36:602–609

Kempton R, McNicol J (1990) Graphical methods for multivariate data. 'Teaching notes' Scottish Agricultural Statistics Service. Invergowrie, Dundee, Scotland

King LH, Schaal BA (1989) Ribosomal-DNA variation and distribution in Rudbeckia missouriensis. Evolution 42:1117–1119

Kochert G, Halward T, Branch WD, Simpson CE (1991) RFLP variability in peanut (Arachis hypogaea L.) cultivars and wild species. Theor Appl Genet 81:565–570

Koller B, Lehmann A, McDermott JM, Gessler C (1993) Identification of apple cultivars using RAPD markers. Theor Appl Genet 85:901–904

Korner JS, Livak KJ (1989) Mutation detection using nucleotide analogues that alter electrophoretic mobility. Nucleic Acids Res 17:7779–7783

Lander ES, Green P, Abrhamson J, Barlow A, Daly MJ, Lincoln SE, Newbury L (1987) MAPMAKER and interactive computer package for constructing primary genetic linkage maps of experimental and natural populations. Genomics 1:174–181

Langridge U, Schwall M, Langridge D (1992) Squashes of plant tissue as substrate for PCR. Nucleic Acids Res 19:6954

Liang P, Pardee A (1992) Differential display of eukaryotic mRNA by means of the polymerase chain reaction. Science 257:967–971
Lynch M, Mulligan BG (1994) Analysis of population genetic structure with RAPD markers. Molecular Ecology 3:91–99
McClelland M, Chada K, Welsh J, Ralph D (1993a) Arbitrary primed PCR fingerprinting of RNA applied to mapping differentially expressed genes. In: Pena SDF et al. (eds) DNA fingerprinting state of the science. Birkhävser Basel PP 103–115
McClelland M, Ralph D, Chada C, Dalal SS, Welsh J (1993b) Fingerprinting RNA using arbitrarily primed PCR. FASEB J 7:1304
McGrath JM, Quiros CF (1990) Generation of alien chromosome additon lines from synthetic *Brassica napus*: morphological, cytogenetics, fertility and chromosome transmission. Genome 33:374–383
Martin GB, Williams JGK, Tanksley SD (1991) Rapid identification of markers linked to a *Pseudomonas* resistance gene in tomato by using random primers and near-isogenic lines. Proc Natl Acad Sci USA 88:2336–2340
Martin GB, Devicente MC, Tanksley SD (1993) High resolution linkage analysis and physical characterisation of *pto* bacterial resistance locus in tomato. Mol Plant-Microbe Interact 6:26–34
Michelmore RW, Paran I, Kesseli RV (1991) Identification of markers linked to disease resistance genes by bulked segregant analysis: a rapid method to detect markers in specific genomic regions using segregating populations. Proc Natl Acad Sci USA 88:9828–9832
Miklas PN, Stavely JR, Kelly JD (1993) Identification and potential use of a molecular marker for rust resistance in common bean. Theor Appl Genet 85:745–749
Mulcahy DL, Cresti M, Sansavini S, Douglas GC, Linskens HF, Mulcahy GB, Vignani R, Pancaldi M (1993) The use of random amplified polymorphic DNAs to fingerprint apple genotypes. Sci Hortic 54:89–96
Neale D, Sederoff R (1991) Genome mapping in pine takes shape. Probe 1, 1–3
Nei M (1973) Analysis of gene diversity in sub-divided populations. Proc Natl Acad Sci USA 70:3321–3323
Nei M, Li W-H (1979) Mathematical model for studying genetic variation in terms of restriction endonucleases. Proc Natl Acad Sci USA 74:5267–5273
Newton CR, Graham A, Heptinsta LE, Powell SJ, Summers C, Kalsheker W, Smith JC, Markhan AF (1989) Analysis of any point mutation in DNA. The amplification refractory mutation system (ARMS). Nucleic Acids Res 17:2503–2515
Orozco-Castillo C, Chalmers KJ, Waugh R, Powell W (1994) Detection of genetic diversity and selective gene introgression in coffee using RAPD markers. Theor Appl Genet 87:934–940
Paran I, Michelmore RW (1993) Development of reliable PCR-based markers linked to downy mildew resistance genes in lettuce. Theor Appl Genet 85:985–993
Paran J, Kesseli R, Michelmore R (1991) Identification of restriction fragment length polymorphism and randomly amplified polymorphic DNA markers linked to downy mildew resistance genes in lettuce using near-isogenic lines. Genome 34:1021–1027

Penner GA, Chong J, Levesquelemay M, Molnar SJ, Fedak G (1993) Identification of RAPD marker linked to the oat stem rust gene – PG3. Theor Appl Genet 85:702–705

Quiros CF, Hu J, This P, Chevre AM, Delseny M (1991) Development and chromosomal localization of genome-specific markers by polymerase chain reaction in Brassica. Theor Appl Genet 82:627–632

Ragot M, Hoisington DA (1993) Molecular markers for plant breeding – comparisons of RFLP and RAPD genotyping costs. Theor Appl Genet 86:975–984

Reiter RS, Williams J, Feldmann RA, Rafalski JA, Tingey SV, Scolnik PA (1992) Global and local genome mapping in *Arabidopsis thaliana* by using recombinant inbred lines and random amplified polymorphic DNAs. Proc Natl Acad Sci USA 89:1477–1481

Rogers SO, Bendich AJ (1985) Extraction of milligram amounts of DNA from fresh, herbarium and mummified plant tissues. Plant Mol Biol 5:69–76

Ronald PC, Albano B, Tabien R, Abenes L, Wu KS, McCouch S, Tanksley SD (1992) Genetic and physical analysis of the rice bacterial blight disease resistance locus, XA21. Mol Gen Genet 236:113–120

Saghai-Maroof MA, Soliman KM, Jorgensen RA, Allard RW (1984) Ribosomal DNA spacer length polymorphisms in barley: Mendelian inheritance, chromosomal location and population dynamics. Proc Natl Acad Sci USA 81:8014–8018

Saiki RK, Gelfand DH, Stoffel S, Scharf S, Higuchi RH, Horn GT, Mullis KB, Erlich HA (1988) Primer-directed enzymatic amplification of DNA with a thermostable DNA polymerase. Science 239:487–491

Schierwater B, Ender A (1993) Different thermostable DNA polymerases may amplify different RAPD products. Nucleic Acids Res 21:4647–4648

Sears ER (1954) The aneuploids of common wheat. Mo Agric Exp Stn Res Bull 572:1–59

Sneath PHA, Sokal RR (1973) Numerical taxonomy, the principles and practice of numerical classification. WH Freeman, San Francisco

Sobral BWS, Honeycutt RJ (1993) High output genetic mapping of polyploids using PCR generated markers. Theor Appl Genet 86:105–112

Stam P (1993) Construction of integrated genetic linkage maps by means of a new computer package – JOINMAP. Plant J 3:739–744

Tulsieram LK, Glaubitz JC, Kiss G, Carlson JE (1992) Single tree genetic linkage mapping in conifers using haploid DNA from megagametophytes. Bio/Technology 10:686–690

van de Ven WT (1992) Construction of a genetic linkage map of Vicia faba. PhD Thesis University of Dundee

Waugh R, Powell W (1992) Using RAPD markers for crop improvement. Trends Biotechnol 10:186–191

Waugh R, Baird E, Powell W (1992) The use of RAPD markers for the detection of gene introgression in potato. Plant Cell Rep 11:466–469

Welsh J, McClelland M (1990) Fingerprinting genomes using PCR with arbitrary primers. Nucl Acids Res 18:7213–7219

Welsh J, McClelland (1991) Genomic fingerprinting using arbitrarily primed PCR and a matrix of pairwise combinations of primers. Nucleic Acids Res 19:5275–5279

Welsh J, Honeycutt RJ, McClelland M, Sobral BWS (1991a) Parentage determination in maize hybrids using the arbitrarily primed polymerase chain reaction (AP-PCR). Theor Appl Genet 82:473–476
Welsh J, Petersen C, McClelland M (1991b) Polymorphisms generated by arbitrarily primed PCR in the mouse – application to strain identification and genetic mapping. Nucleic Acids Res 19:303–306
Welsh J, Chada K, Dalal SS, Cheng R, Ralph D, McClelland M (1992) Arbitrarily primed PCR fingerprinting of RNA. Nucleic Acids Res 20:4965–4970
Welsh J, Perucho M, Peinado M, Ralph D, McClelland M (1993) Fingerprinting of DNA and RNA using arbitrary primed PCR. In: Taylor G, McPherson M, Hames D (eds) PCR II: a practical approach. IRL Press, Oxford, pp 197–217
Wilde J, Waugh R, Powell W (1992) Genetic fingerprinting of Theobroma clones using randomly amplified polymorphic markers. Theor Appl Genet 83:871–877
Wilkie SE, Isaac PG, Slater RJ (1993) Random amplified polymorphic DNA (RAPD) markers for genetic analysis in *Allium*. Theor Appl Genet 86:497–504
Wilkinson RC, Parsons TJ, Albright DG, Klein TA, Braun MJ (1993) Random amplified polymorphic DNA (RAPD) markers readily distinguish cryptic mosquito species (Diptera : Culicidae : Anopheles). Insect Mol Biol 1:205–211
Williams CE, St Clair DA (1993) Phenetic relationships and levels of variability detected by restriction fragment length polymorphism and random amplified polymorphic DNA analysis of cultivated and wild accessions of *Lycopersicon esculentum*. Genome 36:619–630
Williams JGK, Kubelik AR, Livak KJ, Rafalski JA, Tingey SV (1990) DNA polymorphisms amplified by arbitrary primers are useful as genetic markers. Nucleic Acids Res 18:6531–6535
Williams JGK, Reiter RS, Young RM, Scholnik PA (1993) Genetic mapping of mutations using phenotypic pools and mapped RAPD markers. Nucleic Acids Res 21:2697–2702
Yang X, Quiros C (1993) Identification and classification of celery cultivars with RAPD markers. Theor Appl Genet 86:205–212
Young ND, Tanksley SD (1989) RFLP analysis of the size of chromosomal segments retained around the *Tm-2* locus of tomato during backcross breeding. Theor Appl Genet 77:353–359
Yu KF, Pauls KP (1992) Optimization of the PCR program for RAPD analysis. Nucleic Acids Res 20:2606

Chapter 7

RFLP Mapping of Plant Nuclear Genomes: Planning of Experiments, Linkage Map Construction, and QTL Mapping

J.H. van den Berg, S.D. Chasalow, and R. Waugh

7.1 Introduction

Mapping of plant genomes began soon after the development of genetic mapping in *Drosophila* in the early 1900s. These efforts were driven in part by the recognition that genetic maps could facilitate indirect selection based on associations between desired traits and more easily determined characters. As early as the 1930s, genetic linkage groups were presented for tomato (MacArthur 1934) and maize (Emerson et al. 1935). These linkage groups consisted entirely of loci that caused discrete and visible changes in morphological characteristics, and until the 1970s genetic mapping was restricted to these types of markers. Because the number of morphological markers segregating within one cross is generally limited, combination of segregation data from multiple crosses was required to construct one complete linkage map (that is, a map covering an entire genome; e.g. Rick 1980). The development of biochemical markers in the form of isozymes and other polymorphic proteins significantly increased the number of markers which could be followed in a single segregating population (Tanksley 1983). Unfortunately, biochemical markers suffer some of the same inherent limitations as genes for morphological characters. Scoring of both of these marker types depends on gene expression, which may be sensitive to environmental influence, genetic background, developmental stage, and tissue type. Both types of markers have limited abundance.

In 1974, Grozdicker et al. (1974) used differences in the lengths of DNA fragments generated after digesting adenovirus DNA with restriction endonucleases to characterize mutant isolates that exhibited a temperature sensitive phenotype. The potential of using these so-called restriction fragment length polymorphisms (RFLPs) as markers for the construction of complete genetic linkage maps was recognised soon after (Kan and Dozy 1978; Botstein et al. 1980). Since then, RFLPs have been the cornerstone of detailed genetic linkage studies and have been used to construct complete genetic maps of a wide range of species (O'Brien 1993). The power of RFLPs resides in their ability to directly reveal differences in the nucleic acid base sequences of homologous chromosomes. Unlike scoring of conventional morphological and biochemical markers, scoring of RFLPs does not depend on gene expression. RFLPs are numerous, insensitive to the influence of environment and genetic background, developmentally stable, often inherited in a simple Mendelian fashion, and alleles are usually co-dominant.

RFLPs have been used for phylogenetic studies (Miller and Tanksley 1990; Waugh et al. 1990); characterizing genetic relationships among crop varieties (Görg et al. 1992; Ishii et al. 1993); identifying crosses or somatic hybrids; localizing chromosomal segments affecting monogenic traits (see review by Gebhardt and Salamini 1992); map based cloning (Rommens et al. 1989; Giraudat et al. 1992; Martin et al. 1993): and the study of quantitative inheritance (Tanksley 1993; deVincente and Tanksley 1993; Van den Berg 1993; Van Eck et al. 1994). In this chapter we cover theoretical and practical aspects of planning, performing, and analyzing experiments to construct RFLP genetic linkage maps of diploid genomes and use such maps for genetic analysis of quantitatively inherited traits. Theoretical aspects apply equally to the previous chapter which describes an alternative DNA-based marker system (Chap. 6).

Concept and Rationale

RFLPs are the product of allelic differences between DNA restriction fragments caused by nucleotide sequence variability. To detect RFLPs, nuclear DNA is extracted and digested with a restriction enzyme. The resulting restriction fragments are separated according to size by electrophoresis, denatured, and transferred onto a solid support such as a nitrocellulose or nylon filter. The DNA fragments are then hybridized with radioactive- or hapten-labelled probe DNAs. Single-copy DNA probes are preferred. Because they hybridize to unique genetic loci, single-copy probes yield more easily interpretable band profiles in Southern blots than do multi-copy probes.

Using this procedure, at a particular locus, restriction fragments from both homologous chromosomes are revealed. If the restriction fragments corresponding to different alleles at the locus are equal in size, a single band will show on the hybridization filter. If the restriction fragments from different alleles are unequal in size, more than one band will be revealed. This difference in band pattern among alleles represents a Restriction Fragment Length Polymorphism. When an RFLP is observed, the probe detecting it can be used to follow the inheritance of the RFLP locus in a segregating (mapping) population. The process of following RFLP alleles may be repeated with many probes to cover a large proportion of the genome. In the mapping population, alleles at each RFLP locus segregate according to the rules of Mendelian genetics (except for possible non-random segregation distortion; see Sect. 7.5.1 and 7.5.6), so RFLP markers may be used for genetic analysis in the same way as conventional genes. Loci may be assigned to linkage groups, so that every locus in a group is linked to at least one other locus in the group. Within linkage groups the order of markers, and genetic distances between them, can be estimated, yielding an RFLP genetic linkage map.

To localize on an RFLP linkage map genes associated with a particular morphological trait, the trait first must be measured for the individuals in the mapping population. Then one may look for associations between RFLP marker loci and the trait. For example, suppose a double-banded (heterozygous) RFLP pattern at a given marker occurs more frequently in individuals with red flowers than in individuals with white flowers, and a single-banded (homozygous) RFLP pattern at the same marker tends to co-occur more frequently with white flowers. If the differences in frequency are greater than those expected by chance alone, we interpret this observation as evidence that a locus affecting flower color is in genomic proximity to that marker. Markers strongly associated with a trait can be used to test whether a genotype possesses a chromosomal segment associated with a desired value of the trait, without waiting for phenotypic expression of the segment. This application of the RFLP linkage technique is known as Marker Assisted Selection (MAS). A marker tightly linked to a gene for a monogenic trait can also provide a basis for map-based cloning.

RFLP genetic linkage maps are especially useful for the study of traits exhibiting quantitative variation. Such a trait is commonly, though not necessarily, a character measured on a continuous scale –

for example, plant height or yield. It may also be measured on a discrete, categorical scale – for example, disease resistance scored on a five-point scale. A character shows quantitative variation if its value is related to genotype with uncertainty, i.e. the genotype at loci controlling the trait determines not the actual trait value, but rather probabilities of the trait taking on different values. The actual value observed will depend also on additional factors such as environment, measurement techniques and genetic background. Various statistical procedures have been developed for using a complete RFLP linkage map, combined with trait scores for the individuals in a mapping population, to estimate the location and effects of genetic loci underlying a quantitatively inherited trait (see Sect. 7.6).

7.2 Planning of Experiments

The principles of marker selection, map construction, and QTL mapping outlined in Sections 7.3–7.6 apply to most mapping projects regardless of species or objectives. However, the selection of a mapping population is highly dependent on both the species and the objectives of the study and can significantly influence its success. In choosing the size, type, and parents of the mapping population, the segregation of both RFLPs and other traits of interest needs to be considered.

7.2.1 Homozygous Versus Heterozygous Parents

Before starting a mapping project in a new species, pilot studies to evaluate the amount of DNA polymorphism in the species are routinely performed (e.g. McCouch et al. 1988; Eldredge et al. 1992). From these investigations a general distinction regarding the number of polymorphic loci in plants has become apparent. Mainly autogamous (self-fertilizing) species such as tomato, soybean, lentil and wheat are usually highly homozygous, as a consequence of inbreeding. In contrast, mainly allogamous (cross-fertilizing) species such as citrus, leek, and potato tend to outbreed and are usually more heterozygous. Outbreeding species have been observed to contain more polymorphic DNA than inbreeding species (Gebhardt and Salamini 1992). Because outbreeding species contain more DNA sequence variability, mapping populations in these species typically can be created by crossing intraspecific parents. Mapping populations used for inbreeding species typically are derived from crosses between species (intrageneric).

The difference in levels of DNA polymorphism detected within inbreeding and outbreeding species also has consequences for the segregation patterns observed in a mapping population. Because RFLP alleles can segregate in the meioses of heterozygous parents, a cross of these parents can produce up to four different allelic patterns at one particular marker. However, the hybrid progeny of completely homozygous parents are all identical. Thus an F_1 hybrid population can be used as a mapping population if derived from heterozygous parents but not if derived from homozygous parents. Likewise, F_2 and backcross populations will show different segregation patterns for crosses originating from homozygous parents than for those originating from heterozygous parents (Fig. 7.1). Loci in a pair of highly heterozygous parents constitute a mosaic of many different genotypic configurations, with two, three, or four alleles segregating at each locus. Genotypic class frequencies for a given locus in the F_1 may occur in a 1:1:1:1 ratio, a 1:2:1 ratio, or a 1:1 ratio, depending on the parental genotypes at that locus, and a given pair of loci may be linked in either coupling or repulsion phases. This complicates linkage analysis for both marker-loci and QTLs (Sects 7.5 and 7.6). For QTL mapping from heterozygous parents, markers such as RFLPs that exhibit codominant inheritance are preferred, since they provide more information for distinguishing QTL genotypes when more than two QTL alleles are segregating (Van Eck et al. 1994)

7.2.2
Selecting a Segregating Population

7.2.2.1
Population Type

Intercross (F_2). F_2 progenies are probably the most widely used populations for the creation of linkage maps in inbreeding species. When starting with homozygous parents, all three genotypes at a locus (A_1A_1, A_1A_2, and A_2A_2) segregate in the progeny (Fig. 7.1A), al-

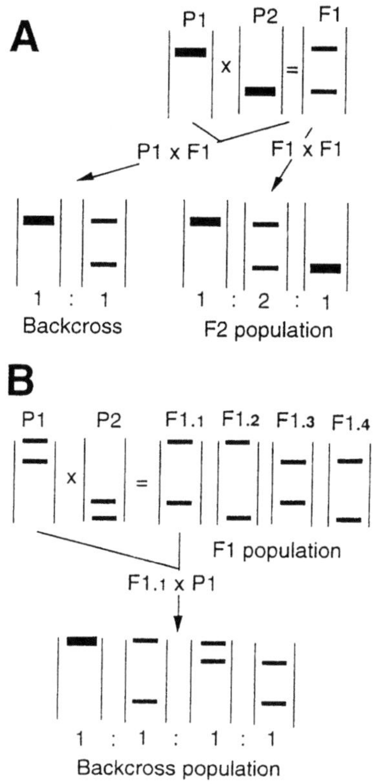

Fig. 7.1A,B. Hypothetical scheme for restriction fragment segregation at one marker. **A** Segregation of RFLP alleles from homozygous parents (inbreeding species) in an F_2 and BC population. **B** Segregation of RFLP alleles from heterozygous parents (outbreeding species) in an F_1 and BC population

lowing the calculation of dominance effects (Tanksley 1993). In addition, recombination may be observed in both gametes forming an F_2 individual. Thus a given population size represents twice as many potentially informative meioses as would be the case for a backcross with the same homozygous parents (Mather 1938; Allard 1956; Ritter et al. 1990).

Backcross (BC). In BC populations derived from inbred parents only two genotypes per locus are represented (Fig. 7.1A): the heterozygous genotype (A_1A_2) and the homozygous genotype (A_1A_1 or A_2A_2) of the recurrent parent. Because one of the homozygous genotypes is missing in the BC population, the calculation of dominance effects is not possible and only dominant or additive alleles can be detected. However, information about gene action may be obtained by studying both reciprocal BCs, because then all three genotypes are represented. In outbreeding species no more than three alleles segregate per locus

(Fig. 7.1B) instead of four as is possible in F_2 crosses. BC populations are used commonly for mapping studies in outbreeding species.

Doubled haploids (DHs). DHs are developed by doubling the chromosomes in gametes, for example by anther culture, to yield homozygous individuals. They have been used most commonly with inbreeding species, but also provide a means of applying mapping techniques developed for inbreeding species to outbreeding species, by providing a homozygous line which may be used as a parent to design a segregating mapping population. In sexually propagated species DHs are also a means of fixing the genotypes in a mapping population in the homozygous state, allowing tests in different years and locations, and use by different research groups to integrate mapping data (Gebhardt and Salamini 1992). DH mapping populations have been used extensively in, for example, barley genome mapping (Heun et al. 1991; Graner et al. 1991).

Recombinant inbred lines (RIs). RIs can be obtained by repeated selfing (inbreeding) of heterozygous individuals, to produce highly homozygous lines. The production of RIs is time consuming and difficult for self-incompatible species. Due to increased recombination and chromosome pairing during inbreeding, however, RI mapping populations allow higher resolution mapping of quantitative trait loci than do F_2 populations (Burr and Burr 1991). Furthermore, QTLs may be detected by comparing the phenotypes of the two homozygotes for alternative marker alleles, rather than by comparing one homozygote to the heterozygote as for a BC population. The difference between the two homozygotes is likely to be greater than that between a homozygote and a heterozygote, except for a QTL with a completely dominant allele. Thus a mapping population consisting of RIs generally requires fewer individuals than does a BC population.

Hybrid population (F_1). Sufficient numbers of RFLP alleles segregate in F_1 mapping populations only for parents that are highly heterozygous (Fig. 7.1B), such as potato (Gebhardt et al. 1989). F_1 populations are frequently used for mapping the genomes of outbreeding species.

7.2.2.2
Parent Selection and Population Size

Monogenic traits. If the objective of a study is to search for RFLPs tightly linked to a monogenic trait, use of two near isogenic lines (NILs; Young et al. 1988) that differ for the trait under study offers an alternative to performing segregation analysis on an entire plant population. NILs, made by repetitive backcrossing with a recurrent parent combined with repetitive selection for the desired characteristic, allow the smallest possible mapping population. After seven backcrosses, the expected proportion of recurrent parent genome in a nearly isogenic line is more than 99% ($1-0.5^n$; n = number of backcrossing cycles) and the line will still contain the allele of interest. Hybridization patterns for digested DNA at almost all marker-loci are expected to be the same for both the inbred and the recurrent parent lines. RFLPs will more probably occur at marker-loci tightly linked to the target gene.

An alternative to the use of NILs is bulk segregant analysis (Michelmore et al. 1991; Pineda et al. 1992). In bulk segregant analysis (see Chap. 6), DNA from a number of individuals possessing the desired trait value is combined into one batch and the DNA from plants lacking that trait value is combined into another. The two bulked DNAs are then digested, Southern blotted, and screened with the marker probes. DNA restriction patterns that differ between the two bulks will most likely originate from marker-loci tightly linked to the target gene associated with the trait of interest. Patterns for all other marker-loci will most likely be isomorphic between the two bulks. For mapping with NILs the maximal distance between the target gene and linked marker-loci depends on the number of backcross cycles. For bulk DNA analysis, the size of the population determines the accuracy of mapping.

Quantitative traits. QTLs (see Sect. 7.6) have been studied in many species, including maize (e.g. Edwards et al. 1987, 1992; Stuber et al. 1987, 1992), tomato (e.g. Paterson et al. 1988, 1991; Tanksley and Hewitt 1988; Garvey and Hewitt 1992; de Vincente and Tanksley 1993), and potato (e.g. Van den Berg 1993; Bonierbale et al. 1994; Leonard-Schippers et al. 1994; Van Eck et al. 1994). The segregation of QTLs can be influenced by the degree to which the parents differ for the trait under study. The selection of parents that are extremes for the quantitative trait may increase the number of detectable QTLs. In general, increasing the population size increases the statistical power for QTL detection and increases the precision of QTL effect and location estimates. For inbreeding species, the number of progeny needed for a particular project may be estimated using formulae and graphs such as those presented by Lander and Botstein (1989) and Van Ooijen (1992). The number of progeny needed to detect a given number of QTLs depends on marker spacing, population type, size of QTL effects, and the number of segregating QTLs (e.g. Darvasi et al. 1993). The number of QTLs segregating in a cross, and the size of their effects, influences the phenotypic differences of the parents, and the genotypic variance of the mapping population. The genotypic variance can be estimated by subtracting the environmental variance from the total experimental variance. The environmental variance can be estimated from measurements of replicate genotypes in the mapping experiment (Van Eck et al. 1994). Methods presented by Lander and Botstein (1989) to estimate the number of segregating QTLs and the number of progeny needed to map them depend on a series of assumptions, including (1) homozygous parental lines; (2) no epistatic interactions among QTLs; (3) normally distributed environmental variance; (4) unlinked QTLs; and (5) alleles in the high strain all increase the phenotypic value. Although these assumptions usually are not satisfied in practice, the formulae and graphs presented can give an indication of the number of segregating QTLs and the size of mapping population needed to map them. Mapping populations for QTL studies typically contain from 150 to 400 individuals.

Selective genotyping. The number of individuals genotyped can be drastically reduced without losing statistical power for the detection of QTLs, using a procedure called selective genotyping (Lander and Botstein 1989). In this approach, only individuals with phenotypes in the tails of the phenotypic distribution of the mapping population are genotyped. Although selective genotyping is highly efficient for the mapping of QTLs, it does have its limitations. First, when only the extremes of the phenotypes are genotyped, it is more difficult to estimate the effects of individual QTLs. This is because the sub-populations that are genotyped are likely to consist of those strains that contain almost all positive alleles or almost all negative alleles of multiple QTLs. These extreme sub-populations are thus deficient in individuals with a mixture of positive and negative alleles (Tanksley 1993). Furthermore, it can be impractical to use selective genotyping for more than one trait, because the individuals that show extreme phenotypes for one trait are not likely to represent the extreme subpopulations for another trait (Tanksley 1993). When different traits are being studied in the same mapping population, it is possible to use a modified version of selective genotyping, in which one genotypes individuals that have extreme phenotypes for one or more of the traits of interest. This will also introduce individuals with mixed sets of favourable and unfavourable alleles for any particular trait (Van den Berg 1993), facilitating the estimation of individual QTL effects.

7.3 Generating and Selecting RFLP Probes

The nuclear DNA of most crop plant species consists largely of repetitive nucleotide sequences. In general, the larger the genome the more repetitive DNA. To generate useful mapping information, it is preferable that each RFLP originates from only one, or at most a few genetic loci. To satisfy this preference, single-copy DNA probes are generally used as RFLP markers. Two sources of such probes are frequently employed: cDNAs and *Pst* I genomic clones (Tanksley et al 1988b).

7.3.1 cDNA Probes

Most plant genes occur in single- or low-copy number in the genome. This makes cDNA libraries a good source of RFLP markers (Tanksley et al. 1988b). The procedures for constructing cDNA libraries are covered in detail in Chapter 3. When using cDNA probes it is important to consider that cDNA is usually relatively short ($\leq$ 1 kb) and the amount of incorporated label is proportional to the size of the probe. Therefore the required autoradiography time of the filters can be long. In addition cDNA is derived from processed gene transcripts, which means they lack introns. When the restriction enzyme cuts within the structural gene associated with the cDNA, the cDNA probe could pick up several fragments of the gene which have only limited homology to the probe. This frequently leads to complex hybridization signals on Southern blots. Occasionally, the lack of continuous homology impedes recognition between the cDNA probe and its homologous DNA in the restriction digest. However, most cDNA probes are excellent genetic markers and are widely used in genome mapping.

7.3.2
Genomic DNA Probes

Genomic DNA (gDNA) probes are usually larger and of better continuous homology with target loci than cDNA probes (Tanksley et al. 1988b). Unfortunately, gDNA contains a high proportion of repeated nucleotide sequences which, if simply cut and cloned randomly, would generate many clones containing repetitive sequences. These are unsuitable as genetic markers. However, evidence indicates that regions associated with single-copy DNA (Tanksley et al. 1988b), including genes (Antequera and Bird 1989), are hypomethylated. Therefore, cutting DNA with a restriction enzyme, such as *Pst* I, that has a methylation-sensitive recognition site, increases the relative abundance of short DNA fragments derived from single copy DNA. In tomato, digestion with *Pst* I resulted in 92% single copy DNA fragments, whereas digestion with *Eco* RI resulted in only 36% single copy clones (Miller and Tanksley 1990). To date, in plants most gDNA probes used in genetic mapping are derived from *Pst* I digests.The production of *Pst* I probes via plasmid libraries is covered in Section 2.2.

7.3.3
Probes from Other Mapping Populations

Within one genus, probes used in one cross can readily be used for another mapping project and within a family, probes can often be used across genus boundaries. For example, maps of potato (Bonierbale et al. 1988) and pepper (Tanksley et al. 1988a) were derived from tomato cDNA and gDNA probes; wheat and oat markers were used to create a barley map (Heun et al. 1991); and maize genomic clones were used for the development of the sorghum map (Whitkus et al. 1992). The use of markers developed for previous studies can speed up the development of new maps considerably and allows comparison of the newly mapped genome with all other genomes mapped with the same probes (e.g. Bonierbale et al. 1988; Whitkus et al. 1992).

7.4
Laboratory Procedures for RFLP Analysis

Having constructed or collected a representative set of cloned DNAs to use as probes, and selected an appropriate segregating population, a detailed study can proceed. Usually, RFLPs between the parental lines are catalogued prior to the analysis of the segregating population. The laboratory techniques employed in standard RFLP analysis are carried out sequentially as follows:

- DNA isolation
- Southern hybridization (radioactive or non-radioactive)
- Autoradiography

These techniques have all been extensively covered earlier in the book (see Chap. 1).

7.5
Linkage Map Construction

Suppose that we have obtained a suitable mapping population and measured genotypic scores for a set of RFLP markers in each individual in the population. In this section we describe the process of constructing a genetic linkage map of these marker loci; that is, using estimated recombination frequencies between loci to estimate the linear arrangement of markers on chromosomes. Available computer programmes for linkage analysis include Linkage (Suiter et al. 1983; Cottingham et al. 1993; Schaffer et al. 1994), Mapmaker (Lander et al. 1987), Map (Morton and Andrews 1989), GMendel (Echt et al. 1992), Surveyor (Shoemaker et al. 1992), RI Manager (Manly and Elliott 1992), Cprop (Letovsky and Berlyn 1992), and JoinMap (Stam 1993). We focus here on general principles, leaving discussion of details specific to a given computer package to the appropriate reference and program documentation.

Construction of a linkage map may be accomplished by five main steps:

1. analysis of segregation at individual loci;
2. detection of linkage between loci;
3. estimation of recombination frequencies and map distances between loci;
4. assembly of loci into linkage groups; and
5. estimation of linear orders of loci within linkage groups.

In practice, these five steps are not necessarily well separated and need not always be followed in strict order. For example, recombination frequency estimates may be revised after the order of loci within linkage groups has been inferred, and often an interactive iterative approach is adopted. We discuss each of these steps in turn below, but first we consider briefly the process of scoring and recording genotypes.

Data Collection

RFLP phenotypes take the form of the presence or absence of particular DNA restriction fragments (bands) on a gel. The phenotype scores for a population may be recorded in a matrix, with one row for each band and one column for each individual. The $(i,j)^{th}$ entry in this matrix gives the score of band i for the jth individual. Absence/presence of a band may be coded as 0/1, −/+, etc., with an additional code for missing values. Data quality is important to consider at this point. Subsequent analysis is only as good as the data you start with. Given many loci scored for many individuals, there are many opportunities for errors to arise in the course of translating raw autoradiograms into a computer file ready for analysis. The use of gel scanners and computerized scoring of bands (e.g. Karaman 1994; Luckenbach et al. 1994) may reduce scoring errors. Misclassification also may be reduced by duplicate scoring (using either duplicate gels, or duplicate reading of a single gel by different observers), followed by computerized comparison of data files and rechecking of gels for any discrepant data points (e.g. Rohrer et al. 1994). Any phenotypes that remain ambiguous are best coded as missing values. Analytic procedures for identifying likely scoring errors are available (see Sect. 7.5.6). Such methods are valuable for identifying some of the errors that inevitably remain in any large data set even after sensible quality assurance techniques.

The exact relationship between RFLP phenotype – bands on a gel – and RFLP genotype – allelic constitution at a set of loci – at first may be incompletely known. A given pair of bands may represent two alternative alleles at a single locus, two redundant components of a single allele, or alleles at two distinct (linked or unlinked) loci. Often these different states may be distinguished by parental phenotypes. However, if a given probe/enzyme combination produces more than one polymorphic band in each parent, the possible relationships between RFLP phenotype and genotype may need to be distinguished by linkage analysis (see Sect. 7.5.2). Complete linkage in repulsion is diagnostic of two alternative alleles at a single locus, and complete linkage in coupling phase is diagnostic of redundant bands representing a single allele. The distinction between RFLP phenotype and genotype may be ignored at first. After preliminary linkage analysis, bands identified as being allelic may be recoded to indicate this (e.g. Stam 1993). Scores for all but one of a set of redundant bands may be eliminated. Usually the band with the fewest missing values should be retained, and scores for extraneous bands may be used to impute missing values for the retained band where possible.

7.5.1 Single-Locus Segregation Analysis

In a given sample of gametes from diploid individuals, representing the product of many meioses, Mendelian segregation predicts that, on average, one-half will carry one allele at a locus and one-half will carry the alternative allele. If a given allele is present in only one parent of a cross, a 1:1 ratio is expected for presence:absence of that allele in the progeny. If a given allele is present in both parents and fertilization is random, a 3:1 ratio is expected. If two codominant alleles have been identified at a locus, a 1:2:1 ratio of diallelic genotypes is expected. Departures from expected Mendelian segregation ratios are known as segregation distortion. Distorted segregation may occur by chance alone. We will call segregation distorted by factors other than chance fluctuation "non-Mendelian segregation". Such factors include differential viability of gamete or zygote genotypes, and genotype misclassification (typing errors). Distortion resulting from different factors need not affect linkage analysis in the same way (cf. Heun and Gregorius 1987). Note also that distorted segregation at one locus may induce segregation distortion at other loci linked to it.

The aim of single-locus segregation analysis is to identify distortion larger than that likely to occur by random fluctuation alone. Such analysis commonly is done before applying methods for detecting linkage and estimating the order of and map distances between loci, which may be sensitive to distorted single-locus segregation (see Sect. 7.5.2 and 7.5.6). We review single-locus tests because they traditionally have been used and because they provide a relatively simple introduction to statistical concepts useful in subsequent sections.

However, we believe they have only limited relevance to most genetic mapping studies, for reasons outlined in Sect. 7.5.2.4.

Departure of a single-locus segregation ratio from Mendelian expectation is commonly tested using a χ^2 (Chi-square) goodness-of-fit test (e.g. Mather 1938; Suzuki et al. 1981). For example, consider 100 progeny from the cross, AB/ab × ab/ab (Table 7.1b), ignoring locus B for the moment. Under the null hypothesis of Mendelian segregation, genotypes Aa and aa each are expected to occur with frequency 100/2 = 50. The χ^2 statistic summarizes the amount of deviation between these expected frequencies and the observed single-locus genotype frequencies, 65 and 35. It is calculated as

$$\chi^2 = \frac{(0_1 - E_1)^2}{E_1} + \frac{(0_2 - E_2)^2}{E_2} ,$$

or in general for s phenotypic classes,

$$\chi^2 = \sum_{i=1}^{S} \frac{(0_i - E_i)^2}{E_i} ,$$

Table 7.1. Two-locus genotype frequencies for 100 progeny simulated from a backcross, AB/ab × ab/ab. (a) No linkage, Mendelian segregation. (b) No linkage, genotype aa has 55% viability. (c) Linkage at recombination frequency θ = 0.25, Mendelian segregation. (d) Linkage at θ = 0.25, genotype aa has 55% viability

a)

	Bb	bb	
Aa	25	25	50
aa	27	23	50
	52	48	100

b)

	Bb	bb	
Aa	32	33	65
aa	19	16	35
	51	49	100

c)

	Bb	bb	
Aa	42	8	50
aa	15	35	50
	57	43	100

d)

	Bb	bb	
Aa	53	10	63
aa	11	26	37
	64	36	100

where Σ indicates summation over all phenotypic classes, and O_i and E_i are the observed and expected frequencies for class i, respectively. Larger values of the χ^2 statistic indicate larger discrepancies between observed and expected frequencies.

If the null hypothesis is true, the χ^2 statistic has approximately a χ^2 distribution with s-1 degrees of freedom. This approximation is generally adequate when each expected value is not too small (roughly speaking, at least 1.5; DeGroot 1975; Smith 1986). A P-value may be obtained for a given χ^2 statistic from a table of the χ^2 distribution, found in most introductory statistics and genetics texts (e.g. Suzuki et al. 1981); alternatively, it may be computed in any statistical computer package. The P-value gives the chance that a statistic at least as great as that observed would be obtained if the null hypothesis were true. Commonly an arbitrary threshold α, such as 0.05 or 0.01, is chosen beforehand and a P-value less than that threshold leads to rejection of the null hypothesis, i.e. is interpreted as evidence for non-Mendelian segregation. Such a threshold defines the type-I error rate; that is, the probability of falsely rejecting a true null hypothesis. For locus A in our example (Table 7.1b), we obtain $\chi^2 = 9.00$ with one degree of freedom and a P-value of about 0.003, indicating strong evidence for non-Mendelian segregation. For locus B in the same example, we compare the observed frequencies, 51 and 49, to the expected, 50 and 50, to obtain $\chi^2 = 0.040$ with one degree of freedom and a P-value of about 0.841, indicating a lack of evidence for non-Mendelian segregation at locus B.

An alternative to the χ^2 test for distortion of single-locus segregation ratios is available. Sometimes called a G test (Garcia-Dorado and Gallego 1992), it is a likelihood ratio test comparing a log-linear model that fixes the class frequencies at their Mendelian expectation to a log-linear model that allows the class frequencies to equal their observed values. The G statistic is calculated as

$$G = 2\sum_{i=1}^{S} O_i(\ln O_i - \ln E_i) ,$$

where ln is the natural logarithm and s, O_i , and E_i are as defined above. This statistic also has approximately a χ^2 distribution with s-1 degrees of freedom, and for large populations is approximately equivalent to the χ^2 statistic. In the example in Table 7.1b, for segregation of locus A we obtain G = 9.14 with 1 degree of freedom, giving a P-value of about 0.003. For locus B we obtain G = 0.040 with 1 degree of freedom, giving a P-value of about 0.841. In this case, the χ^2 and G tests yield almost identical results. A possible advantage of the G test over the χ^2 test is that the single-locus G statistics and the G statistic for linkage between the same two loci (see Sect. 7.5.2) are additive (cf. McCullagh and Nelder 1989, p. 39).

If the segregation ratios of many individual loci are to be tested, a lower threshold α should be used for each test. To see why, consider testing distortion at 100 loci. From the definition of a P-value, on average 5 of the 100 loci would exhibit segregation ratios different enough from expectation by chance alone to yield a P-value less than 0.05. If a threshold of 0.05 were used for each test, on average we would make five false inferences of non-Mendelian segregation. To compensate for multiple testing, instead of controlling the error rate for each test individually we control the global type-I error rate, which is the probability of making one or more type-I errors among k tests. If performing k tests, use of a significance level $\alpha_k = \alpha_0/k$ for each test will result in a global type-I error rate not greater than α_0 (Miller 1966).

7.5.2
Detecting Linkage Between Loci

Testing for genetic linkage between all pairs of loci is usually a precursor to the remaining steps in linkage map construction. In single-locus segregation analysis we considered the frequencies of phenotypic classes determined by a single locus (e.g. the marginal frequencies in Table 7.1). In pairwise linkage analysis, we now consider the frequencies of phenotypic classes formed by the product of the classes for two individual loci (e.g. the individual cell frequencies in Table 7.1). Detection of genetic linkage between two loci is possible if the phenotypic class frequencies expected if the loci were linked differ from those expected if the loci were unlinked. Any test for linkage in one way or another compares the observed class frequencies to one or both of these sets of expected frequencies. A variety of statistical tests for linkage have been proposed and applied in the literature. These differ in the exact null hypotheses they test (in particular, whether or not Mendelian single-locus segregation is assumed) and in their power to detect linkage.

7.5.2.1
Tests with Fixed Expected Single-Locus Segregation Ratios

The χ^2 test or the G test described in Section 7.5.1 for testing single-locus segregation could be applied to test two-locus segregation. If locus 1 has s phenotypic classes and locus 2 has t phenotypic classes, there are st two-locus phenotypic classes and the test has st-1 degrees of freedom. The null hypothesis for such a test states that (1) the two loci exhibit perfect Mendelian segregation (so that any departures from Mendelian expectations are due to chance alone); and (2) the loci are unlinked. In this context we will denote the statistics by χ^2_T and G_T, where the subscript T indicates a test of the total hypothesis consisting of both parts (1) and (2). If this null hypothesis is true, the expected two-locus phenotypic class proportions are the product of the expected Mendelian single-locus proportions. For example (Table 7.1a), given Mendelian segregation the expected proportions of the single-locus classes Aa and Bb each are 1/2, so the expected proportion of class AaBb is 1/4. Thus, two unlinked loci each with 1:1 expected single-locus segregation ratios have an expected two-locus segregation ratio of 1:1:1:1, giving expected two-locus frequencies of 25 in each cell. Rejection of the null hypothesis indicates either that the loci are linked, that one or the other of them exhibits non-Mendelian segregation, or both of these. The frequencies in Table 7.1b were simulated with no linkage between loci A and B but with only 55% viability for individuals of genotype aa. Both the χ^2_T and G_T tests lead to rejection of the null hypothesis at $\alpha = 0.05$ (Table 7.2), leading to a false inference of linkage if we were not aware of the non-Mendelian single-locus segregation. Because the χ^2_T and G_T statistics test the two-part null hypothesis that includes an assumption of Mendelian segregation at individual loci, tests based on χ^2_T and G_T are not, strictly speaking, tests for linkage, and cannot be recommended.

Mather (1946) showed that the χ^2_T statistic may be partitioned into a sum of three χ^2 statistics: $\chi^2_T = \chi^2_A + \chi^2_B + \chi^2_L$. The first two components are test statistics for the single-locus segregation ratios (as described in Sect. 7.5.1). The final one, χ^2_L , may be used to test for linkage. The degrees of freedom corresponding to each statistic are obtained by an analogous partition: st - 1 = (s-1) + (t-1) + (s-1)(t-1). The χ^2_T and χ^2_L statistics test different hypotheses. Let p_{ij} be the probability that a randomly selected progeny is in class i for locus 1 and class j for locus 2. For backcrosses such as the examples in Table 1, the null hypothesis for the test based on χ^2_T is

Table 7.2. Linkage tests for examples in Table 7.1. Examples (b) and (d) were generated with reduced viability of genotype aa recombination frequency; P: P-value. See text for definitions of test statistics.

		a: $\theta=0.5$;		b: $\theta=0.5$;		c: $\theta=0.25$;		d: $\theta=0.25$	
Test	DF[a]	Stat.	P	Stat.	P	Stat.	P	Stat.	P
χ_T^2	3	0.320	0.956	9.20	0.027	31.1	8.02×10^{-7}	48.2	1.89×10^{-10}
χ_L^2	1	0.160	0.689	0.160	0.689	29.2	6.66×10^{-8}	33.6	6.66×10^{-9}
G_T	3	0.320	0.956	9.40	0.024	33.6	2.43×10^{-7}	45.3	7.98×10^{-10}
G_L	1	0.160	0.689	0.233	0.629	31.6	1.88×10^{-8}	30.5	3.31×10^{-8}
χ_U^2	1	0.160	0.689	0.233	0.630	29.7	4.93×10^{-8}	29.9	4.46×10^{-8}
LOD[b]	-	0.035	0.923	0.035	0.923	6.68	2.08×10^{-7}	7.78	1.65×10^{-8}
2 ln (10) LOD	-	0.160	-	0.160	-	30.8	-	35.8	-

[a]Degrees of freedom for χ^2 distribution used to compute p values.
[b]Approximate upper bound, 10^{-LOD}, for p values is given.

$p_{11} = p_{12} = p_{21} = p_{22} = 1/4$. The null hypothesis for the test based on χ^2_L is $p_{11} + p_{22} = p_{12} + p_{21}$ (i.e. the expected frequencies of recombinant and parental phenotypes are equal). The test based on χ^2_L is preferable to that based on χ^2_T, because *random* deviations of the single-locus ratios from Mendelian expectation do not on average inflate the statistic. However, as pointed out by Garcia-Dorado and Gallego (1992), the χ^2_L test is still based on an assumption of Mendelian single-locus segregation, and non-random distortion of single-locus ratios may lead to false inferences regarding linkage (see Sect. 7.5.2.2).

7.5.2.2
Tests for Association

In a manner analogous to Mather's partition of the χ^2_T statistic, the G_T statistic may be partitioned into three components: $G_T = G_A + G_B + G_L$ (Garcia-Dorado and Gallego 1992). Again, a corresponding partition of the degrees of freedom holds: st-1 = (s-1) + (t-1) + (s-1)(t-1). The statistics G_A and G_B are simply those for testing single-locus segregation ratios as described in Section 7.5.1. The statistic G_L , in contrast to χ^2_L , does not depend at all on single-locus segregation ratios. It is used to test for association between the segregation of two loci (i.e. for linkage), against the null hypothesis H_0: $p_{ij} = p_{i+} p_{+j}$, where $p_{i+} = \sum_j p_{ij}$ and $p_{+j} = \sum_i p_{ij}$ are the expected single-locus proportions. This hypothesis implies that the expected two-locus frequencies $E_{ij} = np_{ij} = n(r_i/n)(c_j/n) = r_i c_j /n$, where r_i is the *observed* frequency of individuals with genotype i for locus 1 and c_j is defined similarly for locus 2 (e.g. Kendall and Stuart 1979). Thus, under H_0 the expected two-locus frequencies depend only on the *observed* single-locus segregation ratios. For example, the expected frequency of individuals with phenotype AaBb in Table 7.1b is $E_{11} = (65*51)/100 = 33.15$. Once the expected phenotypic class frequencies are calculated in this manner, the test statistic G_L may be computed using the formula for G given in Section 7.5.1.

For the four examples in Table 7.1, both the χ^2_L and G_L statistics yield correct inferences: neither indicates evidence of linkage in its absence, and both detect linkage when it is present (Table 7.2). Unlike the χ^2_T and G_T tests, when applied to a backcross the χ^2_L and G_L tests are not sensitive to distorted segregation at only one locus. This result holds in general for the G_L test but not for the χ^2_L test. Garcia-Dorado and Gallego (1992) demonstrated by simulation that when segregation at *both* loci was distorted by differential viability, and for crosses other than the backcross, the type-I error rate for the χ^2_L test (but not that for the G_L test) may be inflated above the nominal level (e.g. 0.05). On the other hand, the χ^2_L and G_L tests had similar power to detect linkage when single-locus segregation was Mendelian. Thus use of the G_L statistic clearly is preferable.

Yet another χ^2 test for linkage may be constructed by substituting the expected class frequencies E_{ij} used to construct the G_L statistic into the formula for the χ^2 statistic given in Section 7.5.1. The null hypothesis for a test based on this χ^2 statistic is the same as that for the G_L statistic. We will refer to this statistic as χ^2_U, to emphasize that it is purely a test that the two loci are unassociated. Unlike the G_L statistic, χ^2_U does not form part of an orthogonal (additive) partition of a test statistic for the total hypothesis. While this may result in a slight preference for the use of G_L over χ^2_U on purely theoretical grounds, the two statistics are approximately equivalent for large populations (Kendall and Stuart 1979), and in practice the choice between the two ordinarily makes little difference. For the examples in Table 7.1, the χ^2_U and G_L statistics are identical to three significant digits in the absence of linkage, and are in good agreement in the presence of linkage, regardless of whether or not non-Mendelian segregation at one locus occurs (Table 7.2).

7.5.2.3
The LOD Score Test

The tests for linkage described thus far are goodness-of-fit tests: they compare a given set of expected frequencies to the actual observed frequencies. In contrast, the final test we consider, the LOD score test for linkage (Morton 1955), compares two sets of expected frequencies: those obtained if two loci are linked with recombination frequency θ equal to its maximum-likelihood estimate $\hat{\theta}$, and those obtained if the two loci are unlinked ($\theta = 0.5$). The maximum likelihood estimate (MLE) of θ, $\hat{\theta}$, is obtained by selecting from among all possible values in the interval (0,1) the value that maximizes the probability that the observed data would have occurred. The LOD score is the decimal logarithm ($\log_{10}$) of a likelihood ratio: the likelihood that the observed data would have occurred if $\theta = \hat{\theta}$ divided by the likelihood that the observed data would have occurred if $\theta = 0.5$. (These likelihoods are obtained assuming a multinomial distribution for the phenotypic class frequencies).

LOD stands for log of odds, odds in this case being a synonym for likelihood ratio. There has been some disagreement in the literature over the use of the term LOD score versus use of the term log-likelihood ratio, and over the use of the decimal logarithm versus the natural logarithm (Edwards 1988; Morton 1988a). Regardless of one's position on this issue, we find the use of the term odds ratio as a synonym for odds (e.g. Morris 1992; Lincoln and Lander 1992; Gerber and Rodolphe 1994) unfortunate. An odds is itself a ratio of two probabilities; calling this quantity an odds ratio is redundant. We prefer to restrict the term odds ratio to its well-established usage, meaning the ratio of two odds (e.g. Bishop et al. 1975; Breslow and Day 1980; Agresti 1990), which involves *four* probabilities, not two.

LOD score values greater than zero (i.e. likelihood ratios greater than one) dispose us to prefer $\theta = \hat{\theta}$ to $\theta = 0.5$. This preference "arises from our disposition to prefer that hypothesis which makes more probable what we know to be true" (Barnard 1967). However, "what we know to be true" in this case are the observed phenotype frequencies, which are subject to random variation. As a consequence, an LOD score could be greater than zero by chance alone. Therefore we typically select a predetermined threshold greater than zero, and reject the null hypothesis of no linkage only if the observed LOD score exceeds this threshold. Thresholds for the LOD score test are not clearly connected to a type-I error rate α (Chotai 1984; Gerber and Rodolphe 1994), but approximately the type-I error rate for a LOD score test with threshold C is no greater than $1/10^C$ (Chotai 1984). LOD score thresholds typically seen in the literature range from about 2 to 6.

For the backcross examples in Table 7.1, the LOD score test agrees closely with the χ^2_L, G_L, and χ^2_U tests (Table 7.2). (Here, we show also LOD scores scaled by the constant factor 2ln(10), which converts from the decimal to the natural logarithm and multiplies by two, for ease of comparison to the G_L statistic). For a backcross,

$$\mathrm{LOD} = (x_{11} + x_{22}) \log_{10} (x_{11} + x_{22}) + (x_{12} + x_{21}) \log_{10} (x_{12} + x_{21}) + n\log_{10} (2/n),$$

where x_{ij} is the observed frequency of the phenotypic class having class i for locus 1 and class j for locus 2, and n is the total population size. From inspection of this formula, we see that the LOD score depends only on the total number of recombinant type and total number of parental type progeny. On average, in the absence of linkage, non-Mendelian segregation at a single locus affects the frequency of recombinant and parental classes equally. Thus for a backcross the LOD score is not sensitive to non-Mendelian segregation at a single locus in the absence of linkage. This result does not hold in general if there is non-Mendelian segregation at both loci. It also does not hold in general for analysis of an F_2 generation, although for an F_2 the LOD score appears not to be particularly sensitive to non-Mendelian segregation when two loci are unlinked.

Gerber and Rodolphe (1994) compared the LOD score test to a χ^2 test for linkage applied to the study of haploid megagametophytes of 18 unrelated pine trees. This comparison is not necessarily directly applicable to other mating designs, such as analysis of a single full-sib family, but is nonetheless instructive. They found that the LOD score test was generally more conservative than the χ^2 test. However, for LOD scores greater than about 3, the approximate type-I error rate, $1/10^{\mathrm{LOD}}$, for the LOD score test and that for the χ^2 test showed good agreement. It is just this range of χ^2 p-values (and LOD scores) that is of most interest, because low per-test type-I error rates are required when testing for linkage between many pairs of loci.

7.5.2.4
Analysis Strategies for Detecting Linkage

In general, we prefer to use the LOD score, G_L , or χ^2_U statistics to test for linkage. It is, of course, possible to use more than one test for linkage and perhaps to be reassured by general agreement among the results (e.g. Ellis et al. 1992; Gerber and Rodolphe 1994; Grattapaglia and Sederoff 1994). A completely general statement of an appropriate LOD threshold, or per-test significance level for a χ^2 or G test, is not possible. The choice depends in part on the relative costs of false positive and false negative indications of linkage in a particular situation. The issues of multiple comparisons discussed in Section 7.5.1 in the context of testing single-locus segregation ratios apply equally to testing multiple pairs of loci for linkage. When many pairs are to be tested (e.g. 100 loci yield 4950 pairs), use of a threshold severe enough to control the global type-I error rate below 0.05 may result in an unacceptably high rate of false negatives. Nonetheless, it is helpful to keep in mind the approximate global error rate implicit in a particular choice of threshold or significance level.

In a traditional approach to linkage analysis, we would first test distortion of the individual segregation ratios for a pair of loci. If non-Mendelian segregation is indicated, we might first recheck the scoring of those loci for typing errors. If apparent non-Mendelian segregation remains, we may use a test and estimation procedure in linkage analysis that can compensate for it. Alternatively, loci exhibiting evidence of non-random distortion after single-locus segregation tests may be deleted from initial analyses to construct linkage groups and order loci, and added to the map at their most likely positions afterwards (e.g. Grattapaglia and Sederoff 1994).

There are difficulties with such approaches. If a single-locus test fails to detect non-Mendelian segregation, the likelihood that the locus in question truly exhibits Mendelian segregation depends on the power of the test to detect deviations from it. Failure to detect non-Mendelian segregation does not mean it does not exist. Testing segregation at many loci exacerbates this problem, since the use of lower per-locus false positive rates increases the probability of false negatives. (In fact, it is often advisable to allow a relatively large false positive rate, α, for single-locus tests, in order to reduce the rate of false negatives). Conversely, given a large enough sample size, a single-locus test may detect even very minor departures from Mendelian expectation.

The single-locus tests do not address the critical question: is the subsequent linkage analysis sensitive to segregation distortion, whether random or not, of the magnitude present? As an alternative strategy, we recommend to *always* test for pairwise linkage between loci using a procedure insensitive to single-locus segregation ratios. Ideally, recombination frequencies between each pair of linked loci should be estimated by both a standard method based on an assumption of Mendelian segregation and a method designed to be insensitive to departures from Mendelian segregation (see Section 7.5.6). Comparison of the results of the two analyses allows an assessment of the sensitivity of the estimates to the single-locus segregation ratios. If the two methods produce substantially different estimates, the estimate obtained by the method insensitive to distortion is preferred. Evidence of non-Mendelian segregation from a single-locus test may be taken as a warning to recheck genotype scores and pay extra attention to the results of linkage analyses involving that locus. Single-locus tests should not, however, be used by themselves in an attempt to justify subsequent linkage analyses.

7.5.3 Estimating Recombination Frequencies and Map Distances

Meiotic recombination occurs when an individual produces a gamete with a genotype different from either of the two gamete genotypes that combined to form the individual. Recombination between loci on different chromosomes occurs by independent assortment of chromosomes; between loci on the same chromosome, recombination results from crossing-over (physical exchange of DNA) between nonsister chromatids during meiosis. Recombination takes place between two loci on the chromatid received by a particular gamete if an odd number of crossovers involving that chromatid occur between the loci. The "recombination frequency" θ is defined as the probability of an odd number of crossovers on the same chromatid in the interval between two loci, or more generally, the probability of recombination between two loci. Values of θ are constrained to be in the interval (0, 0.5); $\theta = 0.5$ corresponds to independent segregation of two loci.

The positions of multiple crossover events are not necessarily independent, nor are the chromatids that are involved. Chromatid interference describes the condition in which all pairs of nonsister chromatids are not equally likely to be involved in a crossover; the probabilities instead depend on which were involved in other crossovers (Owen 1950). Crossover interference (also known as chiasma interference) occurs if the probability of a crossover event at one point on a chromosome depends on the occurrence of crossovers at other points. Positive crossover interference holds if one crossover inhibits the occurrence of others nearby. Negative crossover interference holds if the probability that two neighbouring crossovers occur is greater than if they occurred independently.

Recombination frequencies for contiguous chromosomal segments are not in general additive, because of the phenomenon of multiple crossovers. For example, given three loci, L_1, L_2 and L_3 in that order, a gamete recombinant for both the pairs (L_1, L_2) and (L_2, L_3) carries a chromatid involved in an odd number of crossovers in each interval. The sum of two odd numbers is even, so such a gamete is not recombinant for the pair (L_1, L_3). Hence the proportion of gametes recombinant for L_1 and L_3 is less than the sum of the proportions recombinant for the other two locus pairs.

A map function corrects for the unobserved multiple crossover events in an interval. It converts a recombination frequency θ_{ij} to a map distance w_{ij}, defined as the average number of crossovers per chromatid in the interval between loci i and j, over a large number of meioses. Map distances, conventionally expressed in centimorgan (cM) units, are additive over contiguous chromosomal segments. In a 100 cM interval, an average of one crossover occurs per chromatid. A number of alternative mapping functions have been proposed (e.g. Owen 1950; Bailey 1961; Liberman and Karlin 1984; Ott 1991). The most commonly used are those of Haldane (1919) and Kosambi (1944). In general, different mapping functions correspond to different assumed degrees of crossover interference. The Haldane mapping function assumes no crossover interference. The Kosambi mapping function allows for positive interference.

Methods for estimating recombination frequencies may be divided into two categories: pairwise estimation and multipoint estimation. Pairwise estimation uses genotypes for a single pair of loci at a time; multipoint estimation uses genotypes for three or more loci simultaneously. In either case, estimates are obtained most commonly by the method of maximum likelihood (Mather 1938; Haldane and Smith 1947; Morton 1955; Bailey 1961; Ott 1991). The LOD score test for linkage is popular in part because of its intimate relationship to maximum likelihood estimation: computations required for both testing and estimation are performed in a single unified procedure.

The following assumptions are commonly made when estimating recombination frequencies:

- The true recombination frequency θ is the same in both parents of the mapping population. This assumption is made whenever meioses in both parents are informative for estimating θ (e.g. both parents are heterozygous for two loci), and a single estimate (rather than two sex-specific estimates) is to be obtained.
- The true θ is the same for all meioses that yield observed products.
- Gamete genotypes are classified without error. Partition of gametes into genotypic classes based on phenotype may be incomplete, for example due to dominance or unknown linkage phase. However, a particular correspondence between genotypic classes and phenotypic classes is assumed, so that to the maximum extent possible for a given situation, phenotype predicts genotype without error. In particular, genotypes are completely penetrant and phenotypes are correctly scored.
- All genotypes are equally viable.
- Chromatid interference is absent.

In addition, an absence of crossover interference is often assumed. We mention some of the consequences of departures from these assumptions in Section 7.5.6.

7.5.3.1
Pairwise Estimation

When parental and recombinant gametes may be distinguished unambiguously, the recombination frequency θ between a given pair of loci may be estimated simply and intuitively by the observed proportion of recombinant gametes in a sample. This is the case, for example, when (1) two loci each have an allele segregating in one parent only, as in a backcross originating from inbred lines; or (2) haploid genotypes of the meiotic products from individuals with known linkage phases are directly observable, as with analysis of doubled haploids derived from F_1 progeny of a cross between highly homozygous parents, and with analysis of megagametophytes from a known pedigree (e.g. Guries et al. 1978; Tulsieram et al. 1992). In the example in Table 7.1c, for which $\theta = 0.25$, we estimate θ by $\hat{\theta} = (x_{12} + x_{21})/n = (15 + 8)/100 = 0.23$. It may be shown that such an estimator is the maximum likelihood estimator (MLE) of θ: the value of θ that maximizes the probability that the observed data would have occurred, assuming a suitable multinomial distribution for the phenotypic class frequencies (e.g. Bailey 1961). The multinomial probability distribution is a function that assigns a probability to each possible set of phenotypic frequencies. It depends on the set of phenotypic class probabilities, p_{ij}, each of which in turn depends on θ. The likelihood for θ is simply the multinomial probability distribution function considered as a function of θ instead of as a function of the phenotypic frequencies.

When gamete genotypes are incompletely known, so that recombinant gametes may not be counted directly, an MLE of θ must be obtained by analytic or numeric maximisation of the likelihood function over values of θ. Formulae required to obtain the MLE of θ for all types of two-locus allelic configurations – combinations of loci with dominant or codominant alleles, linked in coupling or repulsion, and with two to four alleles segregating – are given by Allard (1956) and Ritter et al. (1990).

7.5.3.2
Multipoint Estimation

Multipoint estimation of θ is based on a partition of progeny phenotypes into classes based on the phenotype at $m \geq 3$ loci. In this case, recombination frequencies $\underline{\theta} = (\theta_{12}, \ldots, \theta_{(m-1)m})$ for multiple contiguous intervals between loci are estimated simultaneously, conditional on the order of loci. The likelihood for the vector $\underline{\theta}$ is now a function of the multilocus phenotypic class frequencies. Maximization of this likelihood over $\underline{\theta}$ yields multipoint MLEs of the components of $\underline{\theta}$.

Multipoint estimation of θ_{ij} is more efficient than pairwise estimation if the linkage phase for loci i and j is unknown in some products of meioses (Morton et al. 1986; Lander et al. 1987). It makes fuller use of available linkage information to produce more precise estimates. For example, in an intercross AB/ab × AB/ab, with two codominant alleles segregating at each locus, progeny of genotypes AB/ab and Ab/aB are not distinguishable without further experiment (i.e. progeny testing). Multipoint estimation, but not pairwise estimation, may make use of information from additional loci linked to the two loci in question to help predict the phase for a particular gamete, and thus predict whether or not it is recombinant. However, the multipoint approach requires prior specification of a locus order, which usually is not known in advance. For this reason, preliminary estimation of recombination frequencies, prior to assembly of linkage groups and initial ordering of loci, usually is done using the pairwise approach. Map distance estimates may be obtained from recombination frequencies, estimated by either the pairwise or multipoint methods, by application of a selected mapping function.

7.5.4 Assembling Linkage Groups

To partition a set of loci into linkage groups, statistical tests for linkage (Sect. 7.5.2), and possibly estimates of recombination frequencies (Sect. 7.5.3), are obtained for all possible pairs of loci. The criterion for linkage of a particular pair of loci is whether the selected test yields a statistic exceeding a specified threshold (or a corresponding p-value less than a specified threshold). Sometimes, an estimated recombination frequency $\hat{\theta}$ less than a specified threshold also is required for a declaration of linkage (e.g. Lander et al. 1987). Use of both criteria may be advisable because the values of test statistics and p-values depend not only on the true value of θ but also on sample size, and because departures from assumptions may influence tests and estimates differently.

A pair of loci for which $\hat{\theta} > 0.5$ may be linked in repulsion. As a consequence, a linkage criterion based on θ is actually that θ not be in a specified window centered at 0.5. In practice, one may simply reverse the coding for one of the loci involved and re-estimate θ. Alternatively, some linkage analysis programmes can directly estimate θ for loci linked in either phase (e.g. JoinMap, Stam 1993).

Linkage groups are formed such that every locus within a given group is linked to at least one other locus in that group. It is a good idea to examine the robustness of linkage group assignments by assembling groups using a range of different threshold values. Such informal sensitivity analysis may highlight subsets of loci for which linkage information is weak. Similarly, one may compare the linkage groups obtained using two or more different tests for linkage. One may wish to estimate initial locus orders within linkage groups (see

Sect. 7.5.5) based only on those loci exhibiting relatively strong evidence of linkage.

7.5.5
Ordering Loci

Once loci are assigned to linkage groups, the final stage of linkage map construction consists of estimating the linear order of loci within linkage groups. This task requires (1) a measure by which to compare alternative possible orders; and (2) a method for searching the set of possible orders. This second requirement arises because an exhaustive search of all possible orders is computationally impracticable unless the number of loci is quite small. For m loci, m!/2 orders are possible. For m = 5, there are 60 possible orders and each possibility may be investigated. However, as m increases, the number of possible orders increases very rapidly. For example, there are 1 814 400 orders for m = 10; about 6.5×10^{11} for m = 15; and about 4.7×10^{157} for m = 100. In general, there is not a unique best-estimated map for a given set of data. The order (and map distances) obtained may depend on the criterion used to assess the "goodness" of different orders, and unless an exhaustive search is performed, may depend as well on the technique used to select orders for comparison.

7.5.5.1
Criteria for Comparing Locus Orders

A satisfactory criterion for assessing a putative order of loci should be a measure of the distance of a given order from the true order. Several criteria have been used for ranking the "goodness" of orders. One criterion is the sum of the pairwise LOD scores (see Sect. 7.5.2) for each pair of loci adjacent in a given order (e.g. MacLean et al. 1985; Olson and Boehnke 1990). Putative orders that maximize (or nearly maximize) this criterion are preferred. For example, suppose we wish to compare the two orders ABC and ACB for three loci, and let $LOD(\theta_{ij})$ be the LOD score for a linkage test between loci i and j. The LOD sum for order ABC is $S_{ABC} = LOD(\theta_{AB}) + LOD(\theta_{BC})$; that for order ACB is $S_{ACB} = LOD(\theta_{AC}) + LOD(\theta_{BC})$. If the true order is ABC, $\theta_{AB} < \theta_{AC}$, $LOD(\theta_{AB}) > LOD(\theta_{AC})$ (all else, e.g. the number of informative meioses, being equal), and consequently $S_{ABC} > S_{ACB}$. In general, maximization of the LOD sum will yield orders that place adjacent to each other those pairs of loci with the strongest evidence of linkage.

An alternative criterion is the maximum multipoint likelihood (or equivalently, log-likelihood) conditional on locus order (see Sect. 7.5.3). For a given order, the multipoint likelihood of the set of recombination frequencies θ between adjacent loci is maximized. Recall that the value of θ at this maximum is the multipoint MLE of θ, and this process yields a "best map" for the given order (e.g. Lander et al. 1987). The orders being considered are ranked according to their maximum log-likelihood; that order yielding the largest maximum log-likelihood (i.e. a best map with largest log-likelihood) is the most likely to have given rise to the observed data. If the difference between the maximum log-likelihoods for orders 1 and 2 is d (where log is the decimal logarithm), the data are 10^d times more likely to arise if order 1 were true than if order 2 were true. The maximum log-likelihood criterion has been implemented in the linkage analysis computer package, MAPMAKER (Lander et al. 1987).

Yet another possible ordering criterion is based on a weighted least-squares (WLS) method for combining estimated map distances for a subset of all possible pairs of loci into a reduced set of estimated map distances for only those pairs of loci adjacent in a given order (Jensen and Jorgensen 1975; Lalouel 1977; Weeks and Lange 1987; Stam 1993). Here, we treat the initial, larger set of estimated map distances as observed distances, and the weights are proportional to the precision of these distance estimates. In this approach, a WLS estimate of the set of map distances between loci adjacent in a given order is found, as the set of values that minimizes a weighted sum of squared differences between observed and expected map distances (a so-called weighted residual sum-of-squares, or wRSS). The criterion associated with this method is the minimum wRSS for a given order; the "best" order has the smallest value of this criterion. The minimum wRSS criterion is implemented in the linkage analysis package JoinMap (Stam 1993).

7.5.5.2
Methods for Searching Possible Orders

Various ad hoc procedures have been proposed for choosing a subset of possible orders to examine when an exhaustive search is impracticable. For example, JoinMap (Stam 1993) automatically builds a map for a given linkage group sequentially. The two loci with the highest LOD score are added first. Then the locus with the highest LOD sum for linkage to the already mapped loci is added to the map. Holding the order of the previously mapped loci fixed, the best fitting position of the new locus is found (as judged by the criterion described above). Next, for each triplet of adjacent loci, the possible orders are compared while holding the order of other loci fixed, and the best fitting triplet order is selected (again, based on the ordering criterion). Once all triplet shuffles have been compared in this way, the next locus is selected for addition to the map, and the entire process is repeated.

MAPMAKER (Lander et al. 1987; Lincoln et al. 1993a) allows the user to specify interactively the orders for which she or he would like to calculate the ordering criterion. Alternatively, an automatic sequential procedure may be used, starting with comparisons of three-point orders and finally computing ordering criteria for all m-point orders that do not include unlikely three-point orders. Again, local shuffling of subsets of loci to look for improvements is possible.

The task of finding a best order of loci is a problem in combinatorial optimization. Ad hoc search procedures such as those described above are heuristic ways of focussing in on subsets of more likely orders and then searching in the neighbourhoods of these relatively likely orders for local improvements. They are approximate optimization methods, which do not guarantee that the order optimizing a given criterion will be found. They have, however, been used with success to order loci based on both real and simulated data (for which the true order is known). Other automatic approximate methods for combinatorial optimization are available, including genetic algorithms and simulated annealing (e.g. Press et al. 1988). Simulated annealing in particular is implemented as the search technique in GMendel (Echt et al. 1992) and Surveyor (Shoemaker et al. 1992). A variety of other techniques for ordering loci within linkage groups have been investigated, involving different ordering criteria and different search methods (e.g. Buetow and Chakravarti 1987; Kammerer and MacCluer 1988; Olson and Boehnke 1990; Rogatko and Zacks 1993).

Different combinations of recombination frequency and map distance estimation methods, ordering criterion, and order search technique have been proposed and successfully applied. None seems to be obviously superior to the rest in all applications; further work may help to clarify this situation.

7.5.6
Sensitivity of Map Construction to Departures from Common Assumptions

The bias of a recombination frequency estimator $\hat{\theta}$ is defined as $E(\hat{\theta}) - \theta$; that is, the difference between the expected value of the estimator and the true (unknown) value. If an assumed model is true, maximum likelihood estimates of θ are asymptotically unbiased: as the number of observed meioses n increases, the bias of $\hat{\theta}$ becomes smaller and smaller (e.g. Bailey 1961). For loci with dominant alleles linked in coupling phase, the bias of $\hat{\theta}$ is negligible relative to its variance for $n \geq 20$ (e.g. Säll and Nilsson 1994). For loci with dominant alleles linked in repulsion, however, the bias of $\hat{\theta}$ can be quite large even for an n of a few hundred, especially for small true values of θ. One would expect the bias to be less severe for loci (such as RFLPs) with codominant alleles.

Violations of the assumptions underlying estimation of distances between loci and orders of loci may introduce additional bias in $\hat{\theta}$, so that the total bias no longer goes to zero as sample size increases, and lead to an increased probability of inferring incorrect locus orders. Different methods of analysis have, in general, different sensitivities to departures from assumptions.

7.5.6.1
Heterogeneity of Recombination Frequencies

The parents of a mapping population may have different rates of recombination because of sex-specific differences, species-specific differences (for interspecific crosses), or differences in genetic background (e.g. when crossing different varieties of a single species; Burt et al. 1991). In this case, a single true recombination frequency does not exist. If only one estimate is obtained for an interval, it may be considered an estimate of the average of the two parental frequencies. Säll and Nilsson (1994) investigated the effects of different recombination frequencies in the male and female parents on pairwise recombination frequency estimates in intercrosses with dominant markers (e.g. RAPDs). They found relatively minor increases in bias for loci linked in coupling phase, but potentially large effects when loci are linked in repulsion.

Differences in recombination frequency among meioses also may induce large biases in recombination frequency estimates for dominant markers linked in repulsion, but have only minor effects for markers linked in coupling (Säll and Nilsson 1994). In either case, effects of differences between parental recombination rates were more serious than effects of differences among meioses within a parent.

Effects of heterogeneous recombination frequencies when codominant alleles are segregating are likely to be smaller than those for dominant alleles. We are not aware of results indicating the sensitivity of multipoint estimation to recombination frequency heterogeneity. Some computer packages (e.g. Mapmaker) do, if desired, provide separate multipoint estimates for each parent.

7.5.6.2
Genotype Misclassification

Misclassification of genotypes, resulting from mis-scoring of phenotypes or from incomplete penetrance, may cause large positive biases in both pairwise and multipoint recombination frequency estimates (Morton and Andrews 1989; Buetow 1991; Shields et al. 1991; Lincoln and Lander 1992; Brzustowicz et al. 1993; Säll and Nilsson 1994). This is true even for low misclassification rates, which may produce only weak segregation distortion not likely to be detected by single-locus tests. Misclassification may also reduce apparent crossover interference and lead to support of incorrect ordering of loci. The effects of misclassification grow more serious as map density increases (i.e. average spacing of loci decreases). Random classification errors tend to inflate the frequency of rare events, such as multiple crossovers, proportionally more than the frequency of more common events. For a dense enough map, the recombination frequency in an interval and the error frequency may be comparable, and a large proportion of apparent crossovers will be spurious. Multiple pairwise estimation of map distances and order is in general less sensitive than multipoint analysis to classification errors, apparently because of the inclusion of pairs of non-adjacent loci, at larger distances (Morton 1988b; Buetow 1991; Shields et al. 1991).

Shields et al. (1991) proposed a method for correcting map distance estimates for misclassification, which they found also improved the recovery of correct locus orders inferred by multiple pairwise analysis. Lincoln and Lander (1992) incorporate the possibility of misclassification into the usual multipoint likelihood model. Using this approach, implemented in Mapmaker, it is possible to construct maps less affected by classification errors, and to identify the genotype scores most likely to be in error. Brzustowicz et al. (1993) also describe several approaches for detecting classification errors and allowing for such errors in linkage analyses.

7.5.6.3
Differential Viability

Reduced viability for a genotype at a single locus does not increase the bias of a pairwise recombination frequency estimate, but does affect the estimated variance of that estimate. Reduced viability for genotypes at two loci does inflate the bias (Bailey 1961). Alternative pairwise estimators of recombination frequency, less sensitive to differential viability, are available (Bailey 1961; Heun and Gregorius 1987; Ritter et al. 1990; Lorieux et al. 1995a, b). Säll and Nilsson (1994) found that, at least for dominant markers, a large increase in the bias of the pairwise recombination frequency estimator required differences in zygote viability large enough to be likely to induce detectable non-random deviations from expected single-locus segregation ratios.

7.5.6.4
Crossover Interference

Multiple pairwise estimation of map distances is less sensitive than multipoint estimation to crossover interference (Morton 1988b; Shields et al. 1991). When estimating a map distance for a single interval, interference is accommodated by application of an appropriate map function to convert recombination frequency into map distance. Pairwise estimation is not insensitive to crossover interference, however, since results depend on the choice of a map function. Some authors recommend comparing likelihoods obtained under different map functions to indicate sensitivity of results to crossover interference (e.g. Shields et al. 1991; Stam 1993).

Multipoint estimates of recombination frequencies and locus orders usually are obtained under an assumption of no crossover interference. After inference of an order, recombination frequencies are converted to map distances by application of a map function, which may allow for interference. However, sensitivity of order determination to interference is not considered (Stam 1993). Speed et al. (1992)

proved that, assuming no chromatid interference and a sufficiently large number of informative meioses, the true order of loci will maximize the multipoint likelihood under the no crossover-interference assumption even when there is crossover interference. However, in practice, the number of informative meioses required is unknown and may be very large.

7.6 Mapping Quantitative Trait Loci

In this section we describe methods of analysis for identifying genetic loci associated with quantitative traits: so-called polygenes or quantitative trait loci (QTLs). A QTL is a gene or set of closely linked genes at which allelic substitutions contribute to the quantitative variation in a specified trait. But what exactly is quantitative variation? Thompson and Thoday (1974) give an illuminating explanation of the key distinction between quantitative and non-quantitative variation:

- "A character shows quantitative or continuous variation when the phenotype is affected by a number of factors, each of which makes a contribution which is small relative to other sources of variation. When these factors include allelic variation at one or more genetic loci, the other sources of variation, which may be environmental, genetic, or both, make it difficult to classify individual genotypes. Thus, segregation at individual loci cannot be studied directly."
- "A character shows qualitative, discontinuous, or Mendelian variation when genetic variation at one or a small number of loci makes large individual contributions to phenotypic variance relative to other genetic and environmental factors... variation in the character allows individuals to be classified into distinct phenotypic classes."

The existence of distinct phenotypic classes allows (usually) unambiguous inference of individual genotypes. Consequently, segregation at individual loci affecting qualitative traits can be studied directly by the methods of classical Mendelian genetics.

Note the emphasis on the *relative* size of contributions to phenotypic variation, rather than the absolute size of effects. Whether or not a trait shows quantitative variation depends on the size of the effects of all factors contributing to variation in the trait. On the one hand are the effects of loci that affect the trait of interest and are segregating in a particular cross. For a given trait, these may be major genes with large effects relative to total variation, minor genes with small effects, or a combination of the two. Whether major, minor or both types of genes control the inheritance of a particular trait, and consequently, whether or not a trait shows quantitative variation, is not an intrinsic property of that trait alone, but depends as well on which loci happen to segregate in a particular cross. On the other hand are the effects of all other factors that contribute to variation in the trait. These include environmental conditions, the resolution of and error in the techniques by which the trait is measured, and the genetic background of the mapping population. Environmental conditions and measurement techniques influence the size of the non-genetic component of variation. Such factors set the baseline level of variation against which the size of the effects of segregating loci are judged. In addition, both environment and genetic background may directly influence the contribution to variation of the segregating loci of interest. Environment may do so by genotype-environment interaction, whereby the effects of particular loci on a trait depend on the environment in which the trait is measured. Genetic background may do so by epistatic interactions amongst genes, whereby the effects of particular loci on a trait depend on the allelic constitution at other loci (which may or may not have alleles segregating in the cross).

From the above perspective, the term quantitative trait may be seen to be a convenient but potentially misleading shorthand for the phrase, a trait showing quantitative variation in a given experiment. The distinction between a quantitative and a qualitative trait is an operational one, valid for a particular set of loci segregating for a given trait in a particular genetic background, under specified environmental conditions and measured by a particular technique. All of the above factors influence whether or not allelic substitution at segregating loci produces the discrete phenotypic classes that define a qualitative trait or the overlapping distributions characteristic of a quantitative trait. This, in turn, determines whether or not the types of analyses described in this section are required.

Given that we have identified a trait showing quantitative variation, and have measured trait values and marker genotypes for all individuals in a mapping population, our goal is to use an RFLP linkage map to help localize regions of the genome associated with the trait. Various analytical techniques have been, and continue to be, developed for this purpose (for reviews, see Weller 1992; Tanksley 1993;

Bradshaw 1994; Jansen 1996b). The most commonly used of these at present are analysis of variance (ANOVA) on single-marker genotype classes (Soller et al. 1976); and interval mapping (Lander and Botstein 1989). We describe these two techniques, discuss their advantages and disadvantages, and then briefly describe some alternative techniques and improvements.

7.6.1 Analysis of Variance on Marker Classes

One-way analysis of variance (ANOVA) on marker phenotype classes is perhaps the simplest way to look for QTLs (e.g. Soller et al. 1976). In this approach, each individual in a population is assigned to a sub-group according to its phenotype at a particular marker locus. Then the mean trait values amongst these sub-groups are tested for differences by ANOVA, which asks whether the differences among sub-groups are large compared to differences amongst individuals within a sub-group. A significant difference among sub-groups is evidence of an association between the marker and the quantitative trait, assumed to result from linkage between the marker and a QTL.

For example, consider a backcross, MQ/mq × mq/mq, where M and m represent two alleles at a marker locus, and Q and q represent alleles at a putative linked QTL. The marker phenotype and a trait value are observed for each individual in the progeny; the genotype at locus Q is unobserved. Ordinarily, in a backcross marker genotype can be inferred unambiquously from marker phenotype. Except for recombination, individuals in the backcross progeny with marker genotype Mm will have QTL genotype Qq; mm individuals will have QTL genotype qq. The observed marker phenotype thus serves as a surrogate for the unobserved QTL genotype. If substitution of a Q allele for a q allele at the QTL affects the trait, we expect the trait values for Mm individuals to differ from the values for mm individuals, on average.

More specifically, let $\bar{\chi}_{Mm}$ and $\bar{\chi}_{mm}$ be the observed mean trait values for marker genotype classes Mm and mm, respectively. Similarly, let μ_{Mm} and μ_{mm} be the unknown true means for genotypes Mm and mm (for example, the mean trait values if an infinite number of individuals were observed in each class). We wish to test the null hypothesis, H_0: $\mu_{Mm} = \mu_{mm}$ against the alternative, H_1: $\mu_{Mm} \neq \mu_{mm}$. This may be done by a two-sample t test. The t statistic gives the difference in the two observed means relative to the amount of variability in the quantitative trait data. A large deviation of the observed statistic, t_{obs}, from zero is more likely if H_1 is true than if H_0 is true. We therefore interpret a suitably large deviation of t_{obs} from zero as evidence that H_0 is false, i.e. that the true means of classes Mm and mm differ. To make more precise the meaning of "suitably large", we may calculate the probability of observing a statistic t as far from or further from zero as is t_{obs} when H_0 is true. This probability is the so-called p-value of the test. A small p-value means we were unlikely to have observed such an extreme t statistic if H_0 were true. We would interpret this as evidence that there is, in fact, a difference in the true means, i.e. we reject H_0. We may, if we wish, compare the p-value to an arbitrary threshold, α, commonly 5%, and reject H_0 only if the p-value is less than α. Following this procedure, α gives the Type I error rate: the probability of rejecting a true null hypothesis. Evidence for a difference in the mean trait values for marker genotype classes Mm and mm is taken as an indication that the unobserved locus Q is a QTL linked to the marker locus. Then α gives the chance that a QTL will be indicated where none actually exists.

The t test just described is equivalent to a corresponding ANOVA F test (in fact, $F = t^2$). The F test is more general, though, being applicable also to a comparison of more than two marker genotype classes, as might arise, for example, in an F_2 generation. In this case, we test the null hypothesis, H_0: $\mu_1 = \mu_2 = ... = \mu_k$ versus the alternative, H_1: at least one pair of means differ. The F statistic is a ratio of the amount of trait variation resulting from differences among the marker classes to the amount of trait variation within marker classes (due to environment, segregation at other loci, etc). It is scaled to have expected value one if H_0 is true; values larger than one by more than is likely to have occurred by chance alone lead us to reject H_0, i.e. the data indicate the presence of a QTL linked to the marker locus. Such a test is performed successively for each marker scored in the mapping population.

The ANOVA F test is based on the following assumptions: (1) the "residual error" in the trait (i.e. unexplained variation, not due to segregation of the marker being tested) is normally distributed; (2) the trait values derive from distributions having a common variance; and (3) the trait values are statistically independent of each other. It is worth a moment to consider the consequences of departures from these assumptions.

The ANOVA F test for inferences about the means of marker genotype classes is quite insensitive to violations of assumption (1) (Scheffé 1959). The trait values are sometimes transformed, for example by taking their logarithm, in an attempt to reduce nonnormality of the residual errors (Scheffé 1959; Atkinson 1987). While this ordinarily does no harm, it is not often essential, especially for the sample sizes generally used for mapping studies.

Inequality of trait variances among the marker-genotype classes (violation of assumption 2) will have little effect on the type I error rate of the F test if class sizes are nearly equal (as would usually be the case for a backcross), but can seriously invalidate inferences about the means if class sizes are unequal. For example, a test with a nominal type I error rate $\alpha = 0.05$ may have an actual type I error rate much larger, the size of the difference depending on the size of the differences in variance and those in class sizes. Diagnosis of unequal variances among marker genotype classes may be aided by simple graphical examination of the deviations of trait values from their marker class mean, for example by histograms or box-plots. Standard statistical tests for equality of variances among classes are not recommended, as they typically are quite sensitive to departures from normality, and tend to lack power to detect non-constant variance (Scheffé 1959). One should be particularly alert for evidence of unequal variances when applying the ANOVA method to an F_2 population (Asins and Carbonell 1988). In this case, the three marker-genotype classes, MM, Mm and mm are (1) expected to have different variances if there is recombination between that marker and a QTL and the two QTL alleles exhibit dominance (i.e. the QTL heterozygote mean is not the average of the two QTL homozygote means); and (2) expected to have different class sizes, having a 1:2:1 ratio for a marker with codominant alleles.

Nonconstant variance of the trait values *within* a marker genotype class may be quite difficult to diagnose in the common case of only a single measurement per individual. However, it may be manifest as lack of normality. Transformation of the trait values may reduce apparent inequality of variances, whether it be among or within marker classes. For example, standard variance-stabilizing transformations exist for binomial and Poisson distributed traits (Atkinson 1987).

Violation of the assumption of independent observations, assumption (3), can have a very serious effect on the ANOVA F test for inferences about means. Fortunately, it is generally safe to assume that this assumption is satisfied for a single full-sib family.

If the validity of the ANOVA F test is in question in a particular application, and data transformation appears inadequate, a variety of alternative tests may be used. For example, data resampling (bootstrap) techniques (Manly 1991; Efron and Tibshirani 1993) may be used to estimate for an F or t statistic p-values that are less dependent on assumptions (1) to (3). When comparing two marker-genotype classes with apparently unequal variances, for large enough class sizes the usual t statistic may be replaced by a Z statistic (having a standard normal distribution under H_0) in which the denominator is calculated using a separate variance estimate for each class (Asins and Carbonell 1988). Rank-based (non-parametric) methods also may be used, for example the Wilcoxon rank-sum test for comparing two classes, or the Kruskal-Wallis test for comparing more than two classes (Lehmann 1975). These have no dependence on the form (e.g. normality) of the trait distributions, and are less sensitive to outliers in the data than are the parametric t and F tests. The cost of these advantages, however, is power: the nonparametric tests in general are less likely to detect differences among means when they exist. The Kruskal-Wallis test is implemented in MapQTL, a computer programme for QTL mapping (Van Ooijen and Maliepaard 1996).

The ANOVA method for QTL detection has the advantage of simplicity; it may be carried out using standard procedures in any statistical computer package. It can be applied to unmapped markers, and may be used when only very few (e.g. 1–3) markers per linkage group, spaced at wide intervals, are available (cf. Darvasi et al. 1993). This makes the ANOVA method particularly well suited for detecting QTLs in preliminary experiments using few markers. It does, however, suffer from a variety of serious deficiencies (cf. Lander and Botstein 1989):

- If a QTL is not located exactly at a marker locus, recombination between the QTL and marker locus may result in a serious underestimation of the QTL effect. In our backcross example, we assumed that Mm individuals had QTL genotype Qq, whereas mm individuals had genotype qq. If recombination occurs, however, some Mm individuals will have QTL genotype qq while some mm individuals will have QTL genotype Qq. The difference between the marker classes is diluted by the presence of the recombinant individuals in each class, and so the difference in marker class means is a biased estimator of the difference in QTL class means. More precisely, the expected value of $\bar{\chi}_{Mm} - \bar{\chi}_{mm}$ is $(1 - 2\theta)(\mu_{Qq} - \mu_{qq})$, where μ_{Qq} and μ_{qq} are the true mean trait values for the QTL genotype classes and θ is the recombination fraction between the marker and QTL.

- Recombination between a marker and a QTL also reduces the power of the method to detect the QTL. Not only is the expected value of the mean difference between marker classes reduced by recombination, but the within-class variation may be increased by the presence of multiple QTL genotypes within each marker class. The net result is a smaller difference on average that is tested relative to greater variability, and thus is more difficult to distinguish statistically from random fluctuation about zero.
- The method does not yield an estimate for the map position of the QTL. In particular, the expected value of the mean difference, $(1 - 2\theta)(\mu_{Qq} - \mu_{qq})$, may take on the same value for a small θ and large QTL effect as it would for a larger θ and smaller QTL effect. Thus the method cannot distinguish a tightly linked QTL with small effect from a loosely linked QTL with large effect.
- The method gives no information on the number of QTLs associated with a marker.
- The presence of segregating QTLs unlinked to a marker may inflate the residual variation against which the size of the effect of a putative linked QTL is measured, resulting in a lower probability of QTL detection.
- A separate test is performed for each marker. While α gives the probability of a false positive for any one test, the probability of at least one false positive somewhere in the genome will be much higher if many markers are tested. This difficulty may be avoided by the use of multiple comparison techniques (see Sect. 7.5.1). Unfortunately, the use of a smaller Type I error rate for each test, while controlling the chance of a false positive for the whole genome, decreases the power of the tests to detect QTL effects (i.e. increases the chance of false negatives). This trade-off is unavoidable.

Additional problems with the ANOVA method for QTL detection arise if the parents of a crossing scheme are not homozygous for alternative QTL alleles (e.g. Soller et al. 1976; Van Eck et al. 1994). For example, suppose the parents of a backcross scheme, MQ/Mq and mQ/mq, are heterozygous for a QTL, carry the same two QTL alleles, Q and q, and are homozygous for alternative marker alleles, M and m. A randomly chosen F_1 individual will be heterozygous for the QTL with probability 0.5, yet only if it is heterozygous will the QTL segregate when that individual is backcrossed to one of its parents. Now consider the backcross to parent 2, MQ/mq×mQ/mq. Assuming no recombination, the Mm progeny class will consist of both QQ and Qq individuals; similarly the mm progeny class will contain both qq and Qq genotypes. As a result, the mean trait difference, $\bar{\chi}_{Mm} - \bar{\chi}_{mm}$ has expected value $\frac{1}{2}(\mu_{Qq} + \mu_{QQ}) - \frac{1}{2}(\mu_{Qq} + \mu_{qq}) = \frac{1}{2}(\mu_{QQ} - \mu_{qq})$ rather than $\mu_{Qq} - \mu_{qq}$. These expectations differ in general, unless Q and q exhibit complete codominance [i.e. $\mu_{Qq} = \frac{1}{2}(\mu_{QQ} + \mu_{qq})$. As a result, the difference in marker-class means is a biased estimator of the QTL effect even if recombination is absent. In addition, the within-class trait variance is likely to be inflated by the variation of QTL genotype within marker classes, and thus an ANOVA F-test (or t-test) will be less powerful: a mean difference of a given size will be harder to distinguish from chance fluctuation.

More generally, when parents are not known to be homozygous at all loci each marker-genotype class in the mapping population may consist of more than one QTL genotype, even in the absence of recombination (Chasalow 1994; Maliepaard and Van Ooijen 1994). This includes the possibility that more than two alleles at a particular QTL are segregating in a cross. When parents are heterozygous for a QTL, the ANOVA method using markers with only two alleles segregating may yield biased estimators of QTL effects, and have decreased detection power due to inflated variances within marker classes. This bias and loss of power may be reduced by using markers with more than two segregating alleles to subdivide the mapping population into four marker-genotype classes. The ANOVA method for QTL detection then may be applied to compare trait means among the four marker classes (e.g. Van Eck et al. 1994).

7.6.2
Interval Mapping

In the ANOVA method, a single marker at a time is analyzed and we assume no recombination between the marker and a putative

QTL. What happens if we wish to allow for recombination? We have seen that each marker-genotype class is composed of two or more QTL genotypes if there is, in fact, recombination. In this case the observed trait measurements within each marker-genotype class take on values with frequencies given by a so-called mixture distribution: a weighted sum of component distributions, one component for each QTL genotype, with weights determined by QTL segregation ratios. Such a distribution may be represented by a mixture model (Titterington et al. 1985; McLachlan and Basford 1988). A variety of techniques have been developed to map QTLs by fitting a mixture model to genotypic and trait data. Some map a putative QTL relative to a single marker at a time (Weller 1986; Hackett et al. 1992). Like the ANOVA method, methods based on single-marker mixture models do not require a pre-existing marker map, and may be used when only one or a few markers per linkage group are available. They do, however, have the disadvantage of decreasing detection power as the distance between the marker and QTL increases. Interval mapping (Lander and Botstein 1989; see also Van Ooijen 1992), in contrast, uses genotypes at two neighbouring markers simultaneously to test for the presence of a QTL in the genomic region flanked by the two markers. It requires a known, accurate marker linkage map (positions and order are assumed known without error). Unlike the ANOVA method, both single-marker mixture models and interval mapping provide estimates of the position (distance from a marker or markers) and effect of a QTL. Knott and Haley (1992) compared the performance of these two mixture-model-based approaches for QTL mapping in simulations of 1000 F_2 progeny. Interval mapping had substantially better power for QTL detection than the single-marker mixture model method only when the markers were widely spaced (at 50 centiMorgan intervals) and the QTL effect was large (accounting for at least 3% of total phenotypic variance). In general, though, interval mapping gave more accurate, less biased estimators of QTL effect and position, and was less sensitive to violations of assumptions such as normality. Since interval mapping, based on flanking-marker mixture models, appears to perform better than or at least as well as methods based on single-marker mixture models, and requires little extra computation time, it is generally preferred. The interval mapping method has been implemented in computer packages, such as MAPMAKER/QTL (Lincoln et al. 1993b) and MapQTL (Van Ooijen and Maliepaard 1996).

Lander and Botstein (1989) discuss the application of interval mapping to a backcross only, but the principles of interval mapping are generally applicable to other types of segregating populations. Details for an F_2 population are given by Paterson et al. (1991), Luo and Kearsey (1992) and Van Ooijen (1992). For a pair of linked markers flanking a single QTL, each two-marker genotype class is composed of individuals with one of two (for a BC), three (for an F_2), or possibly up to four (for an F_1 from heterozygous parents) different QTL genotypes. The trait distribution for each two-marker genotype class is represented by a mixture model with one component for each QTL genotype present in the marker class. Usually, the components are assumed to be normal distributions, with unknown means and a common variance, but the use of other types of distributions, such as binomial, Poisson, exponential etc. also is possible (Lander and Botstein 1989; Jansen 1992; Hackett and Weller 1995; Kruglyak and Lander 1995). Given the genotypes at markers flanking a particular interval and the recombination frequency between those markers, we can obtain a likelihood of the entire mapping population. The likelihood is a function of the unknown parameters of the mixture components, and of the QTL position (i.e. the recombination frequency between the QTL and one of the flanking markers). For any particular set of parameter values, the likelihood gives the probability of observing a particular specified set of trait values for individuals with the observed marker genotypes. The maximum likelihood estimates (MLEs) of the parameters are those parameter values that give the largest probability of observing the set of trait values that actually occurred.

While it is possible to leave the QTL position within a given marker interval as an unknown parameter and obtain its MLE by maximizing the likelihood function (e.g. Jansen 1992), Lander and Botstein (1989) fix the QTL position and then estimate the remaining mixture parameters. If there is no QTL segregating in that interval, the trait distribution within each two-marker genotype class is assumed to consist of only one component. We may also compute the maximum value of the likelihood function under this assumption. The likelihood ratio (LR) for a particular genomic position is the maximum likelihood assuming a QTL at the given position divided by the maximum value of the likelihood assuming no segregating QTL in the interval. Like the LOD score test for linkage (Sect. 7.5.2.3), the LOD score for presence of a QTL is simply $\log_{10}(LR)$.

Thus the LOD score gives a measure of how much more likely are the observed data if there is a QTL at the position than if there is not. It is computed for every position on a discrete grid (e.g. every one centiMorgan) in an interval, and for all intervals between adjacent markers. A plot of the LOD score against map position gives the likelihood ratio profile, or QTL likelihood map (Fig. 7.2), and indicates the relative strength of evidence for a QTL at each position in the genome.

The LOD score may be used as a test statistic to test the hypothesis that a QTL exists at a particular position. For this purpose, a critical value, or LOD threshold, must be specified, corresponding to a desired false positive rate. The null hypothesis of no segregating QTL is rejected only if the LOD score at the tested position exceeds the threshold value. When scanning an entire genome for the presence of QTLs, many hypothesis tests are performed, and as with ANOVA, the problem of multiple comparisons

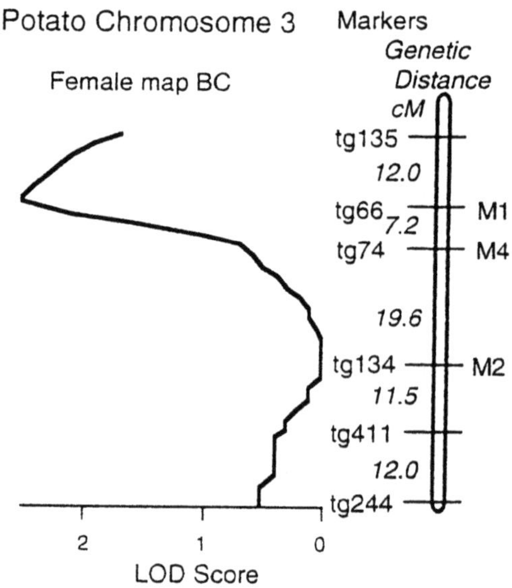

Trait	% explained variance	Mean values aa	Mean values A-
at tg66 (=M1)			
DORMANCY	7	95 d	81 d

Fig. 7.2. QTL likelihood map of potato chromosome 3. The marker order was based on RFLP alleles segregating from the female parent in a potato BC (Van den Berg et al. 1996a). The *peak of the curve* indicates the most likely position of a QTL affecting DORMANCY. The *table below the graph* gives the average DORMANCY (in days) of potato tubers in storage for the two genotypic classes at tg66 (=M1). For an explanation of the allele symbols see Table 7.3

arises. To compensate, the false positive rate for the test at any one position must be quite small; equivalently, the threshold must be relatively large. The global false positive rate α_0 is the chance that the LOD score will exceed the threshold somewhere in the genome, if no QTLs are segregating. Typically, the threshold is chosen to give no more than a 5% global false positive rate, though a threshold corresponding to any desired value of α_0 may be used. The appropriate choice of a value for α_0 depends on the purposes of a particular mapping experiment. It should be made after considering the relative costs of the two types of errors: a false positive (apparent presence of a QTL that does not actually exist) and a false negative (failure to detect a QTL that does exist). Recall that, all else being equal, the smaller one type of error rate is, the larger the other.

Once a false positive rate α_0 has been specified, the threshold required to yield that α_0 depends on the size of the genome, the number of linkage groups, the number and arrangement of markers within linkage groups, and the type of mapping population (e.g. BC, F_2, etc.). For a backcross starting with inbred parents, Lander and Botstein (1989) provide equations to approximate the threshold in the limiting cases of a sparse map, with widely spaced markers, and a dense map, with very closely spaced markers. They also determined, by simulation, threshold values given regularly spaced markers at 0 to 20 centiMorgan intervals, a global false positive rate $\alpha_0 = 5\%$, a normally-distributed trait, and 1, 5, 10, 15 or 20 chromosomes each of length 100 centiMorgans. Threshold values for this set of cases range from about 1.4 to 3.4.

Threshold values for BC and F_2 populations were computed by simulation by Van Ooijen (1992). He gives the null cumulative distribution function (cdf) of the maximum LOD score when no QTL is segregating, for a single 120-centiMorgan chromosome, 25 regularly spaced codominant markers at 5-centiMorgan intervals, no crossover interference, and a normally distributed trait. The null cdf gives the probability of observing on the single chromosome a maximum LOD score less than or equal to a particular value if no QTL is segregating. The value of the cdf at a particular LOD threshold gives $1 - \alpha_k$, where α_k is the probability of a false positive on one chromosome. If c linkage groups, each of length 120 centiMorgans, are to be tested for the presence of QTLs, a false positive rate per chromosome, $\alpha_k = 1 - (1 - \alpha_0)^{1/c}$, will yield a global false positive rate α_0 for the entire genome.

Rebaï et al. (1994) have developed methods for determining approximate thresholds for QTL detection applicable in more general circumstances than those described above. Their methods may be used to obtain appropriate thresholds for any number of codominant markers with any (not necessarily regular) spacing, a genome composed of any number of linkage groups of possibly varying sizes, and any desired global false positive rate α_0. They assume no double recombination (i.e. complete crossover interference) and a normally distributed trait. Formulae are available for BC, F_2 and RI populations, and they present results also for an F_3 progeny from a di-allelic cross. Thresholds obtained by these authors were slightly larger than those obtained by Lander and Botstein (1989) for equivalent cases. A computer programme implementing their methods is available (Rebaï et al. 1994). Nonparametric detection thresholds, more robust to departures from the usual assumptions of QTL mapping models, also have been developed (Churchill and Doerge 1994; Doerge and Churchill 1996).

In regions of the QTL likelihood map where the LOD scores exceed the chosen threshold (and hence the presence of a QTL is indicated), one may wish to estimate its likely position. The MLE of a QTL position is the map location at which the LOD score has its maximum. A one-LOD support interval may be constructed with endpoints at the positions, to the left and right of the MLE, at which the LOD score is one unit less than the maximum (Lander and Botstein 1989; Van Ooijen 1992). A two-LOD support interval for QTL position may be obtained in a similar fashion. One LOD less than the maximum corresponds to a likelihood one-tenth as great as the maximum likelihood, two LOD less corresponds to a likelihood one one-hundredth as great, and so on. The length of a support interval (with fixed level, e.g. one-LOD) gives an indication of the uncertainty in the MLE of QTL position: a short interval indicates a precise estimate, a longer interval indicates a less precise estimate.

If one wishes, such support intervals may be used as confidence intervals, allowing more explicit statements about the probability that a support interval covers the true QTL location. However, the method suggested by Lander and Botstein (1989) for obtaining the confidence level of a given support interval has been shown to yield incorrect results for a population size typically used for QTL mapping (200 backcross progeny), except for a QTL with quite large effect (on the order of three to four times the residual error; Mangin et al. 1994). For example, in some cases a one-LOD support interval has a probability of covering the true QTL position much less than the nominal level of 96.8% calculated by the Lander and Botstein method. In general, the method of Lander and Botstein yields overly optimistic confidence intervals that are shorter than required for the claimed level of confidence. Mangin et al. (1994) describe techniques for obtaining more accurate confidence intervals for QTL location.

When a cross between heterozygous (non-inbred) lines is being studied, interval mapping may be used to locate QTLs on the male and female maps separately (e.g. Bonierbale et al. 1994; Leonard-Schippers et al. 1994). To do so, markers showing a backcross-type (1:1) segregation and those showing an F_2-type (1:2:1) segregation must be analyzed separately, for each parental chromosome. Note that QTL alleles from both parents may be segregating, with as many as four QTL alleles segregating in the cross (e.g. Van Eck et al. 1994; Chasalow 1994). Consequently, any difference between the phenotypic effects of the two QTL alleles from one parent is averaged over the effects of the QTL alleles from the other parent. Interval mapping in this case has problems similar to those seen with the ANOVA method: it yields possibly biased estimators of QTL effects and suffers from decreased power to detect QTLs (relative to crosses originating from inbred lines). Extensions of the interval mapping method designed to avoid such difficulties are under development (Chasalow 1994; Haley et al. 1994; Maliepaard and Van Ooijen 1994; Jansen 1996a; Van Ooijen and Maliepaard 1996).

Interval mapping has several advantages over the ANOVA method for QTL detection and mapping:

- By explicitly allowing for recombination between a QTL and markers, interval mapping provides less biased estimators of QTL effects. It separates the effect of a QTL on the expected trait values from the effect of recombination between the QTL and its flanking markers, providing individual estimates of both QTL effect and position, so recombination does not cause underestimation of the QTL effect.

- Interval mapping has improved power to detect QTLs. With the ANOVA method, putative QTL effects are tested relative to estimated residual variation within marker genotype classes. Recall that variation within a marker class may be inflated by the presence of multiple QTL genotypes resulting from recombination between a

QTL and a marker. In interval mapping, putative QTL effects are tested relative to the residual variation within predicted QTL-genotype classes. Each of these classes is generally more homogeneous with respect to QTL genotype than is a marker-genotype class, and so has less within-class phenotypic variation. This results in more powerful statistical tests for QTL detection. However, one simulation study of large backcross populations (500 or 1000 individuals) found the difference in power between the two methods to be only slight (Darvasi et al. 1993). Differences in power may be greater for smaller populations more commonly used for mapping QTLs.

- Interval mapping provides an estimate of the map position of the QTL, and uncertainty in this estimate may be assessed using support intervals.

  There are still problems, however, with the interval mapping method. The mixture model on which interval mapping is based assumes that no more than one segregating QTL is linked to an interval, and the method tends to work well if this assumption is valid. But if two or more linked QTLs are segregating, tests for QTL presence may result in incorrect inferences, and the method may yield biased estimators of QTL effect and position (Stam 1991; Knott and Haley 1992; Martinez and Curnow 1992). For example, two linked QTLs with effects of the same sign may result in a QTL likelihood map indicating the presence of a so-called ghost QTL at a position intermediate to the two. Two linked QTLs of opposite effect may cancel each other, so that neither is detected. (Note that two or more QTLs showing complete linkage would not be distinguishable from a single segregating QTL, but otherwise would not present a problem).

Multiple *unlinked* QTLs also may have undesirable effects on the interval mapping method (as they do on the ANOVA method). Segregating QTLs not linked to a particular interval will inflate the residual unexplained variation in the trait, reducing the power of tests for QTL detection. An unlinked QTL with sufficiently large effect may even result in false positive QTL detection in the interval (see Appendix A4 of Lander and Botstein 1989). Knott and Haley (1992) investigated the use of an alternative likelihood ratio test for QTL detection designed to avoid such false positives.

7.6.3
Multiple-QTL Methods

Although the traditional ANOVA method may detect more than one QTL, it tests markers one at a time for linkage to a QTL, while ignoring possible effects of additional QTLs not linked to that marker. The ANOVA method for QTL detection may readily be extended, however, by subdividing a mapping population based on genotypes at two or more markers simultaneously. The trait variation among such classes then may be analyzed by multi-way ANOVA, or equivalently, multiple regression (Cowen 1989; Stam 1991; Rodolphe and Lefort 1993; Zeng 1993). Such an approach explicitly allows for simultaneous segregation of multiple QTL, and thus avoids some of the problems this presents for the one-marker-at-a-time ANOVA method and for interval mapping. It continues, however, to suffer from the limitations imposed by assuming no recombination between the markers and QTLs.

Interval mapping also may be extended to account simultaneously for segregation of multiple QTLs (Lander and Botstein 1989; Knapp 1991; Jansen 1992; Jansen 1993). For example, MAPMAKER/QTL can model additive effects (i.e. assuming no epistasis) of multiple unlinked QTLs (Lincoln et al. 1993b). If more than one QTL is segregating in a cross, such methods based on multiple-QTL models have greater power for QTL detection and produce less biased estimators of QTL position and effects compared to single-QTL interval mapping (e.g. Van Ooijen 1994). They can, for example, resolve linked QTL into distinct loci in cases for which single-QTL interval mapping fails.

Unfortunately, models that incorporate the simultaneous effects of multiple QTLs tend to become computationally problematic for more than two or three QTLs. To avoid such difficulties while retaining the advantages of modeling simultaneous effects of multiple QTLs, approximate multiple-QTL models have been introduced, forming the basis for a hybrid, or composite, interval mapping method (Jansen 1993; Jansen 1994; Jansen and Stam 1994; Zeng 1994). This approach combines interval mapping for QTL detection in a particular interval with multiple regression on marker genotypes to control for possible additive effects of QTLs outside that interval. Regression on markers unlinked to the interval reduces residual unexplained variation, leading to more powerful tests for QTL detection. Regression on markers linked to the interval may control for the effects of QTLs outside of but linked to the interval, allowing more accurate estimation of the individual effect and position of a QTL in the interval. The method of Jansen and Stam (1994) has the additional feature of coping with marker genotypes missing for some individuals in a mapping population, and with markers having either dominant or codominant alleles. It is available in the package, MapQTL (Van Ooijen and Maliepaard 1996).

Further research is required to understand more fully the behaviour of multiple-QTL methods under a wide range of circumstances, and unresolved issues remain in such areas as model selection (e.g. choosing which markers to include in the regression; Hackett 1994) and thresholds for QTL detection (Jansen 1994). Nonetheless, because of their generally superior performance, it seems likely that we will witness an increasing use of multiple-QTL methods as computer implementations of them become more widely available.

7.6.4 Epistasis

The methods described in the previous section (7.6.3) assume that multiple QTLs have additive effects on a given trait: the effect of substituting one allele for another at a given QTL is the same regardless of the allelic constitution at other QTLs for the same trait, and the expected trait value is determined by the sum of individual QTL effects. However, epistatic interactions among two or more QTLs may control the value of a given trait. In this case, the effect of allelic substitution at one QTL depends on the allelic constitution at other QTLs, and in general the expected trait value depends in a nonadditive manner on the combined allelic constitution for the entire set of epistatic QTLs.

For example, a study of dormancy in potato (Van den Berg et al. 1996a) revealed evidence of epistasis between two marker loci, M1 and M3 (Table 7.3). Here, there is a large difference in average dormancy between genotype classes A- and aa for marker M1 only if marker M3 has genotype bb. The presence of epistasis can have important consequences for marker-aided selection. For example, to achieve long dormancy, favourable alleles at both M1 and M3 should be present (row 1, Table 7.3). However, if a breeder were selecting for short dormancy a favourable allele at only one of the interacting loci would be sufficient (row 2,3 or 4, Table 7.3). Kreike et al. (1996) have also reported evidence of epistasis between QTLs in potato.

Table 7.3. Average DORMANCY in each of the four genotypic classes associated with loci M1 and M3

Genotype Class		Average DORMANCY
M1	M3	(days)
aa	bb	102
aa	B-	85
A-	bb	80
A-	B-	82

Allele symbol a means not A; b means not B; and - indicates that the allele is unknown

To test for epistasis, one may include interaction terms in the multi-way ANOVA model (e.g. Fatokun et al. 1992). Two-way ANOVA with interaction could be performed for all possible pairs of loci (with significance-level adjustment for multiple comparisons). To reduce the computational burden, and problems of multiple comparisons, one may test for epistasis among only those markers with statistically significant main effects in one-way ANOVA (Bonierbale et al. 1994), or among only those loci with LOD scores greater than one (Van den Berg et al. 1996a,b). Note, however, that it is possible in principle for a locus to have no apparent effect on the trait when considered in isolation by a one-way analysis, but to have a significant main effect when considered jointly with an epistatic locus. Moreno-Gonzalez (1992) has described modeling of epistasis among QTLs using multiple regression techniques. The inclusion of interaction effects in multiple-QTL extensions of interval mapping (both exact and approximate) requires further methodological research.

7.6.5 Pleiotropic Effects

QTL analysis of related traits provides the opportunity to study the similarities in genetic control of these traits. In a study in potato (Van den Berg 1993; Van den Berg et al. 1993), at least three of the QTLs detected for branching and two of the QTLs for leaf size were either pleiotropic for tuberization or else tightly linked to loci affecting tuberization. The relationship between the traits controlled by pleiotropic loci were as might be predicted by theory: early tuberization was associated with reduced branching and longer leaves (Ewing and Struik 1992). The associations among these traits were not detected by conventional (non-genetic) correlation analyses. Study of pleiotropic QTLs may prove useful in unravelling the different physiological components of quantitative traits. Recent developments of multivariate QTL mapping methods (Jiang and Zeng 1995; Korol et al. 1995; Ronin et al. 1995), which use information on the segregation of two or more traits simultaneously, should stimulate and facilitate such studies on pleiotropy.

7.7 Conclusion

The discovery of biochemical genetic markers has revolutionized the field of genome mapping. One group of such markers are those based on DNA RFLPs. The large number of RFLPs segregating in specific crosses allows screening of marker-loci closely associated with single gene traits. This facilitates map based cloning (Martin et al. 1993), marker-assisted progeny selection, and the construction of complete RFLP maps of plant genomes. These RFLP genome maps, in turn, facilitate the mapping of genes associated with quantitative traits.

Since the recognition of DNA RFLPs as tools in genetic mapping (Botstein et al. 1980) several other DNA based markers have been developed (e.g. RAPD markers; Chap. 6). The sections in this chapter regarding the selection of a mapping population, map construction, and QTL mapping are also applicable to other genetic markers. In selecting a specific marker for mapping, one should consider the purpose of the study. For example, if inferences are desired about effects of multiple QTL alleles in outbreeding species, marker-loci with co-dominant alleles are preferable (van Eck et al. 1994); for map-based cloning co-dominance is not necessary.

Various computer software packages are available to aid in the use of RFLP markers to map QTLs. Furthermore, methods for more accurate mapping of QTLs continue to be developed. New methods are especially needed for diploid outbreeding species and for genetic mapping in polyploids (e.g. Wu et al. 1992). The mapping of QTLs and analysis of the effects of QTL alleles, including epistatic interactions among alleles at different loci, has resulted in a better understanding of quantitative genetics in general. In particular, QTL analysis of physiologically related characters may help to unravel complex polygenic traits into more comprehensible components, much as classical genetic analysis of major genes has aided the dissection of metabolic pathways.

Acknowledgements. We are grateful to C.A. Hackett, J.W. McNicol and T. Connolly for helpful discussions, and J.E. Bradshaw and J.W. McNicol for critical review of the manuscript. Preparation of this manuscript was supported in part by the Scottish Office Agriculture and Fisheries Department.

References

Agresti A (1990) Categorical data analysis. John Wiley & Sons, New York, 558 pp
Allard RW (1956) Formulas and tables to facilitate the calculation of recombination values in heredity. Hilgardia 24:235–278
Antequera F, Bird AP (1989) Unmethylated CpG islands associated with genes in higher plant DNA. EMBO J 7:2295–2299
Asins MJ, Carbonell (1988) Detection of linkage between restriction fragment length polymorphism markers and quantitative traits. Theor Appl Genet 76:623–626
Atkinson AC (1987) Plots, transformations, and regression. An introduction to graphical methods of diagnostic regression analysis. Clarendon Press, Oxford, 282 pp
Bailey NTJ (1961) Introduction to the mathematical theory of genetic linkage. Oxford University Press, Oxford, 298 pp
Barnard GA (1967) The use of the likelihood function in statistical practice. In: Proc 5th Berkeley Symp on Math Statist and Prob, University of California Press, Berkeley
Bishop YMM, Fienberg SE, Holland PW (1975) Discrete multivariate analysis: theory and practice. The MIT Press, Cambridge, 557 pp
Bonierbale MW, Plaisted RL, Tanksley SD (1988) RFLP maps based on a common set of clones reveal modes of chromosomal evolution in potato and tomato. Genetics 120:1095–1103
Bonierbale MW, Plaisted RL, Pineda O, Tanksley SD (1994) QTL analysis of trichome-mediated insect resistance in potato. Theor Appl Genet 87:973–987
Botstein D, White RL, Skolnick M, Davis RW (1980) Construction of a genetic linkage map in man using restriction fragment length polymorphisms. Am J Hum Genet 32:314–331
Bradshaw JE (1994) Theory for locating quantitative trait loci. In: Bradshaw JE, Mackay GE (eds) Potato genetics, CAB International, Wallingford, pp 101–107
Breslow NE, Day NE (1980) Statistical methods in cancer research, Vol I. The analysis of case-control studies. IARC, Lyon, 338 pp

Brzustowicz LM, Mérette C, Xie X, Townsend L, Gilliam TC, Ott J (1993) Molecular and statistical approaches to the detection and correction of errors in genotype databases. Am J Hum Genet 53:1137–1145
Buetow KH, Chakravarti A (1987) Multipoint gene mapping using seriation. I. General methods. Am J Hum Genet 41:180–188
Buetow KH (1991) Influence of aberrant observations on high-resolution linkage analysis outcomes. Am J Hum Genet 49:985–994
Burr B, Burr FA (1991) Recombinant inbreds for molecular mapping in maize: theoretical and practical considerations. Trends Genet 7:55–60
Burt A, Bell G, Harvey PH (1991) Sex difference in recombination. J Evol Biol 4:259–277
Chasalow SD (1994) Mixture models for mapping QTL without inbred lines. In: Van Ooijen JW, Jansen J (eds) Biometrics in plant breeding: Applications of molecular markers. Proc 9th Meet of the Eucarpia section, Biometrics in plant breeding, 6–8 July 1994, Wageningen, The Netherlands
Chotai J (1984) On the LOD score method in linkage analysis. Ann Hum Genet 48:359–378
Churchill GA, Doerge RW (1994) Empirical threshold values for quantitative trait mapping. Genetics 138:963–971
Cottingham RW, Idury RM, Schaffer AA (1993) Faster sequential genetic linkage computations. Am J Hum Genet 53:252–263
Cowen NM (1989) Multiple linear regression analysis of RFLP data sets used in mapping QTLs. In: Helentjaris T, Burr B (eds) Development and application of molecular markers to problems in plant genetics. Cold Spring Harbor Laboratory Press, Cold Spring Harbor, pp 113–116
Darvasi A, Weinreb A, Minke V, Weller JI, Soller M (1993) Detecting marker-QTL linkage and estimating QTL gene effect and map location using a saturated genetic map. Genetics 134:943–951
DeGroot MH (1975) Probability and statistics. Addison-Wesley, Reading, 607 pp
DeVincente MC, Tanksley SD (1993) QTL analysis of transgressive segregation in an interspecific tomato cross. Genetics 134:585–596
Doerge RW, Churchill GA (1996) Permutation tests for multiple loci affecting a quantitative character. Genetics 142:285–294
Echt C, Knapp S, Liu B-H (1992) Genome mapping with non-inbred crosses using GMendel 2.0. Maize Genet Coop Newslett 66:27–29
Edwards AWF (1988) Lod score or log-likelihood. Nature 333:308
Edwards MD, Helentjaris T, Wright S, Stuber CW (1992) Molecular-marker-facilitated investigations of quantitative trait loci in maize. 4. Analysis based on genome saturation with isozyme and restriction fragment length polymorphism markers. Theor Appl Genet 83:765–774
Edwards MD, Stuber CW, Wendel JF (1987) Marker-facilitated investigations of quantitative-trait loci in maize. I. Numbers, genomic distribution and types of gene action. Genetics 116:113–125
Efron B, Tibshirani RJ (1993) An introduction to the bootstrap. Chapman and Hall, London

Eldredge L, Ballard R, Baird WV, Abbott A, Morgens P, Callahan A, Scorza R, Monet R (1992) Application of RFLP analysis to genetic linkage mapping in peaches. HortScience 27:160–163

Ellis THN, Turner L, Hellens RP, Lee D, Harker CL, Enard C, Domoney, Davies DR (1992) Linkage maps in pea. Genetics 130:649–663

Emerson RA, Beadle GW, Fraser AC (1935) A summary of linkage studies in maize. Cornell Univ Agric Exp Stn Mem 180

Ewing EE, Struik PC (1992) Tuber formation in potato: induction, nitiation, and growth. Hortic Rev 14:89–198

Fatokun CA, Menacio-Hautea DI, Danesch D, Yound ND (1992) Evidence for orthologous seed weight genes in cow pea and mung bean based on RFLP mapping. Genetics 132:841–846

Garcia-Dorado A, Gallego A (1992) On the use of the classical tests for detecting linkage. J Hered 83:143–146

Garvey TC, Hewitt JD (1992) Use of molecular markers to locate quantitative trait loci linked to high soluble solids content in a hybrid of *Lycopersicon cheesmanii*. J Am Soc Hortic Sci 117:497–499

Gebhardt C, Salamini F (1992) Restriction fragment length polymorphism analysis of plant genomes and its application to plant breeding. Int Rev Cytol 135:201–237

Gebhardt C, Ritter E, Debener T, Schachtschnabel U, Walkemeier B, Uhrig H, Salamini F (1989) RFLP analysis and linkage mapping in *Solanum tuberosum*. Theor Appl Genet 78:65–75

Gerber S, Rodolphe F (1994) Estimation and test for linkage between markers: a comparison of lod scores and χ^2 test in a linkage study of maritime pine (*Pinus pinaster* Ait.). Theor Appl Genet 88:293–297

Giraudat J, Hauge BM, Valon C, Smalle J, Parcy F, Goodman HM (1992) Isolation of the *Arabidopsis ABI3* gene by positional cloning. Plant Cell 4:1251–1261

Görg R, Schachtschnabel U, Ritter E, Salamini F, Gebhardt C (1992) Discrimination among 136 tetraploid potato varieties by fingerprints using highly polymorphic DNA markers. Crop Sci 32:815–819

Graner A, Jahoor A, Schondelmaier J, Siedler H, Pillen K, Fischbeck G, Wenzel G, Herrmann RG (1991) Construction of an RFLP map of barley. Theor Appl Genet 83:250–256

Grattapaglia D, Sederoff R (1994) Genetic linkage maps of *Eucalyptus grandis* and *Eucalyptus urophylla* using a pseudo-testcross: mapping strategy and RAPD markers. Genetics 137:1121–1137

Grodzicker T, Williams J, Sharps P, Sambrook J (1974) Physical mapping of temperature-sensitive mutations of adenovirus. Cold Spring Harbor Symp Quant Biol 39:439–446

Guries RP, Friedman ST, Ledig FT (1978) A megagametophyte analysis of genetic linkage in pitch pine (*Pinus rigida* Mill.) Heredity 40:309–314

Hackett CA (1994) Selection of markers linked to quantitative trait loci by regression techniques. In: Van Ooijen JW, Jansen J (eds) Biometrics in plant breeding: applications of molecular markers. Proc 9th Meet of the Eucarpia section, Biometrics in plant breeding, 6–8 July 1994, Wageningen, The Netherlands

Hackett CA, Weller JI (1995) Genetic mapping of quantitative trait loci for traits with ordinal distributions. Biometrics 51:1252–1263

Hackett CA, Ellis RP, Foster BP, McNicol JW, Macaulay M (1992) Statistical analysis of a linkage experiment in barley involving quantitative trait loci for height and ear-emergence time and two genetic markers on chromosome 4. Theor Appl Genet 85:120–126

Haldane JBS (1919) The combination of linkage values, and the calculation of distance between the loci of linked factors. J Genet 8:299–309

Haldane JBS, Smith CAB (1947) A new estimate for the linkage between the genes for color-blindness and haemophilia in man. Ann Eugen 14:10–31

Haley CS, Knott SA, Elsen J-M (1994) Mapping quantitative trait loci in crosses between outbred lines using least squares. Genetics 136:1195–1207

Heun M, Gregorius H-R (1987) A theoretical model for estimating linkage in F_2 populations with distorted single gene segregation. Biom J 29:397–406

Heun M, Kennedy AE, Anderson JA, Lapitan NLV, Sorrells ME, Tanksley SD (1991) Construction of a restriction fragment length polymorphism map for barley (*Hordeum vulgare*). Proc Natl Acad Sci USA 88:3691-3695

Ishii T, Terachi T, Mori N, Tsunewaki K (1993) Comparative study on the chloroplast, mitochondral and nuclear genome differentiation in two cultivated rice species, *Oryza sativa* and *O. glaberrima*, by RFLP analyses. Theor Appl Genet 86:88–96

Jansen RC (1992) A general mixture model for mapping quantitative trait loci by using molecular markers. Theor Appl Genet 85:252–260

Jansen RC (1993) Interval mapping of multiple quantitative trait loci. Genetics 135:205–211

Jansen RC (1994) Controlling the type I and type II errors in mapping quantitative trait loci. Genetics 138:871–881

Jansen RC, Stam P (1994) High resolution of quantitative traits into multiple loci via interval mapping. Genetics 136:1447–1455

Jansen RC (1996a) A general monte carlo method for mapping multiple quantitative trait loci. Genetics 142:305–311

Jansen RC (1996b) Complex plant traits: time for polygenic analysis. Trends in Plant Science 1:89–94

Jensen J, Jørgensen JH (1975) The barley chromosome 5 linkage map. I. Literature survey and map estimation procedure. Hereditas 80:5–16

Jiang C, Zeng Z-B (1995) Multiple trait analysis of genetic mapping for quantitative trait loci. Genetics 140:1111–1127

Kammerer CM, MacCluer JW (1988) Empirical power of three preliminary methods for ordering loci. Am J Hum Genet 43:964–970

Kan YW, Dozy AM (1978) Polymorphism of DNA sequence adjacent to human β-globin structural gene: relationship to sickle mutation. Proc Natl Acad Sci USA 75:724–732

Karaman Z (1994) Towards an automated interpretation of gel electrophoresis. In: Van Ooijen JW, Jansen J (eds) Biometrics in plant breeding: applications of molecular markers. Proc 9th Meet of the Eucarpia section, Biometrics in plant breeding, 6–8 July 1994, Wageningen, The Netherlands

Kendall M, Stuart A (1979) The advanced theory of statistics, vol II. Inference and relationship, 4th edn. Charles Griffin, London, 748 pp
Knapp SJ (1991) Using molecular markers to map multiple quantitative trait loci: models for backcross, recombinant inbred, and doubled haploid progeny. Theor Appl Genet 81:333–338
Knott SA, Haley CS (1992) Aspects of maximum likelihood methods for the mapping of quantitative trait loci. Genet Res Camb 60:139–151
Korol AB, Ronin YI, Kirzhner VM (1995) Interval mapping of quantitative trait loci employing correlated trait complexes. Genetics 140:1137–1147
Kosambi DD (1944) The estimation of map distances from recombination values. Ann Eugen 12:172–175
Kreike CM, Kok-Westeneng AA, Vinke JH, Stiekema WJ (1996) Mapping of QTLs involved in nematode resistance, tuber yield and root development in *Solanum* sp. Theor Appl Genet 92:463–470
Kruglyak L, Lander ES (1995) A nonparametric approach for mapping quantitative trait loci. Genetics 139:1421–1428
Lalouel JM (1977) Linkage mapping from pair-wise recombination data. Heredity 38:61–77
Lander ES, Botstein D (1989) Mapping Mendelian factors underlying quantitative traits using RFLP linkage maps. Genetics 121:185–199
Lander ES, Green P, Abrahamson J, Barlow A, Daly MJ, Lincoln SE, Newburg L (1987) MAPMAKER: an interactive computer package for constructing primary genetic linkage maps of experimental and natural populations. Genomics 1:174–181
Lehmann EL (1975) Nonparametrics. Statistical methods based on ranks. Holden-Day, Oakland, and McGraw-Hill, New York, 457 pp
Leonard-Schippers C, Gieffers W, Schäfer-Pregl R, Ritter E, Knapp SJ, Salamini F, Gebhardt C (1994) Quantitative resistance to *Phytophtora infestans* in potato: a case study for QTL mapping in an allogamous plant species. Genetics 137:67–77
Letovsky S, Berlyn MB (1992) Cprop – a rule-based program for constructing genetic maps. Genomics 12:435–446
Liberman V, Karlin S (1984) Theoretical models of genetic map functions. Theor Pop Biol 25:331–346
Lincoln SE, Lander ES (1992) Systematic detection of errors in genetic linkage data. Genomics 14:604–610
Lincoln SE, Daly MJ, Lander ES (1993a) Constructing genetic linkage maps with MAPMAKER/EXP Version 3.0: a tutorial and reference manual. Whitehead Institute for Biomedical Research Technical Report, 3rd edn Cambridge, Massachusetts
Lincoln SE, Daly MJ, Lander ES (1993b) Mapping genes controlling quantitative traits with MAPMAKER/QTL Version 1.1: a tutorial and reference manual. Whitehead Institute for Biomedical Research Technical Report 2nd edn. Cambridge, Massachusetts
Lorieux M, Goffinet B, Perrier X, González de León D, Lanaud C (1995a) Maximum-likelihood models for mapping genetic markers showing segregation distortion. 1. Backcross populations. Theor Appl Genet 90:73–80

Lorieux M, Perrier X, Goffinet B, Lanaud C, González de León D (1995b) Maximum-likelihood models for mapping genetic markers showing segregation distortion. 2. F_2 populations. Theor Appl Genet 90:81–89
Luckenbach C, Luckenbach A, Ritter H (1994) Restriction fragment length polymorphism: molecular weight analysis and calculation with a scanner-based computer system. Electrophoresis 15:149–152
Luo ZW, Kearsey MJ (1992) Interval mapping of quantitative trait loci in an F_2 population. Heredity 69:236–242
MacArthur JW (1934) Linkage groups in tomato. J Genet 29:123–133
MacLean CJ, Morton NE, Lew R (1985) Efficiency of LOD scores for representing multiple locus linkage data. Genet Epidemiol 2:145–154
Maliepaard C, Van Ooijen JW (1994) QTL mapping in a full-sib family of an outcrossing species. In: Biometrics in plant breeding. Proc 9th Meet of the Eucarpia section, Biometrics in plant breeding, 6–8 July 1994, Wageningen, The Netherlands
Mangin B, Goffinet B, Rebaï A (1994) Constructing confidence intervals for QTL location. Genetics 138:1301–1308
Manly BFJ (1991) Randomization and Monte Carlo methods in biology. Chapman and Hall, London
Manly KF, Elliott RW (1991) RI Manager, a microcomputer program for analysis of data from recombinant inbred strains. Mamm Genome 1:123–126
Martin GB, Brommonschenkel SH, Chunwongse J, Frary A, Ganal MW, Spivey R, Wu T, Earle ED, Tanksley SD (1993) Map-based cloning of a protein kinase gene conferring disease resistance in tomato. Science 262:1432–1436
Martinez O, Curnow RN (1992) Estimating the locations and the sizes of the effects of quantitative trait loci using flanking markers. Theor Appl Genet 85:480–488
Mather K (1938) The measurement of linkage in heredity. Methuen, London, 132 pp
Mather K (1946) Statistical analysis in biology. Methuen, London, 267 pp
McCouch SR, Kochert G, Yu ZH, Wang ZY, Khush GS, Coffman WR, Tanksley SD (1988) Molecular mapping of rice chromosomes. Theor Appl Genet 76:815–829
McCullagh P, Nelder JA (1989) Monographs on statistics and applied probability 37. Generalized linear models, 2nd edn. Chapman and Hall, London, 511 pp
McLachlan GJ, Basford KE (1987) Mixture models. Inference and applications to clustering. Marcel Dekker, New York, 253 pp
Michelmore RW, Para I, Kesseli RV (1991) Identification of markers linked to disease resistance genes by bulked segregant analysis: a rapid method to detect markers in specific genomic regions by using segregating populations. Proc Natl Acad Sci USA 88:9828–9832
Miller JC, Tanksley SD (1990) RFLP analysis of phylogenetic relationships and genetic variation in the genus *Lycopersicon*. Theor Appl Genet 80:437–448
Miller RG (1966) Simultaneous statistical inference. McGraw-Hill, New York, 272 pp
Moreno-Gonzalez J (1992) Genetic models to estimate additive and non-additive effects of marker-associated QTL using multiple regression techniques. Theor Appl Genet 85:435–444

Morris C (1992) Academic Press dictionary of science and technology. Academic Press, San Diego
Morton NE (1955) Sequential tests for the detection of linkage. Am J Hum Genet 7:277–318
Morton NE (1988a) Lod score *Redivivus*. Nature 334:477–478
Morton NE (1988b) Multipoint mapping and the emperor's clothes. Ann Hum Genet 52:309–318
Morton NE, Andrews V (1989) MAP, an expert system for multiple pairwise linkage analysis. Ann Hum Genet 53:263–269
Morton NE, MacLean CJ, Lew R, Yee S (1986) Multipoint linkage analysis. Am J Hum Genet 38:868–883
O'Brien SJ (ed) (1993) Genetic maps. Locus maps of complex genomes, 6th edn. Cold Spring Harbor Laboratory Press, Cold Spring Harbor
Olson JM, Boehnke M (1990) Monte Carlo comparison of preliminary methods for ordering multiple genetic loci. Am J Hum Genet 47:470–482
Ott J (1991) Analysis of human genetic linkage, rev edn. Johns Hopkins University Press, Baltimore, 302 pp
Owen ARG (1950) The theory of genetical recombination. Adv Genet 3:117–157
Paterson AH, Lander ES, Hewitt JD, Peterson S, Lincoln SE, Tanksley SD (1988) Resolution of quantitative traits into Mendelian factors by using a complete linkage map of restriction fragment length polymorphisms. Nature 335:721–726
Paterson AH, Damon S, Hewitt JD, Zamir D, Rabinowitch HD, Lincoln SE, Lander ES, Tanksley SD (1991) Mendelian factors underlying quantitative traits in tomato: comparison across species, generations, and environments. Genetics 127:181–197
Pineda O, Bonierbale MW, Plaisted RL, Brodie BB, Tanksley SD (1992) Identification of RFLP markers linked to the *H1* gene conferring resistance to the potato cyst nematode *Globodera rostochiensis*. Genome 36:152–156
Press WH, Flannery BP, Teukolsky SA, Vetterling WT (1988) Numerical recipes in C. The art of scientific computing. Cambridge University Press, Cambridge
Rebaï A, Goffinet B, Mangin B (1994) Approximate thresholds of interval mapping tests for QTL detection. Genetics 138:235–340
Rick CM (1980) Linkage report: tomato linkage survey. Rep Tomato Genet Coop 30:1–17
Ritter E, Gebhardt C, Salamini F (1990) Estimation of recombination frequencies and construction of RFLP linkage maps in plants from crosses between heterozygous parents. Genetics 125:645–654
Rodolphe F, Lefort M (1993) A multi-marker model for detecting chromosomal segments displaying QTL activity. Genetics 134:1277–1288
Rogatko A, Zacks S (1993) Ordering genes: controlling the decision-error probabilities. Am J Hum Genet 52:947–957
Rohrer GA, Behrens DW, Keele JW (1994) A simplified procedure for entry of raw genotypic data. J Comp Biol 1:111–119
Rommens JM, Iannuzzi MC, Kerem B, Drumm ML, Melmer G, Dean M, Rozmahel R, Cole JL, Kennedy D, Hidaka N, Zsiga M, Buchwald M, Riordan JR, Tsui L-C, Collins FS (1989) Identification of the cystic fibrosis gene: chromosome walking and jumping. Science 245:1059–1065

Ronin YI, Kirzhner VM, Korol AB (1995) Linkage between loci of quantitative traits and marker loci: multi-trait analysis with a single marker. Theor Appl Genet 90:776–786
Säll T, Nilsson N-O (1994) The robustness of recombination frequency estimates in intercrosses with dominant markers. Genetics 137:589–596
Schaffer AA, Gupta SK, Shriram K, Cottingham RW (1994) Avoiding recomputation in linkage analysis. Hum Hered 44:225–237
Scheffé H (1959) The analysis of variance. John Wiley & Sons, New York, 477 pp
Shields DC, Collins A, Buetow KH, Morton NE (1991) Error filtration, interference, and the human linkage map. Proc Natl Acad Sci USA 88:6501–6505
Shoemaker J, Zaitlin D, Horn J, DeMars S, Kirschman J, Pitas J (1992) A comparison of three Agrigenetics maize RFLP linkage maps. Maize Genet Coop Newslett 66:65–68
Smith CAB (1986) Chi-squared tests with small numbers. Ann Hum Genet 50:163–167
Soller M, Brody T, Genizi A (1976) On the power of experimental designs for the detection of linkage between marker loci and quantitative loci in crosses between inbred lines. Theor Appl Genet 47:35–39
Speed TP, McPeek MS, Evans SN (1992) Robustness of the no-interference model for ordering genetic markers. Proc Natl Acad Sci USA 89:3103–3106
Stam P (1991) Some aspects of QTL analysis. In: Pesek J, Hartmann J (eds) Biometrics in plant breeding. Proc 8th Meet of the Eucarpia section, Biometrics in plant breeding, 1–6 July 1991, Brno, Czechoslovakia
Stam P (1993) Construction of integrated genetic linkage maps by means of a new computer package: JoinMap. Plant J 3:739–744
Stuber CW, Edwards MD, Wendel JF (1987) Molecular marker-facilitated investigations of quantitative trait loci in maize. II. Factors influencing yield and its component traits. Crop Sci 27:639–648
Stuber CW, Lincoln SE, Wolff DW, Helentjaris T, Lander ES (1992) Identification of genetic factors contributing to heterosis in a hybrid from two elite maize inbred lines using molecular markers. Genetics 132:823–839
Suiter KA, Wendel JF, Case JS (1983) LINKAGE-1: a Pascal computer program for the detection and analysis of genetic linkage. J Hered 74:203–204
Suzuki DT, Griffiths AJF, Lewontin RC (1981) An introduction to genetic analysis, 2nd edn. WH Freeman, San Francisco, 911 pp
Tanksley SD (1983) Gene mapping. In: Tanksley SD, Orton TJ (eds) Isozymes in plant genetics and breeding, Part 1A. Elsevier, Amsterdam, pp 109–138
Tanksley SD (1993) Mapping polygenes. Annu Rev Genet 27:205–233
Tanksley SD, Hewitt J (1988) Use of molecular markers in breeding for soluble solids content in tomato – a re-examination. Theor Appl Genet 75:811–823
Tanksley SD, Bernatzky R, Lapitan NL, Prince JP (1988a) Conservation of gene repertoire but not gene order in pepper and tomato. Proc Natl Acad Sci USA 85:6419–6423
Tanksley SD, Miller J, Paterson A, Bernatzky R (1988b) Molecular mapping of plant chromosomes. In: Gustafson JP, Appels R (eds) Chromosome structure and function. Plenum press, New York, pp 157-173

Thompson JN, Thoday JM (1974) A definition and standard nomenclature for "polygenic loci". Heredity 33:430–437
Titterington DM, Smith AFM, Makov UE (1985) Statistical analysis of finite mixture distributions. John Wiley & Sons, Chichester, 243 pp
Tulsieram LK, Glaubitz JC, Kiss G, Carlson JE (1992) Single tree genetic linkage mapping in conifers using haploid DNA from megagametophytes. Bio/Technology 10:686–690
Van den Berg JH (1993) Identification of chromosomal segments and gibberellins affecting potato (*Solanum tuberosum* L.) development. PhD Diss, Cornell University, Ithaca, 250 pp.
Van den Berg JH, Bonierbale MW, Ewing EE, Plaisted RL, Tanksley SD (1993) Use of RFLP-linkage to study genetics and physiology of potato tuberization. Proc 12th triennial Conf EAPR, Paris, pp 278–279
Van den Berg JH, Ewing EE, Plaisted RL, McMurry S, Bonierbale MW (1996a) QTL analysis of potato tuber dormancy. Theor Appl Genet (in press)
Van den Berg JH, Ewing EE, Plaisted RL, McMurry S, Bonierbale MW (1996b) QTL analysis of potato tuberization. Theor Appl Genet (in press)
Van Ecke HJ, Jacobs JME, Stam P, Ton J, Stiekema WJ, Jacobsen E (1994) Multiple alleles for tuber shape in diploid potato detected by qualitative and quantitative genetic analysis using RFLPs. Genetics 137:303–309
Van Ooijen JW (1992) Accuracy of mapping quantitative trait loci in autogamous species. Theor Appl Genet 84:803–811
Van Ooijen JW (1994) Comparison of a single-QTL model with an approximate multiple-QTL model for QTL mapping. In: Van Ooijen JW, Jansen J (eds) Biometrics in plant breeding: applications of molecular markers. Proc 9th Meet of the Eucarpia section, Biometrics in plant breeding, 6–8 July 1994, Wageningen, The Netherlands
Van Ooijen JW, Maliepaard C (1996) MapQTL (tm) version 3.0: Software for the calculation of QTL positions on genetic maps. CPRO-DLO, Wageningen
Waugh R, Van de Ven WTG, Phillips MS, Powell W (1990) Chloroplast DNA diversity in the genus *Rubus* (Rosaceae) revealed by Southern hybridization. Plant Syst Evol 172:65–75
Weeks DE, Lange K (1987) Preliminary ranking procedures for multilocus ordering. Genomics 1:236–242
Weller JI (1986) Maximum likelihood techniques for the mapping and analysis of quantitative trait loci with the aid of genetic markers. Biometrics 42:627–640
Weller JI (1992) Statistical methodologies for mapping and analysis of quantitative trait loci. In: Beckmann JS, Osborn TC (eds) Plant genomes: methods for genetic and physical mapping. Kluwer, Dordrecht, pp 181–207
Whitkus R, Doebley J, Lee M (1992) Comparative genome mapping of sorghum and maize. Genetics 132:1119–1130
Wu KK, Burnquist W, Sorrells ME, Tew TL, Moore PH, Tanksley SD (1992) The detection and estimation of linkage in polyploids using single-dose restriction fragments. Theor Appl Genet 83:294–300
Young ND, Zamir D, Ganal MW, Tanksley SD (1988) Use of isogenic lines and simultaneous probing to identify DNA markers tightly linked to the Tm-2a gene in tomato. Genetics 120:579–585

Zeng Z-B (1993) Theoretical basis for separation of multiple linked gene effects in mapping quantitative trait loci. Proc Natl Acad Sci USA 90:10972–10976
Zeng Z-B (1994) Precision mapping of quantitative trait loci. Genetics 136:1457–1468

Part III

Genetic Engineering: Methodology and Analysis

Chapter 8

Plant Gene Transfer

R.H. POTTER and M.G.K. JONES

8.1 Introduction

The production of transgenic plants requires the transfer of genetic material into a plant cell, stable integration of the introduced DNA in the genome of the plant and regeneration of whole plants such that the introduced gene is stably maintained in subsequent generations. For all methods of gene transfer, using either vectors or naked DNA, with cloned genes or less specific DNA sequences, the major requirement is for regeneration of fertile transgenic plants from the cells. Thus, the success and efficiency of transformation depends on tissue culture systems, and the methods required vary with species.

The transfer of genetic information falls into two major areas: specific gene transfer, in which one or a few individual genes, which are normally well characterized, may be inserted into plant cells, and transfer of blocks of genes which are usually less well defined, using protoplast fusion technology to generate various types of somatic hybrids.

Specific Gene Transfer

Gene transfer systems can be divided broadly into two types: vector-mediated, in which another organism is used to effect the transfer and/or integration; and direct gene transfer, in which naked DNA is introduced into cells. So far, viral vectors have not yet proved to be useful for plant tranformation and so vector-mediated transformation relies almost exclusively on use of the soil bacteria, *Agrobacterium tumefaciens* and *A. rhizogenes* as vectors. An important point to note in *Agrobacterium*-mediated vector transformation is that the specific DNA is integrated into the host chromosome as a part of the transformation process, thus stable maintenance of the DNA is not usually a problem with this method of transformation.

In contrast, the methods which involve naked DNA (either as whole or partial chromosomes, isolated DNA or engineered plasmids) must enable DNA both to enter the cell and also integrate into the recipient genome. The integration of naked DNA into the genome is an inefficient process; consequently, efficient methods of introducing DNA into the cells are required to produce as many transformed plants as possible. This aspect has considerable bearing on the tissue culture methodology used, both for preparing the cells for uptake/introduction of DNA and for culture and selection later to produce transgenic plants. Because of these problems, *Agrobacterium* is generally regarded as the more efficient method. However, the limited host range of *A. tumefaciens* (almost exclusively to dicotyledonous plants) and more complicated cloning vector construction have resulted in wide use of direct gene transfer systems.

When considering which method of gene transfer to use in a particular case, the host range of *Agrobacterium* is a vital factor, since different strains show differences in virulence for different plant species. There are many different strains of *Agrobacterium* containing different Ti (*A. tumefaciens*) and Ri (*A. rhizogenes*) plasmids. The DNA plasmids carry the genes which are transferred and the genes which effect the transfer. Tobacco has been used as a model plant for much transformation work and so *Agrobacterium* strains suited for tobacco transformation are the most widely available. These strains may be less suitable for some other species.Thus the limited host range associated with a specific strain of *Agrobacterium* may, in fact, only reflect the limited number of strains tested. In general, most dicotyledonous plants seem susceptible to *Agrobacterium* – with varying degrees of virulence. In contrast, few monocots have successfully been transformed in this way. Currently, only asparagus is routinely transformed by *Agrobacterium*, although reports of rice transformation have recently been published (Raineri et al. 1990; Park and Smith 1994).

Direct gene transfer also has "host range" problems. These relate to the ability to regenerate fertile transgenic plants from single, transformed, cells. Regeneration is the limiting factor in most transformation systems. An advantage of direct gene transfer is that any piece of DNA may be transferred without using specialized vectors, or even without isolating DNA at all. Where the gene of interest has been cloned, all that is required is that the gene is ligated to the correct plant expression signals and possibly linked to a selectable marker gene to aid identification of transgenic plants. This contrasts with the specialized vectors required for *Agrobacterium* transformation because of the special mechanism of T-DNA transfer.

In deciding which system of transformation to use, both the plant to be transformed and the trait to be introduced must be considered. With regard to the plant species, the first point is the tissue culture/regeneration system available for the particular plant. The second point is susceptibilty to *Agrobacterium* and, in addition to this, the availability of an efficient strain and easily manipulated vectors. Sometimes these two factors may go together, i.e. only certain tissues can be used for regenation of fertile plants and only some tissues are susceptible to *Agrobacterium*. The efficiency of regeneration of transformed tissues is important. For example, although there appear to be strains of *Agrobacterium* that will infect rice, the efficiency of procedures to introduce DNA into protoplasts and regenerate transgenic plants is so successful that direct gene transfer into protoplasts is most widely used to obtain transgenic plants.

The trait/gene to be transferred also affects the choice of transformation system. If the trait is controlled by one or only very few genes and these genes have been cloned, then the use of specific transformation techniques are clearly more efficient. Linking of the desired gene (driven by a suitable plant promoter to direct expression in the desired tissue) to a separate selectable marker gene enables transformed cells to be selected directly and thus only transformed plants regenerated. For a further discussion on reporter genes and selection systems, see Chapter 9.

Fusion-Mediated Transformation

Some desirable traits are, however, controlled by many genes, and currently little is known at the molecular level about the specific genes involved in many important agronomic characteristics. In such cases, whole or partial genome transfer offers an approach to transfer the trait without any knowledge of the molecular basis. Methods of whole or partial genome transfer (through protoplast fusion or transfer of isolated genomic DNA) require post transfer screening to detect cells or plants with the desired trait (for further details, see Chap. 10). This "shotgun" approach relies on an efficient system of transferring the genetic material such that regeneration of all treated cells will produce many which have integrated foreign DNA. The desired trait can then be selected at the whole-plant level. As a result, efficient regeneration protocols are needed and this emphasizes the need for good tissue culture expertise. If plants can be regenerated from protoplasts then methods of somatic hybridization can be used in which protoplast fusion between intact or irradiated donors are fused to recipient protoplasts. This results in the generation of "symmetric" (i.e. combining both genomes) or "asymmetric" (i.e. partial genome transfer from donor) somatic hybrids.

8.2 *Agrobacterium*-Mediated Gene Transfer

Agrobacterium-mediated gene transfer relies on a natural system for the transfer and integration of DNA which is found in two related soil bacteria, *Agrobacterium tumefaciens* and *A. rhizogenes*. This is a plasmid-encoded trait with two main regions on the plasmid – the T-DNA (the region of the plasmid that is transferred and integrated into the plant) and the *vir* region which codes for the proteins that effect the transfer. A particularly useful finding is that these two regions can work in *trans* as well as *cis*, enabling more efficient vector systems to be developed. The T-DNA is a region defined by two border sequences. The DNA between these border sequences is transferred and integrated into the host cell chromosomes in the transformed cell. Only the border sequences are required for transfer; the DNA in the wild-type Ti and Ri plasmids between the borders is not involved in the transformation process and can be replaced by other genes.

The native Ti (or Ri) plasmid is large (approx. 200 kbp) and so vectors are now usually constructed using smaller plasmids which are easier to handle. The genes to be transferred are cloned between the left and right border sequences contained in a plasmid, using *E.coli* as a host. The plasmid is then transferred to a disarmed *Agrobacterium* host strain (a strain in which the internal T-DNA sequences have been removed from the Ti-plasmid). In earlier work, the cloned genes were integrated into the Ti plasmid by homologous recombination and the

Agrobacterium then used for transformation (co-integrative vectors). This system required a bacterial selectable marker inside the T-DNA in order to select for the recombinants and was not particularly efficient. Once it was found that the *vir* genes worked in *trans* (i.e. they did not need to be on the same molecule of DNA as the transferred genes in order to effect transfer) then "binary" vector systems were developed. Disarmed Ti plasmids were isolated which contained no T-DNA at all and broad host range plasmids were constructed (which could replicate in both *E.coli* and *Agrobacterium*) with a multiple cloning site between the left and right borders. Chosen DNA could now be easily introduced into this binary vector and, when mobilized into the *Agrobacterium* strain, be maintained and subsequently transferred into plants. This advance has made it easier to transfer cloned genes into an *Agrobacterium* vector system and to prepare vectors for transformation.

For efficient plant transformation, a selectable marker gene is required to select transformed plant cells in culture. A readily assayable reporter gene may also be included to enable checking and characterization of inserts. Many vectors have been constructed which contain such marker genes and have restriction sites to insert desired genes. Vector construction involves sub-cloning of the desired genes in *E. coli*, followed by mobilization from *E.coli* into *Agrobacterium*. For this mobilization, tri-parental mating with an *E.coli* helper strain has been widely used (Ditta et al. 1980), alternatively direct transformation of *Agrobacterium* can be achieved using electroporation (Shen and Forde 1989).

8.2.1 Transformation of Tobacco with *Agrobacterium tumefaciens*

Reagents

- GYPC (0.2% glucose, 0.2% yeast extract, 0.2 M sodium phosphate pH 7, 0.2% casein hydrolysate) with suitable antibiotics, 10 ml liquid in 50-ml flasks
- MS30 (Murashige and Skoog, 1963, salts and vitamins with 30 g/l sucrose) 1 mg/l BAP, solidified with 0.5% agar in petri dishes
- MS30, 1 mg/l BAP with 200 mg/l cefotaxime, 50 mg/l kanamycin (or other selective agent), solidified with 0.5% agar in petri dishes

- MS30, 0.1 mg/l BAP with 200 mg/l cefotaxime, 50 mg/l kanamycin (or other selective agent), solidified with 0.5% agar in deep petri dishes
- MS30, 200 mg/l cefotaxime, 50 mg/l kanamycin (or other selective agent), solidifed with 0.8% agar in sterile 60-ml tissue culture jars
- 70% ethanol

Equipment

- Sterile petri dishes
- Sterile cork borer, approx. diameter 0.5 cm
- Forceps (blunt end)
- Sheets of dental wax
- Sterile paper towels
- Nescofilm
- 60-ml pots
- Shaking incubator set at 28 °C
- Growth chamber

Procedure

This protocol is based on that published by Horsch et al. (1985). All procedures must be carried out under sterile conditions, preferably in a class II downward flow microbiological containment cabinet.

1. Grow *A. tumefaciens* containing binary or co-integrate plasmids in GYPC, with appropriate antibiotics to maintain plasmid, overnight at 28 °C. Pour solution into a sterile petri dish.

2. Sterilize a sheet of dental wax in 70% ethanol for 5 min, then place in a sterile petri dish. Remove leaves from an in vitro-grown tobacco plant and cut discs from the leaves using a cork borer on the dental wax sheet.

3. Place discs into *A. tumefaciens* suspension for up to 5 min, then blot off excess *A. tumefaciens* on the sterile paper towel and place the disc onto MS30, 1 mg/l BAP with no antibiotics with the underside of the leaf in contact with the media.

4. Culture for 2 days at 25 °C, 16 h light, and then transfer the discs to MS30, 1 mg/l BAP with antibiotics, maintaining the same orientation, and culture for 2 days without sealing the dishes. Then seal dishes with Nescofilm and culture as before for 4 weeks.

5. Transfer discs to MS30, 0.1 mg/l BAP and antibiotics, seal and culture for a further 4 weeks. Sub-culture shoots into 60 ml pots with MS30 and antibiotics and culture until strong root growth is evident (approximately 4 weeks).
6. Transfer plantlets to pasteurized soil, taking care to wash off all agar from roots. Keep plantlets in high humidity for the first week and then gradually reduce to normal greenhouse conditions.

Tobacco plants will grow in most compost/soil mixtures and under most daylight regimes. If flowering is required, longer daylengths must be provided (minimum 14 h). It must be remembered that transgenic plants may pose an environmental risk and so have to be grown under suitable containment conditions. Requirements for this differ from country to country, and permission must be obtained from the relevant regulatory authority (institutional and national) before this work can be undertaken.

Notes and troubleshooting

- Many different media are used for the growth of different *A. tumefaciens* strains; GYPC is simple to prepare and is suitable for growth of many strains. The particular antibiotic to use and in what amount depends on the plasmid. Binary plasmids are quickly lost in the absence of selection and so appropriate antibiotics must be used. Disarmed *Agrobacterium* strains can be obtained from a variety of sources, as can binary vectors for plant transformation. Clontech (Palo Alto, California) supply a complete kit with bacterial strains and binary vectors for plant transformation which also contain the β- Glucuronidase (GUS) marker gene (Jefferson et al. 1986). A hypervirulent disarmed *Agrobacterium tumefaciens* strain, AGL1, can be obtained from Dr. R A Ludwig (Lazo et al., 1991) and a range of smaller binary vectors is available from Prof. P. Maliga (Hajdukiewicz et al. 1994).

- The usual varieties of tobacco for transformation are Wisconsin 38 and SR1. These are easily maintained in vitro on MS30 with 0.8% agar. The use of in vitro-grown plants removes the need to surface sterilize the plant tissue before transformation. This ensures that tissue is in good condition.

- Cutting the leaf discs on sheets of wax ensures a clean cut with the minimum of damage to the leaf cells at the cut edge of the tissue. *A.*

tumefaciens infects only wounded tissue, and so complete discs should be cut. In order to keep good contact of the cut edges with the media, softer medium is used and the tissue is pushed gently down into the media. For this reason, the leaf mid-rib is avoided when cutting leaf discs. Leaving the plates unsealed for a few days after sub-culture helps to reduce overgrowth of the tissue by the *A. tumefaciens*.

- Reducing the level of BAP induces multiple shoot formation from the callus tissue. Continued growth of shoots, green leaves and healthy root growth are symptoms of resistance to the selective agent and thus indicative of the transformed nature of the plantlet. Further evidence may be obtained by a variety of phenotypic and molecular tests detailed elsewhere in this book.

8.2.2
Transformation of Tomato Using *Agrobacterium tumefaciens*

Reagents

- GYPC with appropriate antibiotics
- 70% ethanol
- 5% w/v sodium hypochlorite solution, with one or two drops of detergent, e.g. Tween 20
- Sterile distilled water
- MS0 (Murashige and Skoog 1963, salts with Gamborg et al. 1968, B5 vitamins, 30 g/l sucrose), 1 mg/l zeatin solidified with 0.8% w/v agar in petri dishes
- MS0 liquid.
- MS0, 1 mg/l zeatin, 200 mg/l cefotaxime, 100 mg/l kanamycin, solidified with 0.8% w/v agar in deep petri dishes
- MS0, 0.1 mg/l zeatin, 200 mg/l cefotaxime, 100 mg/l kanamycin, solidified with 0.8% w/v agar in deep petri dishes
- MS0, 200 mg/l cefotaxime, 100 mg/l kanamycin, solidified with 0.8% w/v agar in sterile 60-ml tissue culture jars

Equipment

- Sterile deep (20–25 mm) petri dishes
- Sterile filter papers (same size as petri dishes)
- Blunt-end forceps
- Scalpel

- Nescofilm
- 60-ml pots
- Growth chamber

Procedure

This method is based on that from McCormick et al. (1991) and is used for varieties of tomato commonly used in research laboratories. We have used it routinely to transform a common cultivated variety, Grosse Lisse. All procedures must be carried out under sterile conditions, preferably in a class II downward flow microbiological containment bench.

1. Surface sterilize tomato seeds by washing briefly (1–2 min) in 70% ethanol and then soaking for 20 min in 5% sodium hypochlorite. Wash at least four times with sterile distilled water and allow to germinate for 8–10 days in deep petri dishes with moist filter papers (sterile) at 25 °C, 16-h days.

2. Grow up *A. tumefaciens* containing binary or co-integrate plasmid in GYPC, with appropriate antibiotics to maintain plasmid, overnight at 28 °C. Dilute 1:20 with liquid MSO.

3. Remove cotyledons from tomato seedlings, cut at least once across the widest part and place pieces with the lower epidermis uppermost on solid MSO with 1 mg/l zeatin. Cover surface of agar with cotyledon pieces.

4. Flood surface of medium with *A. tumefaciens* suspension. Allow to adsorb for 30 min, then pipette off as much of the suspension as possible. Culture for 2 days (25 °C, 16-h days).

5. Transfer pieces onto MSO with 1 mg/l zeatin and antibiotics. Seal with Nescofilm and culture for 3 weeks and then transfer onto MSO with 0.1 mg/l zeatin and antibiotics.

6. Transfer callus/shoots to MSO with 0.1 mg/l zeatin and antibiotics in petri dishes every 3 weeks until shoots emerge.

7. Transfer shoots onto MSO with antibiotics in 60-ml jars for rooting. Once roots are established, transfer plantlets to pasteurized soil, taking care to wash off all agar from roots. Keep plantlets in high humidity for the first week and then gradually reduce to normal greenhouse conditions.

Once established in soil, tomatoes are easy to maintain, and flower and set seed readily. Remember that local restrictions for the containment of genetically manipulated organisms must be complied with.

Notes and troubleshooting

- This procedure was optimized for the cultivar VF36 by McCormick et al. (1991), but we have found it works well for Grosse Lisse. Other cultivars may require optimization of the regeneration protocol, which should be done before transformation.

- Cotyledons provide a ready source of material, but they should only be used while compact and quite young. Once the first true leaves have emerged, the cotyledons are too old and respond poorly. *A. tumefaciens* infects only wounded tissue, so it is important to cut the cotyledon at the widest part in order to maximize the wounded area. Cutting the cotyledons under liquid (liquid MS0 for example) can help the infection process, but is not strictly necessary if the pieces are placed onto media soon after cutting.

- Adding acetosyringone to the *A. tumefaciens* cultures after dilution can increase the infectivity with some strains. A stock in ethanol at 1.48 mM is convenient; 25 μl per ml of culture can be added to give a final concentration of 375 μM. Some workers have reported that the effect is better if the plant tissue is incubated with acetosyringone for 3–4 h before transformation.

- Although it may appear strange to place the cotyledon pieces with the lower epidermis uppermost, better regeneration has been observed if oriented in this way. Callus is usually seen by the time of the first transfer onto reduced zeatin on up to 50% of the cotyledon pieces. One or two sub-cultures are usually necessary to obtain shoots after this. After three sub-cultures, it is unlikely that more shoots will be produced.

- As for tobacco, healthy green shoots with obvious root growth in the selectable agent indicates transformed growth. As rooting can be a problem in tissue culture, healthy-looking shoots which do not produce roots can be cultured further and tested to see if they are transformed. Addition of 0.1 mg/l NAA to the media can be used to promote rooting for these shoots or they can be dipped in a rooting compound before planting into soil directly.

8.3 Direct Gene Transfer

For plants which are not readily transformed with *A. tumefaciens* spp., direct gene transfer offers an alternative approach for transformation. In this case, the strategy is to introduce many copies of the naked DNA into the plant cell and then to select those cells which have incorporated the DNA into their genome. The integration of naked DNA into plant chromosomes is inefficient compared with transformation by *A. tumefaciens*, therefore the tissue culture system must be more efficient, i.e producing more cells which receive DNA and can subsequently be cultured and regenerated to plants.

Transformation of plants with naked DNA has two elements: the transfer of DNA into the plant cell, and the culture of the plant cells to identify and regenerate plants from those cells which have incorporated the introduced DNA. For the first element, the major barrier is the plant cell wall. This limits DNA both physically and chemically, and must either be breached or removed for DNA transfer. By using cell-wall-degrading enzymes the cell wall can be removed from plant cells and protoplasts isolated in an isotonic medium. The plasma membrane can then be permeabilized either by chemicals or by electric pulses to allow the uptake of DNA from the surrounding solution. Breaching the cell wall can also be achieved by micro-injection with fine micropipettes. Alternatively, success has now been achieved using microparticles coated with DNA and accelerated to high velocity (microprojectile bombardment). Protoplast transformation and microprojectile bombardment will be discussed more fully below.

8.3.1 Direct Gene Transfer to Rice Protoplasts

The regeneration of fertile plants from protoplasts of rice (Abdullah et al. 1986) was an important advance. The initial reports of regeneration were quickly followed by reports of transformation using direct gene transfer (e.g. Shimamoto et al. 1989), and now transgenic rice plants can be produced routinely and new improved cultivars are being developed in several countries using this approach. The basis for the successful regeneration of plants from protoplasts was the establishment of embryogenic suspension cultures from which many protoplasts capable of regeneration could be isolated. Many protoplasts are needed because the uptake and integration of naked DNA is not an efficient process and, even with improvements in culture, the regeneration of fertile plants occurs from few of the treated protoplasts. The initiation and maintenance of the suspension cultures is clearly a crucial step in direct gene transfer to obtain transgenic rice plants. In this chapter, the production of transgenic plants from embryogenic suspension cultures is described.

The culture of protoplasts often requires more complex media composition than when starting from explants for callus cultures. Combinations of different sugars are used in initial protoplast culture media to aid the initiation of cell wall synthesis and make up for compounds which may diffuse out of isolated protoplasts. Some additional factors that promote culture of these cells cannot yet be replaced by additions to the media. This problem may be solved by using nurse cell cultures (suspension cultures of either the same source as the protoplasts or a different genotype to enable identification of derived regenerated plants; e.g. Kyozuka et al. 1987), although culture of the protoplasts at a high initial density is more common.

The frequent need for a high initial plating density means that selection for transformed cells is usually delayed until initial divisions have taken place and colony formation is well established. This may cause a problem of chimaerism if the integration of DNA into the host cell chromosome takes place after cell division has occurred. For this reason, regeneration of plants via somatic embryogenesis is preferred to regeneration via organogenesis. It is thought that somatic embryos arise from fewer initial callus cells than spontaneously developing meristems involved in organogensis.

Reagents

- AA2 medium (Abdullah et al. 1986)
- CPW salts (Frearson et al. 1973) with 13% (w/v) mannitol (CPW13M)
- CPW13M, with 1% Cellulase Onozuka RS, 0.1% Pectolyase Y23
- KPR medium (Abdullah et al. 1986)
- Sterile, plasmid DNA
- KPR medium with 2.4% agarose
- Kanamycin

- N6 medium (Chu et al. 1975), 8% sucrose, 0.4% agarose
- MSD4, (MS salts and vitamins, 8% sucrose, 0.8% agar)
- Electroporation medium: 800 mg NaCl, 20 mg KCl, 20 mg KH_2PO_4/l + 10% glucose pH 7.1
- Fluorescein diacetate (FDA) 10 mg/ml in acetone

Equipment

- Sterile pipettes
- Sterile sieves, 500, 250, 64, 45, 30 μm internal diameter
- Sterile petri dishes
- Sterile 10 ml centrifuge tubes
- Bench-top centrifuge capable of providing 100 *g*
- Haemocytometer
- Electroporation apparatus, with chamber/cuvettes
- 3-cm diameter sterile petri dishes
- 5-cm diameter sterile petri dishes
- 27 °C shaking incubator
- 45 °C water bath
- 27 and 25 °C growth rooms

Procedure

This is based on the procedure of Zhang et al. (1988) and requires a good embryogenic suspension culture of rice. Varieties successfully transformed with this method include Taipei 309 and Nipponbare, which both show good tissue culture response.

1. Subculture suspension cultures at no more than 1:3 innoculum to fresh AA2 medium every 7 days with vigourous shaking at 27 °C in the dark.

2. Sieve suspension 4–6 days after subculture through 500 μm sieve placed in pre-weighed 9 cm sterile petri dish. Remove as much of the medium as possible from the small clumps and re-weigh to estimate weight of callus.

3. Add 20 ml CPW13M for every 1 g of callus in the dish and allow to pre-plasmolyze for 1 h at room temperature.

4. Remove the CPW13M and replace with CPW13M with enzymes (10 ml/1 g of callus) and incubate in the dark at 27 °C with gentle shaking (20–30 rpm) for 3–4 h.

5. Sieve the protoplast/enzyme mixture sequentially through 250-, 64-, 45- and 30-μm sieves into another 9-cm petri dish and wash the sieves with CPW13M.

6. Transfer the mixture to sterile centrifuge tubes and centrifuge at 80 *g* for 5 min to pellet the protoplasts.

7. Remove the supernatant and gently resuspend the protoplasts in CPW13M. Centrifuge at 80 *g* for 5 min again.

8. Repeat step **7** two more times and then combine protoplasts in one tube suspended in 10 ml of KPR.

9. Count the number of protoplasts using the haemocytometer and estimate the yield. Add 1 drop of FDA solution to 1 ml of KPR and mix 1 drop of this with an equal volume of protoplast suspension. Allow to settle for 5 min, then observe under epifluorescent microscopy. Viable protoplasts fluoresce bright green with long wavelength UV illumination.

10. Adjust the protoplasts to a density of 2×10^6 ml in electroporation medium, heat shock at 45 °C for 5 min then cool rapidly in ice.

11. Add sterile plasmid DNA to a concentration of 50 μg/ml, allow to stand on ice for at least 5 min and then electroporate in a capacitor discharge electroporator at 450 V/cm (distance between electrodes in cuvette) and a capacitance of 125 μF (Biorad Gene-Pulser apparatus).

12. Add an equal volume of KPR, 2.4% agarose at 45 °C (molten), mix well and plate in 3-cm petri dishes, 1–2 ml/dish.

13. Culture at 27 °C in the dark for 2 weeks, by which time colonies should be visible. Cut the agarose into four blocks and place into 5-cm petri dishes with 5 ml of liquid KPR containing selective agent at an appropriate concentration. One or two blocks can be placed into non-selective medium to check plating efficiency and colony formation.

14. Continue culture for 2 weeks then replace liquid medium and culture for a further 2 weeks. By now colonies should be large enough to be transferred individually to solid N6 in 9-cm petri dishes. Maintain selection for transgenic material.

15. Culture at 27 ° C in the dark, sub-culturing to fresh medium every 4 weeks until shoots are seen. Gently separate these shoots from the callus and place onto MSD4 in deep petri dishes. Try to remove as much callus as possible, but do not cut the shoot/callus piece, use two pairs of blunt forceps to break off any friable callus from the shoot base.

16. Culture at 25 °C in diffuse light until the shoot turns green and good roots emerge. Gently wash off the agar from these roots and pot into very wet compost/potting mix in a pot with no drainage holes, 1-l plastice beakers are a convenient size. Cover with plastic for the first week to keep the humidity high.

17. Grow at 28 °C and high light intensity, with no more than 14-h days. Flood the plants regularly to maintain a saturated water content of the soil. If plants are fertile, seeds will be produced about 4 months after planting and tillers will continue to be produced over several months.

Notes and troubleshooting

- The quality of the suspension culture is critical for plant regeneration from rice protoplasts. Maintain a high innoculum to fresh medium ratio (no more than 1:3) and look at the cultures every week to check morphology. Embryogenic cultures have small clumps of round cells with a dense cytoplasm and almost no vacuole. The presence of long thin cells with extensive vacuolation indicates poor suspension cultures. Even with a good subculture regime, cultures will lose their regenerative capacity over time: new suspension cultures should be started periodically.

- Many protoplasts (up to 2×10^7/g fresh weight) can be isolated with over 90% viability from good suspension cell cultures. If there are many non-viable cells, count only viable cells when adjusting density for electroporation. Plating efficiency (the number of protoplasts dividing to form colonies) depends on plating density. For any division and colony formation to occur these protoplasts must be cultured above 5×10^5/ml. A nurse culture system may be used with an aliquot from the original suspension culture feeding the dividing protoplasts to overcome this problem, but such sytems require specialized culture vessels to keep protoplasts separate from the nurse culture and selection is often difficult.

- Embedding the protoplasts in agarose solidified medium (agarose bead culture) simplifies exchange of media and selection. Plating efficiency can be monitored directly as can the survival and growth during selection. Individual colonies picked from the agarose blocks are daughter cells from a single protoplast and so the transformation frequency can be scored easily.

- Regeneration of plants from the colonies proceeds by somatic embryogenesis and so each shoot should be associated with root primordia. However, the orientation of most somatic embryos, with the shoot primordia on the outside of the callus and the root primordia inside, often makes it difficult to separate plantlets with shoots and roots free of callus. It is much better to leave some callus tissue associated with the plantlet than to damage the delicate root primordia with rough handling.

- Depending on the age and condition of the suspension culture, a proportion of the developing plantlets may be albino and will not become green once placed in the light. For the production of transgenic plants, it is not worth continuing with such material, and if the proportion of albinos becomes too great, it is probably not worth using that suspension culture for protoplast isolation any longer.

8.3.2
Particle Bombardment

The introduction of DNA into cells by mechanical means (often referred to as micro-injection) has been used routinely for transformation of reptile, insect, fish and mammalian tissues. After some initially promising results, this technique has not been widely applied to plant transformation. Major problems include breaching the plant cell wall, the low rate of injection that can be achieved and the low rate of integration of foreign DNA into the host cell genome after injection. A technique for introducing DNA through cell walls into many cells could overcome some of these problems. This was achieved by bombarding plant tissues with very small metal particles coated with DNA. In this technique, referred to as biolistics or particle bombardment, metal particles coated with DNA are accelerated to high velocity (usually under partial vacuum to reduce deceleration of the particles) and then penetrate plant tissues (Klein et al. 1987). Initially, this was achieved using an explosive charge (a blank 0.22 bullet) in a gun barrel driving a large plastic macrocarrier on which was placed a slurry of DNA-coated microparticles. The macrocarrier was subsequently stopped by a plate with a small hole in the centre (stopping plate) to allow passage of the microparticles into an evacuated chamber containing the plant tissues. Various other methods have since been developed to accelerate the particles, for example water vapour produced by capacitor discharge through a small water droplet (Christou et al. 1988), bursts of compressed air or helium produced by various means (e.g. Finer et al. 1992). However, all of these methods are based on the same principle of accelerating tungsten or gold microparticles to high velocity to enable them to penetrate plant cell walls.

The original technology, licensed to DuPont and now available through Biorad, has been copied by many groups, and equipment is generally available. Many machines will produce adequate results for transient expression (expression of non-integrated DNA a few days after bombardment) but the high efficiencies needed to obtain transgenic plants often require optimization of various parameters, and it is significant that the best results have generally been obtained by larger groups working with the most sophisticated technology. This is not to say that only the sophisticated equipment works, but it usually produces more consistent results and conditions can be optimized more easily.

Particle bombardment has been a major breakthrough in the transformation of cereals. Previously, only rice could be transformed routinely via protoplasts to produce fertile transgenic plants. However, the major cereals, maize, wheat and barley, have now all been transformed using particle bombardment (Gordon-Kamm et al. 1990; Vasil et al. 1992; Wan and Lemaux 1994). Two procedures are normally used which differ in the tissue used for bombardment. Where highly regenerable cell cultures can be obtained (for example some lines of maize and wheat), such cell cultures may be used as target tissue for bombardment. These cells are then cultured under strong selection pressure to select transgenic tissues. In contrast, with barley and other lines of maize and wheat, freshly isolated explants (immature embryos or microspore cultures) are bombarded directly, or after only a few days culture. These tissues are good sources of regenerable cell cultures, and callus tissue is induced from the bombarded material under selection and plants subsequently regenerated via somatic embryogenesis. Another approach, used initially with soybean and also successful with barley, is to bombard meristematic regions of shoots or imbibed seeds and, after regeneration without selection, to identify sectors of the plants which are transgenic and which produce transgenic flowers and seed.

The selective agent used is an important factor in the production of transgenic cereals, especially for particle bombardment. Selective agents usually used for dicots (the antibiotics kanamycin and hygromycin) are often not effective for the selection of transgenic monocot callus. This deficiency can be overcome in rice by using protoplasts, where selection by kanamycin is adequate. However, for the chimaeric tissue produced by bombardment of explants, more effective selective agents must be used. The most widely applied of these is the herbicide Basta (Bayer AG), which contains the active ingredient PPT (phosphinothricin) that effectively kills susceptible tissues at low concentrations. A resistance gene, *bar*, which encodes phosphinothricin acetyl transferase (PAT), confers resistance to Basta. This combination of selective agent and resistance gene is now frequently used in experiments on cereal transformation.

8.3.2.1
Transformation of Barley Immature Embryos by Particle Bombardment

Reagents

- Callus induction medium (MS salts, 30 g/l maltose, 1 mg/l thiamine-HCl, 0.25 g/l myo-inositol, 1 g/l casein hydrolysate, 0.69 g/l proline, 2.5 mg/l dicamba) solidified with 3.5 g/l Gelrite.
- Selection medium, callus induction medium with 5 mg/l Basta
- Regeneration medium (MS salts, 30 g/l maltose, 2 g/l glutamine) solidified with 3.5 g/l Gelrite
- 1.1 M $CaCl_2$
- 100 mM spermidine (free base)
- Tungsten particles, 1.1 μm median diameter (M17 grade) washed in ethanol and then rinsed three times in sterile distilled water. Store frozen in sterile distilled water.
- 2% sodium hypochlorite solution
- Sterile distilled water
- Plasmid DNA

Equipment

- Sterile laminar flow hood
- 25 °C incubator/growth room
- Particle bombardment device (DuPont PDS1000 or similar)
- Blank 0.22 cartridges
- Macrocarriers and stopping plates (sterilized by immersion in ethanol)

Procedure

This method is based on Wan and Lemaux (1994) and has been used successfully to obtain transgenic plants of the spring barley cultivar Golden Promise.

1. Surface sterilize immature barley caryopses by immersion in 2% sodium hypochlorite for 5 min followed by four washes in sterile distilled water. Seeds from the middle of the spike should be examined and the whole spike used 10–14 days after fertilization when the embryo is 1–2 mm long.
2. Using forceps and a scalpel, remove the pericarp and separate the embryo from the rest of the caryopsis. Place scutellum side uppermost on callus induction medium. Place 16 embryos together in the centre of a petri dish for bombardment.
3. Prepare the tungsten/DNA slurry as follows. Add 25 µl of plasmid DNA solution (1 mg/ml) to 1.25 mg of tungsten particles, vortex well and add 500 µl of 1.1 M $CaCl_2$. Vortex again and add 50 µl 100 mM spermidine, vortex and keep on ice for 10 min. Centrifuge for 5 min at 500 *g* and decant 550 µl of the supernatant. Prepare just prior to use and keep on ice until used.
4. Remove macrocarriers and stopping plates from ethanol and allow to dry in a sterile laminar air flow bench. Wash the inside of the bombardment apparatus with ethanol in the flow bench.
5. Place petri dish with embryos into apparatus at a suitable distance from the end of the barrel (10 cm is a good place to start). Place stopping plate into holder and insert into apparatus and close door.
6. Agitate slurry of tungsten/DNA either using a sonic probe or by dragging the tube vigourously across a grating and immediately pipette a 1-µl aliquot onto a macroprojectile. Insert macro-

projectile into barrel and place blank cartridge in behind the macroprojectile.

7. Assemble firing mechanism and evacuate chamber to 25–28 mm Hg before detonating cartridge. Release vacuum slowly, open chamber door and remove dish.

8. Culture embryos in the dark at 25 °C. Transfer bombarded embryos onto selection medium 1 day after bombardment.

9. Transfer tissue/callus pieces every 14 days onto fresh selection medium. Break up large pieces of callus and culture separately. Callus pieces showing more vigorous growth should be transferred separately. Resistant callus lines can be maintained by monthly subculture onto fresh selection medium.

10. Transfer embryogenic resistant callus to regeneration medium with 1 mg/l Basta and culture under fluorescent lights (45–55 μ Einsteins/m^2, 16-h days). Green plantlets should be transferred onto hormome-free callus induction medium in deep culture vessels as they reach 2 cm in height (approximately 4 weeks after transfer) and transferred to soil once roots have developed.

CAUTION. Although the apparatus is designed to withstand thc force of the explosive charge, care should be taken to follow the instructions supplied with the apparatus. In particular the cartridges should be handled carefully and under no circumstances should they be forced into the barrel or prised out from the top. The chamber and vacuum trap should be cleaned regularly to prevent the build up of unexploded powder.

Notes and roubleshooting

- Although care should be taken to avoid damaging the embryos during isolation, good results have recently been obtained by bisecting embryos longitudinally. Any surface of the embryo may be bombarded, but it appears that the best results are obtained by bombarding the scutellum and culturing with the scutellum uppermost. Some workers have suggested soaking the tissue to be bombarded in a high osmoticum for several hours prior to bombardment. What effect this may have is uncertain and it may be specific to certain tissues; however, particle bombardment is still developing rapidly and any suggestions should be considered.

- A common problem of particle preparation is clumping of the particles, which can cause severe tissue damage. This can be alleviated by using a sonic probe to disperse the particles. Alternatively a 100-μm mesh placed 2 cm above the tissue not only disperses large clumps of particles but also appears to increase the number of cells transiently expressing introduced genes (Gordon-Kamm et al. 1990). An alternative method of preparation of DNA-coated particles involves an ethanol precipitation of the DNA onto the particles with the resulting slurry being allowed to dry onto the macrocarrier before bombardment. This method is used more often for gold particles than for tungsten and is also recommended for different types of apparatus which use a burst of gas (water vapour of helium) to accelerate a piece of foil carrying the particles.

- Bombardment parameters can be optimized by carrying out experiments on transient expression with DNA encoding reporter genes. The reporter gene β-Glucuronidase (GUS) is usually used. Cells expressing GUS are visualized by staining after 1 or 2 days to determine the number of cells containing the DNA. This technique is often used to determine promoter sequences active in particular tissues (Kalla et al. 1994). See Chapter 9 for protocols.

- Culture and regeneration of barley and wheat from cultured immature embryos is a relatively routine procedure; however, the efficiency of transformation is low and so an efficient regeneration system has to be employed to counteract this. Many different media have been employed with varying results and those given here should not be regarded as definitive. The regeneration medium in particular has been the subject of much analysis with a reduction in inorganic nitrogen (nitrate and ammonium salts) and an increase in organic nitrogen (amino acids, especially glutamine). One important factor is the genotype of the plant used. Some varieties respond much better to culture and produce many more shoots.

8.4 Protoplast Fusion

The ability to regenerate whole plants from isolated protoplasts has enabled a range of fusion techniques to be developed. These include the combination of complete genomes from protoplasts of different species (e.g. Fish et al. 1988), combination of part of a "donor" genome with a recipient genome (asymmetric hybridization; e.g. Li et al. 1993) and transfer of organelles such as chloroplasts and mitochondria to create "cybrids" (e.g. Akagi et al. 1989), see also Chapter 5. In sexual crosses haploid genomic material is combined and segregation of genes occurs. In somatic hybridization the complete genomes may be combined, including those of organelles. In specific gene transfer, the gene of interest must be well characterized at the molecular level, but in asymmetric hybridization, useful traits may be transferred without any detailed knowledge of the genes that underlie the trait.

Examples of this approach include the production of somatic hybrids between the wild potato species *Solanum brevidens* and *S. tuberosum* to transfer resistance to potato virus diseases for introgression into potato breeding lines (Fish et al. 1988). Dihaploid lines of potato have been fused to produce tetraploid lines combining the characters of the parents (Waara et al. 1989). Partial genome transfer has been used to transfer varying amounts of genetic material from donor protoplasts (irradiated) to a recipient strain (Li et al. 1993), and novel combinations of mitochondria have been produced by fusing enucleated and intact protoplasts to produce male sterile lines of *Brassica* species for production of high-yielding F_1 hybrid seed (Barsby et al. 1987).

Plant protoplasts can be fused routinely using approaches based on chemical or physical treatments. Chemical fusogens make the membranes of protoplasts adhere and then fuse to form heterokaryons. In electrofusion, electric fields are used initially to bring protoplasts into close contact and then to induce pore formation and fusion.

Chemical fusogens which have been used include sodium nitrate, treatment with Ca^{2+} ions and high pH, polyethylene glycol (PEG), or a combination of the last two treatments. PEG causes protoplasts to agglutinate and become closely associated. Subsequent elution of the PEG, especially with a high Ca^{2+}/high pH solution, disrupts the plasma membrane and allows the protoplasts to fuse. PEG-based treatments have been used successfully to obtain many somatic hybrids, but there are some disadvantages in using chemical fusogens. The process is not easily controlled, and fusion frequencies are variable. There are also many different published protocols, most derived empirically, which are difficult to compare directly. These problems have led to development of the more recent technique of electrofusion.

When protoplasts are suspended in media of low conductivity, application of high-frequency electric current causes them to become polarized. This polarization causes the protoplasts to migrate to regions of higher field intensity, and when two cells meet during this migration, they are strongly attracted to each other because they act as dipoles. When a series of protoplasts align, they form stable chains parallel to the field lines, and application of one or a series of high voltage direct current pulses can induce pores in contacting membranes which can lead to their fusion. The alignment voltage and frequency and the time allowed for alignment can be controlled, together with the fusion voltage and duration, for precise control of the fusion conditions. This control of the fusion conditions enables a high yield of the desired fusion products to be obtained in mass fusion experiments. If high fusion frequencies are obtained, early identification of heterokaryons from parental protoplasts may not be necessary. Conditions can even be chosen specifically to eliminate unfused protoplasts of one parental type, making recovery of the desired hybrid products simpler.

Whatever the method of fusion, to ensure reproducibility, clean protoplast preparations are required. An efficient protoplast to plant regeneration system is also required and some knowledge of the compatability of hybrid genomes is desirable for the production of new plant varieties with specifically altered characteristics. This approach is now being applied to a range of crop plants, particularly *Solanum* spp., brassicas, forage legumes and rice.

8.4.1 Chemical Fusion

Reagents

- Solution A: 72 g/l glucose, 0.52 g/l $CaCl_2.2H_2O$, 95 mg/l KH_2PO_4
- Solution B: 500 g/l PEG (MW1540), 1.54 g/l $CaCl_2.2H_2O$, 95 mg/l KH_2PO_4, pH 5.5
- Solution C: 54 g/l glucose, 7.36 g/l $CaCl_2.2H_2O$, 3.75 g/l glycine, pH to 10.5 with NaOH. Make fresh before use.

  All solutions should be filter sterilized

Equipment

- Sterile glass centrifuge tubes
- Sterile cover slips
- 6-cm diameter petri dishes

- Sterile Pasteur pipettes
- Centrifuge capable of 200 *g*

Procedure

1. Protoplasts from each fusion partner are mixed at a 1:1 ratio in a glass centrifuge tube. The protoplast mixture is centrifuged at 200 *g* and the pellet resuspended in solution A to give a final density of 1×10^6/ml.
2. Pipette 150 µl of the protoplast solution onto a sterile cover slip placed in a 6-cm Petri dish. Allow the protoplasts to settle out.
3. Add 450 µl of the PEG solution (B) carefully to the protoplast droplet and incubate at room temperature for 40–50 min. This will cause the protoplasts to aggregate.
4. Carefully add 500 µl of the high Ca^{2+} high pH wash (solution C), and after 10 min add a further 500 µl. Washing induces protoplast fusion.
5. After 5 min add 1 ml of protoplast culture medium.
6. The protoplasts will remain adhering to the coverslip. Gently wash the protoplasts five times at 5-min intervals with a total of 10 ml of protoplast culture media.
7. Add culture media to produce the final protoplast plating density.

Notes and troubleshooting

- Different plant species require different culture media for division, colony formation and regeneration. Often, different media are required for the parent protoplasts. Selection against one parental type can be made to increase the proportion of heterokaryons which form colonies. Plating density is critical for division and colony formation, the protoplasts should be counted in a haemocytometer before adjusting to final plating density. If the protoplasts need to be concentrated to achieve this density, they should be collected in a tube and centrifuged gently (no more than 50 *g*), then resuspended in a volume chosen to give the required plating density.

8.4.2 Electrofusion

Reagents

- Electrofusion wash: 9% mannitol, 1 mM $CaCl_2$, filter sterilized

Equipment

- Electrofusion apparatus. Commercial apparatus can be purchased, for example GCA/Precision Scientific group, 3737 W. Cortland Street, Chicago Illinois 60647, USA or Krüss GmbH, Borsteler Chaussee 85–99a, Hamburg, Germany.
- Alternatively, equipment can be made incorporating an oscillator with a range of up to 1.5 MHz and 1–20 V peak-peak amplitude, together with either a pulse generator with stepwise control of pulse width (10 μs-1 ms) and amplitude (40–200 V) or a capacitor capable of delivering up to 800 V/cm with a time constant of around 1 ms.
- Electrofusion chambers. Analytical – parallel wire electrodes (0.1 mm diameter) 20 mm long with a separation of 1 mm fixed to a microscope slide; Preparative – lamellar electrodes (CFA404) Krüss GmbH; "Transferable" electrodes – a series of thin brass plates grouped as parallel sheets separated by 1 mm and dipped into suspensions of protoplasts in the square chambers of multiwell Sterilin culture dishes.

Procedure

Construct a pulse duration-fusion response curve

1. Pipette protoplasts at suitable density (5×10^5 to 2×10^6/ml) in electrofusion wash between electrodes on the glass slide of the analytical chamber.

2. Apply the alternating field at 1 MHz and and 10–20 V (the protoplasts must not rotate). If protoplasts do not align, increase the alignment voltage or wash the protoplasts again by centrifugation and resuspension in electrofusion wash to remove interfering salts.

3. Set pulse characteristics (e.g. 10 μs at 1500 V/cm) and apply the pulse when the average chain length is five to ten protoplasts.

4. Maintain the alternating field, but reduce the amplitude to about 30% of the former value to hold the chains in place while the characteristics are scored over the following 10–15 min. Count the

number of protoplasts aligned in chains that fuse, and the number involved in each fusion event.

5. From these data, calculate
   a) the percentage of protoplasts aligned
   b) the percentage of aligned protoplasts fused
   c) the percentage of 1/1 fusion products
   d) the percentage of total protoplasts involved in fusion events.

6. Collect data from pulses of increasing length (10, 20, 50, 100 µs) and amplitude (1.0, 1.5, 2.0 kV/cm), and from these the parameters giving optimum fusion are chosen – usually those for the shortest pulse duration giving maximum percentage fusion.

Large-scale electrofusions

7. Introduce 0.5 ml of protoplasts into the electrode chamber (Krüss chamber or Sterilin square multiwell culture vessel) and insert the electrodes.

8. Apply the fusion conditions determined above, and after 10 to 15 min dilute the protoplasts to their final culture density with appropriate culture medium.

Notes and troubleshooting

- If protoplasts do not align in the AC field, centrifuge and resuspend in fresh fusion medium. Lack of alignment is always caused by too high a salt concentration in the protoplast preparation. Because high fusion frequencies are possible using electrofusion, harsh conditions for the selection against one or both parental types are not normally required. The proportion of heterokaryons can often be increased by manipulation of the ratio of parental protoplasts. For example, for mixtures of protoplasts with different fusion characteristics, use the settings giving optimum fusion for the more responsive population, but then increase the ratio of the less responsive to the responsive to 1:5 or 1:10. For large-scale electrofusion, the square well culture dishes made by Sterilin can be used with transferable electrodes that can be moved from well to well, allowing many protoplasts to be fused rapidly.

References

Abdullah R, Cocking EC, Thompson JA (1986) Efficient plant regeneration from rice protoplasts through somatic embryogenesis. Biotechnology 4:1087–1090

Akagi H, Sakamoto M, Negishi T, Fujimura T (1989) Construction of rice cybrid plants. Mol Gen Genet 215:501–506

Barsby TL, Chuong PV, Yarrow SA (1987) The combination of Polima CMS and cytoplasmic triazine resistance in *Brassica napus*. Theor Appl Genet 73:809–814

Christou P, McCabe DE, Swain WF (1988) Stable transformation of soybean callus by DNA-coated gold particles. Plant Physiol 87:671–674

Chu CC, Wang CC, Sun CS, Hsu C, Yin KC, Chu CY, Bi FY (1975) Establishment of an efficient medium for anther culture of rice through comparative experiments on the nitrogen sources. Sci Sin 18:659–668

Ditta G, Stanfield S, Corbin D, Helinski D (1980) Broad host range DNA cloning system for gram-negative bacteria: construction of a gene bank of *Rhizobium meliloti*. Proc Natl Acad Sci USA 77:7347–7351

Finer JJ, Vain P, Jones MW, McMullen MD (1992) Development of the particle inflow gun for DNA delivery to plant cells. Plant Cell Rep 11:323–328

Fish F, Karp A, Jones MGK (1988) Production of somatic hybrids by electrofusion in *Solanum*. Theor Appl Genet 76:220–266

Frearson EM, Power JB, Cocking EC (1973) The isolation culture and regeneration of *Petunia* leaf protoplasts. Dev Biol 33:130–137

Gamborg OL, Miller RA, Ojima K (1968) Nutrient requirements of suspension cultures of soybean callus. Exp Cell Res 50:151–158

Gordon-Kamm WJ, Spencer TM, Mangano ML, Adams TR, Daines RJ, Start WG, O'Brien JV, Chambers SA, Adams WR, Willetts NG, Rice TB, Mackey CJ, Krueger RW, Kausch AP, Lemaux PG (1990) Transformation of maize cells and regeneration of fertile transgenic plants. Plant Cell 2:603–618

Hajdukiewicz P, Svab Z, Maliga P (1994) The small, versatile pPZP family of *Agrobacterium* binary vectors for plant transformation. Plant Mol Biol 25:989–994

Horsch RB, Fry J, Hoffman NL, Wallroth M, Eichholtz D, Rogers SG, Fraley RT (1985) A simple and general method for transferring genes into plants. Science 227:1229–1231

Jefferson RA, Burgess SM, Hirsh D (1986) β-Glucuronidase from *Escherichia coli* as a gene-fusion marker. Proc Natl Acad Sci USA 83:8447–8451

Kalla R, Shimamoto K, Potter R, Nielsen PS, Linnestad C, Olsen, O-A (1994) The promoter of the barley aleurone-specific gene encoding a putative 7-kDa lipid transfer protein confers aleurone cell-specific expression in transgenic rice. Plant J 6:849–860

Klein TM, Wolf ED, Wu R, Sanford JC (1987) High-velocity microprojectiles for delivering nucleic acids into living cells. Nature 327:70–73

Kyozuka J, Hayashi Y, Shimamoto K (1987) High frequency plant regeneration from rice protoplats by novel nurse culture methods. Mol Gen Genet 206:408–413

Lazo GR, Stein PA, Ludwig RA (1991) A DNA transformation competent *Arabidopsis* genomic library in *Agrobacterium*. Bio/Technology 9:963–968

Li Y-G, Tanner GJ, Delves AC, Larkin PJ (1993) Asymmetric hybrid plants between *Medicago sativa* L. (alfalfa, lucerne) and *Onobrychis viciifolia* Scop. (sainfoin). Theor Appl Genet 87:455–463

McCormick S (1991) Transformation of tomato. In: Lindsey K (ed) Plant tissue culture manual: fundamentals and applications. Kluwer, Dordrecht, Chap B6

Murashige T, Skoog F (1962) A revised medium for rapid growth and bioassays with tobacco tissue cultures. Physiol Plant 15:473–497

Park SH, Smith RH (1994) Rice transformation using *Agrobacterium* and the shoot apex. Poster No 2064, 4th Int Congr Plant molecular biology, Amsterdam, 19–24 June 1994

Raineri DM, Bottine P, Gordon MP, Nester EW (1990) *Agrobacterium*- mediated transformation of rice (*Oryza sativa L.*). Bio/Technology 8:33–38

Shen W-J, Forde BJ (1989) Efficient transformation of *Agrobacterium* spp. by high voltage electroporation. Nucleic Acids Res 20:8385

Shimamoto K, Terada R, Izawa T, Fujimoto H (1989) Fertile transgenic rice plants regenerated from transformed protoplasts. Nature 338:274–276

Vasil V, Castillo AM, Fromm ME, Vasil IK (1992) Herbicide-resistant fertile transgenic wheat plants obtained by microprojectile bombardment of regenerable callus. Bio/Technology 10:667–674

Waara S, Tegelstrom H, Wallin A, Eriksson T (1989) Somatic hybridisation between anther-derived dihaploid of potato (*Solanum tuberosum* L.) and the identification of hybrid plants by isozyme analysis. Theor Appl Genet 77:49–55

Wan Y, Lemaux PG (1994) Generation of large numbers of independantly transformed fertile barley plants. Plant Physiol 104:37–48

Zhang HM, Yang H, Rech EL, Golds TJ, Davis AS, Mulligan BJ, Cocking EC, Davey MR (1988) Transgenic rice plants produced by electroporation-mediated plasmid uptake into protoplasts. Plant Cell Rep 7:379–384

Chapter 9

Molecular Characterization of Transformed Plants

J.F. Topping and K. Lindsey

9.1 Introduction

The generation of genetically transformed plants is central to, and has indeed revolutionized, plant molecular biology. This is true for studies at both the fundamental and more applied levels of research. For researchers interested in unravelling the roles of specific genes in particular pathways of growth and development, the introduction into plants of foreign genes and gene promoters linked to reporter genes allows the detailed study of the temporal, spatial and quantitative expression of plant genes and the activities of associated regulatory sequences. In our own laboratory, we use these techniques in a programme of insertional mutagenesis to identify developmentally interesting genes (Topping et al. 1991; Lindsey et al. 1993; Topping et al. 1994). In the more applied area of genetic engineering, which is directed towards crop improvement, the introduction of novel genes encoding, for example, resistance to various pests and herbicides into economically important species, is in the long term likely to develop into a major branch of the plant breeding industry (Lindsey 1992). There are several well-characterized and very successful methods which are currently being employed to introduce specific genes and gene regulatory sequences into plants and these are described in Chapter 8 of this book.

In this chapter, we will first introduce the genes which are used most frequently as selectable markers. We will then go on to describe the selection and segregation procedures which identify transformants and allow determination of the number of genetic loci into which the foreign DNA has integrated. The molecular techniques which are used to ascertain beyond doubt that individual plants are transformed will then be described. Moving on from selectable to assayable or screenable genes, we will discuss the merits of the two most commonly used reporter genes, *uid*A (*gus*A, encoding β-glucuronidase; Jefferson et al. 1987) and *luc* (encoding firefly luciferase; Ow et al. 1986), and protocols for the assay of the respective gene products in plant cells will be presented.

9.2 Selectable Marker Genes

If the aim of a transformation experiment is to study the expression of a stably inserted foreign gene, it is necessary to be able to select the transformed plants/cells/protoplasts from a background of non-transformed material. This is achieved by incorporating, into the gene construct to be introduced, a selectable marker gene under the transcriptional regulation of a promoter sequence which is constitutively, or as near as possible constitutively, expressed. The most commonly used promoter is that of the 35S RNA gene of the cauliflower mosaic virus (CaMV) (known as the 35S promoter; Nagy et al. 1985). The selectable marker genes most frequently used are those encoding resistance to antibiotics or herbicides. Examples of genes encoding resistance to antibiotics include neomycin phosphotransferase, *npt*II (which confers resistance to kanamycin and G418 as well as neomycin; Pridmore 1987), hygromycin phosphotransferase, *hpt* (which confers resistance to hygromycin B; van den Elzen et al. 1985) and aminoglycoside adenylytransferase, *aad*A (which confers resistance to streptomycin; Chinault et al. 1986; Jones et al. 1987). Examples of genes encoding resistance to herbicides include phosphinothricin acetyltransferase, *bar* (which confers resistance to phosphinothricin and bialophos; De Block et al. 1987) and mutant 5-enolpyruvylshikimate-3-phosphate synthase, EPSPS (which confers resistance to glyphosate; Shah et al. 1986).

The choice of marker gene is often influenced by the plant species being transformed. For example, certain monocot species are not very sensitive to the more commonly used antibiotics such as kanamycin (Hauptmann et al. 1988). Therefore, prior to embarking on a transformation experiment, it is important to determine the effectiveness of a particular agent in allowing the unequivocal selection of transformed plant cells. Failure to do this can result in the regeneration of untransformed "escapes" and a good deal of wasted time. Many of the selective agents used do not entirely inhibit the growth of non-transformed cells/plants but may either stunt the growth of, and/or bleach the leaves of, non-transformed plants. Table 9.1 lists some commonly used selective agents, the range of concentrations at which they are commonly used, the effect on the wild-type plants and the corresponding gene which confers resistance.

Segregation Analysis

Depending on the nature of the experiment, it may be important to determine the number of transgenes which are present in a transformant. A first step towards this is to use a genetic approach, to determine the segregation of the introduced resistance marker in the progeny of a transformed line. Using such segregation analysis, it is possible to determine the number of non-linked genetic loci into which the transgene has inserted. Detailed molecular analysis is necessary,

Table 9.1. Some commonly used selective agents

Selective agent	Working concentration (mg/l)	Basis of selection	Gene	Reference
Kanamycin	10–750	Inhibits Ct[a] translation bleaching of sensitive tissues	*npt*II	Reynaerts et al. (1988)
G418 (Geneticin)	10–2000	Inhibits Ct[a] translation bleaching of sensitive tissues	*npt*II	Reynaerts et al. (1988)
Hygromycin B	10–400	Inhibits translation browning of sensitive tissues	*hpt*	Reynaerts et al. (1988)
Streptomycin	10–300	Inhibits Ct[a] translation bleaching of sensitive tissues	*aad*A	Jones et al. (1987)
Phosphinothricin and Bialaphos (Basta)	20–50	Competitive inhibitor of glutamine synthase	*bar*	De Block et al. (1987); Wohlleben et al. (1988); D'Halluin et al. (1992)
Chlorsulfuron	2–20	Inhibitor of acetolactate synthase	mutant *ALS*	D'Halluin et al. (1992)

[a]Chloroplastic

however, to estimate the number of transgenes present, since more than one copy of the foreign DNA can integrate at a single genetic locus; this is described later in this chapter. In most instances, it is desirable to study a line which contains a single copy of the foreign DNA. This is true particularly when the aim of the experiment is to study the expression of a native promoter-reporter gene construct in vivo. Multiple copies of the construct may interfere with each other by, for example, competition for trans-acting factors or, by the phenomenon of co-suppression (Matzke et al. 1989; Jorgensen 1990; Hobbs et al. 1993), both of which may affect the expression of the reporter gene. Reporter genes will be discussed in some detail later in this chapter.

Segregation analysis can only be carried out if the primary transformant (T_1) is fertile. The procedure involves collecting T_1 seeds, surface sterilizing the seed in a dilute bleach solution, and plating out on sterile growth media containing the appropriate selective agent. The T_2 generation is then scored for resistant and sensitive individuals. The expected ratios of resistant to sensitive seedlings for different numbers of insertional events, causing embryo- or seedling-lethal and non-lethal mutations, are presented in Table 9.2.

Reagents

- 70% (v/v) ethanol
- Sterilant: 5–10%, v/v (depending on species) commercial bleach solution containing 0.05% Tween 20 as a surfactant
- Sterile double distilled water (SDW)
- Germination medium: half-strength Murashige and Skoog medium

Table 9.2. Segregation frequencies of *npt*II in transgenic plants

Insertion Event	Ratio Kan^r: Kan^s
Insertion at one locus in redundant DNA	3:1
Insertion at two unlinked loci in redundant DNA	>3:1 <15:1
Insertion at two linked loci in redundant DNA	<15:1
Insertion at more than two loci of redundant DNA	>15:1
Insertion causing lethal gametophic mutation or Loss or inactivation of selectable marker gene	0:1
Insertion in one locus, causing an embryo-lethal mutation	2:1 (75% germination)

(Sigma), 10 g/l sucrose, 25–50 mg/l kanamycin sulphate (depending on species), 8 g/l agar.

Equipment

- Pasteur pipettes
- 9 cm Petri dishes
- Micropore tape (3 MM)/Nescofilm
- Culture room/incubator

Procedure

1. To surface sterilize the seeds:
   a) immerse seeds in 70% (v/v) ethanol for 30 s, then remove the ethanol with a Pasteur pipette,
   b) immerse the seeds in sterilant for 15–20 min.
2. Remove the bleach solution with a Pasteur pipette and wash the seeds six times in SDW.
3. Suspend the seeds in a drop of sterile water (if they are small, such as for tobacco or *Arabidopsis*) and, using a pipette, transfer them to the surface of agar plates containing germination medium. Spread out the seeds evenly (approx. 150 per 9-cm Petri dish) and seal the plates with Micropore tape (3 MM) or Nescofilm.
4. Incubate the plates for 2 days in dim light (approx. 22–25 °C) and then for a further 3–4 weeks in a standard light regime for cultures (at least 50 μmol/m^2/s) to allow root growth. Note that some species, such as *Arabidopsis*, require vernalization at 4 °C for 2–7 days to maximize germination frequency.
5. Count the numbers of seedlings that are kanamycin-resistant (full green) and -sensitive (bleached cotyledons, and lacking roots and true leaves).

9.3 Molecular Characterization of Transformants

The presence of foreign DNA within the genome must be demonstrated in order to confirm that a plant has been transformed. This is routinely carried out by Southern analysis. Southern analysis involves the extraction of total genomic DNA from the transformed plant, and digesting it with appropriate restriction endonucleases; it is then size-fractionated by agarose gel electrophoresis. Following the transfer of the fractionated DNA to a nylon membrane by capillary blotting (known as Southern blotting or Southern transfer), a radiolabelled fragment of the foreign DNA is used as a probe to detect, by hybridization, the presence of the homologous sequence within the plant genomic DNA. This technique allows not only the detection of foreign DNA but also an estimation of the number of copies and the arrangement of the foreign gene(s) within the plant genome. Detailed protocols for the extraction of genomic DNA, Southern blotting, radiolabelling DNA and hybridization have all been discussed in detail earlier in the book (see Sects. 1.1, 1.2, 1.3, 1.4).

Polymerase chain reaction (PCR) techniques can also be used to demonstrate the presence of foreign DNA. However, it is the case that, although PCR gives results rapidly, it is necessary to carry out a range of control experiments to avoid potentially false-positive results that may result from contaminating DNA sequences. However, PCR can sometimes be the only option when the amount of tissue available for analysis is limited. We recommend that if there is sufficient material, Southern analysis should be performed on transformed plants wherever possible. Nevertheless, the use of PCR to analyze transgenics is discussed in Section 2.1.

9.3.1 Interpreting the Results of Southern Analysis

By carefully selecting the restriction enzymes with which the plant genomic DNA is cut prior to electrophoresis, it is possible to gain information about the organization of a specific DNA sequence (e.g. foreign DNA) present within a particular transformant. The number of copies of that sequence can be estimated by two commonly used methods. In the first, genomic DNA is digested with restriction enzymes that cut twice within the target sequence, to excise an internal fragment. The copy number is estimated by calibrating the intensity (following autoradiography) of the internal fragment band against samples of known copy numbers of the foreign DNA run on the gel. It is important for accurate quantification that each track of the gel contains equal amounts of genomic DNA, and genomic DNA from untransformed plants must be added to the copy number control samples, to allow the differentiation of non-specific background hybridization. In the second method, the genomic DNA is digested with a restriction enzyme known to cut at only one site within the target DNA. Here, border fragments are generated, whereby the second site of cleavage is in the genomic DNA rather than the target DNA. Since the site of the second cleavage is expected to be different for each independent insertion event, the number of bands observed is a good indication of the number of copies present. Tandem arrays of a gene construct similarly can be diagnosed by cutting with single enzymes that cut once within the construct. The expected result in this case is a relatively intense band, representing artificially generated internal fragments within the tandem repeat, each of the same size, plus the border fragments, of different sizes. This experiment is important in order to differentiate between single and mutiple insertion events when segregation data have indicated a single locus insertion. In all cases, it is important to run a negative control comprising non-transformed genomic DNA, to check that, under the experimental conditions used, there is no background hybridization which could confuse the analysis.

9.4 Analysis of Reporter Gene Expression

A reporter gene encodes an enzyme which is readily assayable in vitro and, ideally, can be localized histochemically in vivo, to provide both quantitative and qualitative information. Over the years several genes have been utilized, including *cat* (chloramphenicol acetyltransferase; e.g. Lindsey and Jones 1987), *lac*Z (β-galactosidase; Helmer et al. 1984) and *npt*II (neomycin phosphostransferase; Reiss et al. 1984). However, with the exception of *npt*II, which is still in common use as a selectable marker, *gus*A (β-glucuronidase; Jefferson et al. 1987) and *luc* (firefly luciferase; Ow et al. 1986) have in recent years become the reporter genes of choice in the majority of laboratories.

It is useful, when studying the expression of a particular promoter-reporter gene fusion in a transient assay, to use the expression of a second reporter gene linked to a constitutively expressed promoter as an internal standard. This makes it possible to correct for variations in transformation efficiencies between the samples. For this reason, *gus*A and *luc* are routinely used together in transient assays when the DNA is introduced by electroporation or microprojectile bombardment (e.g. Twell et al. 1989; see also Chap. 8).

9.4.1
*gus*A (β-Glucuronidase) as a Reporter Gene

The *gus*A gene was isolated from *Escherichia coli* (Jefferson et al. 1987) and a number of gene cassettes are available (e.g. from Clontech) which allow the insertion of promoter regions adjacent to the *gus*A gene generating either transcriptional or translational gene fusions. β-glucuronidase (GUS) can be both assayed fluorimetrically (using 4-methylumbelliferone glucuronide, MUG, as substrate) and localized in situ histochemically (using 5-bromo-4-chloro-3-indolyl glucuronide, X-gluc, as substrate).

The GUS fluorimetric assay has several advantageous features: (1) it is extremely sensitive (expression within a single cell type of an organ can be detected); (2) the assay is simple, rapid and many samples can be processed simultaneously (up to 96 in a microtitre plate); (3) plant background activity is low and where it has been detected (e.g. in pollen of some species) it can be inhibited (Mascarenhas and Hamilton 1992; Wilkinson et al. 1994); and (4) the assay is quantitative and can readily be carried out in conjunction with a rapid protein assay to allow accurate quantification of the data.

9.4.1.1
Assay of β-Glucuronidase (GUS) Activity by Fluorimetry

This protocol can be used for assaying the transient expression of GUS in protoplasts following transient gene expression studies, or in intact cells from stably transformed plants. The only difference in protocol in each case is that, following transformation of protoplasts, thorough washing must be carried out to remove contaminating cell-wall-degrading enzymes, which have GUS-like activity and can interfere with the assay. Plant material can be frozen prior to the assay at –80 °C without any loss in activity. Do not, however, freeze samples in the GUS buffer, as this does result in significant loss of activity.

Reagents

- GUS buffer: 50 mM sodium phosphate buffer (pH 7.5), 10 mM β-mercaptoethanol, 0.1% (v/v) Triton X-100 and 1 mM EDTA
- Acid-washed sand (Sigma)
- 5mM 4-methylumbelliferone glucuronide, MUG (Sigma)
- 4-methylumbelliferone, 4-MU (Sigma)
- 0.2 M Na_2CO_3

Equipment

- 1.5-ml Eppendorf tubes
- Vortex mixer
- Microfuge
- Water bath, 37 °C
- Fluorescence spectrometer
- Micropipettes

Procedure

1. Grind the tissue (ca. 10^6 protoplasts or 10–100 mg tissue) in 200 µl of GUS buffer, in an Eppendorf tube with a small amount of acid-washed sand. Vortex thoroughly and leave on ice until all the samples have been ground.

2. Pellet the cell debris by centrifugation in a microfuge (12 krpm, 5 min).

3. Transfer 100 µl of the supernatant to a fresh Eppendorf tube which contains 0.7 ml of GUS buffer. Save the remainder for a protein assay (see below).

4. Equilibrate the samples at 37 °C for 5 min.

5. To each sample add 200 µl 5 mM MUG made up in GUS buffer (i.e. at a final concentration 1 mM in the reaction mix). The substrate can be made up in advance and stored in the dark at –20 °C.

6. Incubate at 37 °C.

7. Take 100 µl samples at various known time points (e.g. 0, 10, 20, 60 and 90 min) and add directly to 0.9 ml of 0.2 M Na_2CO_3. The Na_2CO_3 stops the reaction and enhances the fluorescence. The samples can be stored at room temperature in the dark until the last time points have been taken.

8. Assay the fluorescence by removing aliquots (200 µl) of the stopped reactions into a 96-well microtitre plate. Measure the fluorescence at an excitation wavelength of 365 nm and an emission wavelength of 455 nm.

9. A standard curve should be constructed by measuring the fluorescence of serial dilutions of the reaction product, 4-MU (Sigma). Thus, in conjunction with the protein estimation detailed below, the enzyme activity per sample can be expressed as pmoles 4-MU produced/mg protein/min.

9.4.1.2
Assay of Protein Content in Tissue Homogenate

Reagents
- Bradford's reagent (Bio-Rad)
- SDW
- BSA

Equipment
- Spectrophotometer
- Micropipettes

Procedure

1. Remove 5 μl of the crude cell extract (step **3** of the protocol above) and add directly to a 1-ml cuvette containing 200 μl of Bradford's reagent plus 795 μl of SDW; mix well.
2. Leave at room temperature for 5–10 min. The samples can be left for up to 1 h but no longer because the protein precipitates in the GUS buffer.
3. Read the absorbance, against a blank, at 595 nm. BSA can be used to construct a calibration curve.

9.4.2
Histochemical Localization of GUS Activity

In stably transformed plants, the expression pattern of *gus*A gene fusions can be analyzed very precisely both temporally and spatially by virtue of a simple histochemical assay. The GUS activity can be localized accurately in vivo using X-Gluc (Biosynth) as a substrate (see Fig. 9.1). The reaction takes place in two steps: first the glucuronic acid moiety is cleaved to generate a colourless indoxyl intermediate, which is in turn oxidized to an insoluble blue precipitate. The advantages in using GUS for this type of experiment include: (1) a low background GUS activity in the majority of plant tissues (but see Wilkinson et al. 1994); (2) the product is non-diffusible and accumulates within the plant cells where the gene is expressed; and (3) the substrate is readily taken up into the plant cells by simple diffusion or following vacuum infiltration of tissues.

There are, however, some disadvantages: (1) the assay is not readily suitable for use on living tissues, although some progress is being made in the development of non-lethal assay techniques (Kirchner et al.

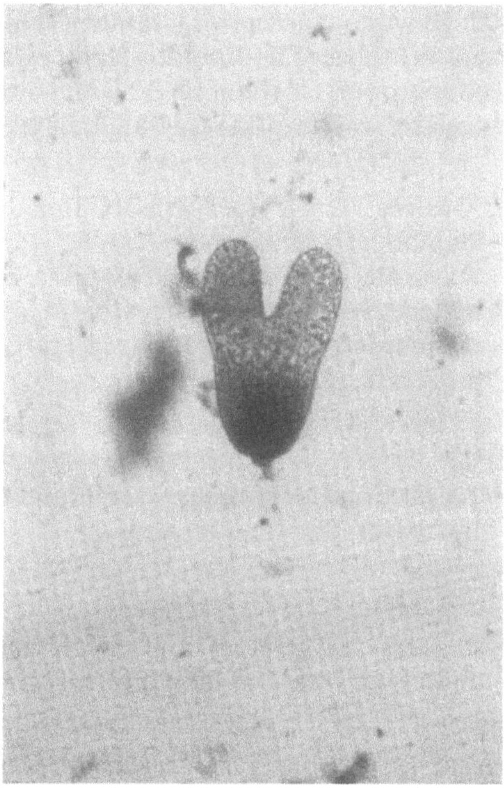

Fig. 9.1. Arabidopsis embryo stained with X-gluc to reveal GUS activity in the basal part of the embryonic axis, following transformation with a promoter trap vector (see Topping et al. 1994 for more details)

1993); (2) the enzyme and product are very stable and can therefore give an unrealistic picture of the steady-state abundance of the native protein for which GUS is being used as a reporter; (3) in tissues where there is a high level of GUS expression some leakage of the colourless reaction intermediate has been shown to occur; however, this can be minimized by the addition of either or both the oxidative agents potassium ferri- and ferrocyanide to a final concentration of up to 5 mM in the substrate solution (Stomp 1990; Mascarenhas and Hamilton 1992); and (4) as with all histochemistry, the results are not quantitative.

The preparation of the plant material for staining depends upon the individual tissues and organs involved. For example the roots, flowers and leaves of *Arabidopsis* and the roots of tobacco seedlings can be stained without any prior treatment. However, stems and leaves of, for example, tobacco and potato plants should be cut into thin (1–3-mm) sections prior to staining. Vacuum infiltration is sometimes useful, especially when working with larger tissue samples.

Reagents

- X-Gluc: 20 mM stock in N-N-dimethylformamide, stored in 100-μl aliquots at –20 °C.
- Substrate solution: 1 mM X-Gluc in 100 mM sodium phosphate buffer (pH 7) containing 10 mM EDTA, 1–5 mM each of potassium ferri- and ferrocyanide and
- 0.1% Triton X-100
- 95% (v/v) ethanol

Equipment

- 1.5-ml Eppendorf tubes/96-well microtitre plate
- Incubator, 37 °C
- Micropipettes

Procedure

1. Place the tissue in an Eppendorf tube or well of a microtitre plate and cover with 100–200 μl of substrate solution.
2. If the incubation is being carried out in a microtitre plate, ensure that the sample is well sealed to prevent the samples from drying out.
3. Incubate at 37 °C overnight (or for shorter periods if appropriate).
4. If necessary, soak the tissue in 95% ethanol to clear the tissue of chlorophyll; the blue precipitate is stable in ethanol.
5. If the tissue is to be fixed for sectioning, begin the fixation process after the overnight incubation.

9.4.3 *luc* (Firefly Luciferase) as a Reporter Gene

Firefly luciferase (LUC) is encoded by the *luc* gene (Ow et al. 1986), which is suitable as a marker in both transiently transfected and stably transformed plant cells. Plasmids containing *luc* are commercially available (e.g. from Promega). The quantitative assay procedure is usually carried out in cell extracts, but the in vivo detection of *luc* expression is also possible (Ow et al. 1986). A particular advantage of using luciferase as a reporter is that plants have essentially no background luciferase activity, providing a high signal-to-noise ratio. In this chapter we will describe only the in vitro assay for LUC activity.

Luciferase (Luc) activity. Firefly luciferase oxidizes the substrate luciferin, a heterocyclic carboxylic acid, in the presence of ATP, to produce of a flash of light, the intensity of which is directly proportional to the enzyme activity over several orders of magnitude and which is detected by a luminometer. Crude cell extracts are placed in the reaction chamber of the luminometer and the substrates luciferin and ATP are injected directly. The light emitted is quantified using a photon counter and a photomultiplier tube. In our laboratory we use a Clinilumat luminometer (Berthold Instruments UK Ltd) and the method given below is suitable for this type of machine.

Reagents

- Extraction buffer: 0.1 M potassium phosphate buffer (pH 7.5) plus 1 mM DTT, added immediately prior to use
- ATP buffer: 50 mM Hepes (pH 7.5), 20 mM $MgCl_2$, 10 mM ATP
- Luciferin solution: 0.1 M D-luciferin (Sigma) in 0.1 M potassium phosphate buffer, pH 7.5

Equipment

- Luminometer
- 1.5-ml Eppendorf tubes
- Microfuge
- Micropipettes

Procedure

1. Grind the tissue (10–100 mg) in 0.5 ml of extraction buffer in an Eppendorf tube with a small amount of acid-washed sand (Sigma). Store the samples on ice.
2. Centifuge the samples at 12 krpm for 2 min in a microfuge.
3. Transfer the supernatant to fresh tubes on ice. The samples can be frozen at –70 °C at this stage with no loss of enzymatic activity.
4. Make up the ATP buffer and place in one of the substrate chambers.
5. Make up the luciferin solution [0.1 M D-luciferin (Sigma) in 0.1 M potassium phosphate buffer, pH 7.5] and place in the second substrate chamber.

6. Dilute the cell extracts by mixing 50 μl of extract with 450 μl of ice cold extraction buffer. This dilution step can be omitted if the activity in the sample is found to be very low.
7. Add 25 μl of the diluted extract to a sample cuvette.
8. Start the machine running. The substrates for the reaction are injected automatically into the cuvette.
9. Total light emission (arbitary light units) is measured over a period of 10 seconds.
10. Repeat the assay using a fresh cuvette for each sample.
11. Express the data as arbitrary light units produced/10 sec/μg protein.

Note It is important to ensure that the plant tissues assayed are free of bacteria, e.g. agrobacteria, which will give false positive results. The presence of agrobacteria can be determined by mashing up a sample of the tissue to be tested and incubating it in Luria-Bertani medium (Sambrook et al. 1989) overnight at 29 °C.

References

Chinault AC, Blakesley VA, Roessler E, Willis DG, Smith CA, Cook RG, Fenwick RG (1986) Characterization of transferable plasmids from *Shigella flexneri* 2A that confers resistance to trimethoprim, streptomycin and sulfonamides. Plasmid 15:119–131

De Block M, Botterman J, Vandewiele M, Dockx J, Thoen C, Gosselé V, Rao Movva N, Thompson C, Van Montagu M, Leemans J (1987) Engineering herbicide resistance in plants by expression of a detoxifying enzyme. EMBO J 6:2513–2518

D'Halluin K, Bossut M, Bonne E, Mazur B, Leemans J, Botterman J (1992) Transformation of sugarbeet (*Beta vulgaris* L.) and evaluation of herbicide resistance in transgenic plants. Bio/Technology 10:309–314

Hauptmann RM, Vasil V, Ozias-Akins P, Tabaeizadeh Z, Rogers SG, Fraley RT, Horsch RB, Vasil IK (1988) Evaluation of selectable markers for obtaining stable transformants in the Gramineae. Plant Physiol 86:602–606

Helmer G, Casadaban M, Bevan MW, Kayes L, Chilton MD (1984) A new chimeric gene as a marker for plant transformation: the expression of *Escherichia coli* β-galactosidase in sunflower and tobacco cells. Bio/Technology 2:520–527

Hobbs SLA, Warkentin TD, DeLong CMO (1993) Transgene copy number can be positively or negatively associated with transgene expression. Plant Mol Biol 21:17–26

Jefferson RA, Kavanagh TA, Bevan MW (1987) GUS fusions: β-glucuronidase as a sensitive and versatile gene fusion marker in higher plants. EMBO J 6:3301–3307

Jones JDG, Svab Z, Harper EC, Hurwitz CD, Maliga P (1987) A dominant nuclear streptomycin resistance marker for plant cell transformation. Mol Gen Genet 210:86–91

Jorgensen R (1990) Altered gene expression in plants due to trans-interactions between homologous genes. Trends Biotechnol 8:340–344

Kirchner G, Kinslow CJ, Bloom GC, Taylor DW (1993) Nonlethal assay system of β-glucuronidase activity in transgenic roots of tobacco. Plant Mol Biol Rep 11:320–325

Lindsey K (1992) Genetic manipulation of crop plants. J Biotechnol 26:1–28

Lindsey K, Jones MGK (1987) Transient gene expression in electroporated protoplasts and intact cells of sugar beet. Plant Mol Biol 10:43–52

Lindsey K, Wei W, Clarke MC, McArdle HF, Rooke LM, Topping JF (1993) Tagging genomic sequences that direct transgene expression by activation of a promoter trap in plants. Transgenic Res 2:33–47

Mascarenhas JP, Hamilton DA (1992) Artifacts in the localization of GUS activity in anthers of petunia transformed with a CaMV 35S-GUS construct. Plant J 2:405–408

Matzke MA, Primig M, Trnovsky J, Matzke AJM (1989) Reversible methylation and inactivation of marker genes in sequentially transformed tobacco plants. EMBO J 8:643–649

Nagy F, Odell JT, Morelli G, Chua N-H (1985) Properties of expression of the 35S promoter from CaMV in transgenic tobacco plants. In: Zaitlin M, Day P, Hollaender A (eds) Biotechnology in plant science: relevance to agriculture in the eighties. Academic Press, Orlando, pp 227–235

Ow DW, Wood KV, DeLuca M, De Wet JR, Helinski DR, Howell SH (1986) Transient and stable expression of the firefly luciferase gene in plant cells and transgenic plants. Science 234:856–859

Pridmore RD (1987) New and versatile cloning vectors with kanamycin-resistance marker. Gene 56:309–312

Reiss B, Sprengel R, Will H, Schaller H (1984) A new and sensitive method for quantitative and qualitative assay of neomycin phosphotransferase in crude cell extracts. Gene 30:211–218

Reynaerts A, De Block M, Hernalsteens J-P, van Montagu M (1988) Selectable and screenable markers. In: Gelvin SB, Schilperoort RA (eds) Plant molecular biology manual Kluwer, Dordrecht A9:1–16

Sambrook J, Fritsch EF, Maniatis T (eds) (1989) Molecular cloning: a laboratory manual 2nd edn. Cold Spring Harbor Laboratory Press, Cold Spring Harbor

Shah DM, Horsch RB, Klee HJ, Kishore GM, Winter JA, Tumer NE, Hironaka CM, Sanders PR, Gasser CS, Aykent S, Siegel NR, Rogers SG, Frayley RT (1986) Engineering herbicide tolerance in transgenic plants. Science 233:478–481

Stomp A-M (1990) Use of X-Gluc for histochemical localisation of glucuronidase. In: Editorial comments. United State Biochemical, Cleveland, p 5

Topping JF, Wei W, Lindsey K (1991) Functional tagging of regulatory elements in the plant genome. Development 112:1009–1019

Topping JF, Agyeman F, Henricot B, Lindsey K (1994) Identification of molecular markers of embryogenesis in Arabidopsis thaliana by promoter trapping. Plant J 5:895–903

Twell D, Klein TM, Fromm ME, McCormick S (1989) Transient expression of chimeric genes delivered into pollen by microprojectile bombardment. Plant Physiol 91:1270–1274

Van den Elzen PJM, Townsend J, Lee KY, Bedbrook JR (1985) A chimaeric hygromycin resistance gene as a selectable marker in plant cells. Plant Mol Biol 5:299–302

Wilkinson JE, Twell D, Lindsey K (1994) Methanol does not specifically inhibit endogenous β-glucuronidase (GUS) activity. Plant Sci 97:61–67

Wohlleben W, Arnold W, Broer I, Hillemann D, Strauch E, Pühler A (1988) Nucleotide sequence of the phosphinothricine N-acetyltransferase gene from Streptomyces viridochromogenes Tü 494 and its expression in Nicotiana tabacum. Gene 70:25–37

Chapter 10

Molecular Characterization of Somatic Hybrids

Y.-S. Xu and E. Pehu

10.1 Introduction

Gene flow between sexually incompatible plant species is restricted by the evolutionary boundaries in sexual crosses. Somatic cell hybridization and DNA transformation have opened new horizons for combining genes from unrelated species (Jain et al. 1992; Lindsey 1992). If the gene of interest has been cloned, transformation systems such as the *Agrobacterium* method (Horsch et al. 1987), direct gene transfer (Jain 1993), biolistic delivery system (Lindsey 1992) and microinjection (Neuhaus et al. 1987) provide the most efficient way of introducing it into a foreign host genome. When the goal is to transfer several genes or when the genes of interest are unidentified, protoplast fusion and related techniques, however, offer an alternative approach for alien gene transfer (see also Chaps. 8 and 9).

The selection of heterokaryons from the mixture of self-fused and unfused protoplasts following protoplast fusion becomes necessary because the portion of desired fusion products is often less than 10% of the total protoplast population (Naton et al. 1992). Several selection strategies have been developed. The most widely used ones include fluorescence labelling of protoplasts prior to fusion and selection of heterocaryons by fluorescence activated cell-sorter (Galbraith 1984; Glimelius et al. 1986; Boromova and Knopf 1991) or micromanipulator (Kao 1977; Sundberg and Glimelius 1986; Waara et al. 1989); selection based on hybrid vigour (Shieder 1978; Debenath and Wenzel 1987; Fish et al. 1987, 1988; Deimling et al. 1988), intermediate morphology (Binding et al. 1982; Gleddie et al. 1986; Kobayash et al. 1991) or recently complementary antibiotic resistance, whereby parental lines transformed with two antibiotic resistance genes are selected after fusion (Komari et al. 1989; Masson et al. 1989; Thomas et al. 1990).

In practice, a combination of the above-listed selection criteria are often used.

Calli or plantlets obtained by one of the selection procedures described above, are only putative somatic hybrids. Therefore, it is essential to confirm their hybrid nature. This is done because parental material may slip through the selection procedure, elimination of chromosomes from the initial fusion product may take place and somaclonal variation may occur. Reliable methods are therefore required to verify nuclear hybridity of the fusion products and to analyze the nuclear and cytoplasmic genome composition of the hybrids.

10.2 General Characterization of Somatic Hybrids

Phenotypic traits can be very helpful in characterizing somatic hybrids. Thus, morphological characters such as leaf shape, trichome types, pigmentation and leaf morphology have often been used to identify somatic hybrids (Austin et al. 1985; Fish et al. 1987, 1988; Jain et al. 1988; Pehu et al. 1989). Morphological assessments are often essential, and important to evaluate the usefulness of the plants obtained.

Another very informative and relatively easy method for characterizing hybrids is cytogenetic analysis. Chromosome counts, together with morphological characterization, can give supporting evidence to the hybrid nature of the fusion products. Detailed studies on the composition of parental chromosomes of somatic hybrids can be carried out based on chromosomal morphology where parental karyotypes are distinct. Cytogenetic studies of somatic hybrids between *Arabidopsis thaliana* and *Brassica campestris* (Gleba and Hoffmann 1978), *Atropa belladonna* and *Nicotiana chinensis* (Gleba et al. 1982) and *A. belladonna* and *N. plumbaginifolia* (Gleba et al. 1988) detected possible introgression of genomic sequences between the two species' chromosomes. However, cytology is handicapped in terms of rapid, large-scale screening of somatic hybrids. Thus, flow cytometric measurement of nuclear DNA content is an invaluable technique for the evaluation of the cytogenetic stability of hybrid plants (Fahlesson et al. 1988; Chaput et al. 1990); see Chapter 12 for cytological characterization.

Before the development of RFLP and RAPD analyses methods, polymorphism shown in electrophoretic banding patterns of isozymes was a regularly used method for the identification of inter- and intraspecific somatic hybrids (Austin et al. 1985; Debnath et al. 1987; Fish et al. 1987, 1988; Deimling et al. 1988; Jain et al. 1988; Famelaer et al. 1989; Waara et al. 1989, 1991).

The focus of the present chapter is on the recently developed molecular methods of characterizing somatic hybrids and thus the reader is referred to the papers listed in the above section for detailed description of morphological, cytogenetic and isozyme analyses.

10.3 Molecular Characterization of Somatic Hybrids

10.3.1 Nuclear Genome

The commonly used methods outlined above (morphology, isozyme patterns, cytogenetic analysis) do not provide much detailed information on the genetic composition of the hybrids, and these methods rely on the use of full-sized plants. It is, for example, difficult to analyze the loss or doubling of specific chromosomes or chromosomal fragments in aneuploid or polyploid somatic hybrids. Recent developments in plant molecular genetics have provided powerful new tools for the characterization, identification and genetic analysis of somatic hybrids. The nuclear constitution of somatic hybrids has been studied by random amplified polymorphic DNA (RAPD) markers (Baird et al. 1992; Xu et al. 1993a), restriction fragment length polymorphism (RFLP) analysis (Pehu et al. 1989; Bates 1992) and highly repetitive DNA sequences such as species-specific DNA probes (Saul and Potrykus 1984; Pehu et al. 1990; Piastuch and Bates 1990; Puite and Schaart 1993) and rDNA (Gleba et al. 1988; Thomas et al. 1990).

Somatic hybrid plants between *Solanum tuberosum* and *S. brevidens* obtained according to the methods of Fish et al. (1987, 1988) and Xu et al. (1993b) have been used to develop the procedures described above in our laboratory. These are discussed in more detail below.

10.3.1.1
RAPDs

The methodology behind producing RAPDs, including trouble-shooting tips, has been extensively discussed in Chapter 6 and so will not be expanded upon here.

In our study, each of the five primers directed the amplification of a genome-specific "fingerprint" for both of the fusion parents. The somatic hybrids showed the sum of the parental banding profiles, whereas regenerants from one of the parents had the same banding pattern or one similar to that of the parent. This method requires only very small amounts of plant material (5–10 mg of leaf material is sufficient for 50 PCR reactions). As with RFLP analysis, RAPDs are DNA-based and directly reflect the hybridity of the nuclear genome. The advantages that RAPDs have over the other genetic tests are that the experiments are simple and quick to perform. The usefulness of this approach in crops will, to a certain extent, be dependent on the level of polymorphism detected with RAPD markers (Fig. 10.1).

Although the banding patterns of symmetric somatic hybrids are expected to be the sum of the bands of both parents, unpredictable banding patterns or disappearance of some of parental weak bands of the symmetric somatic hybrids can be found (Xu et al. 1993a). The reasons might be the primer binding competition (Williams et al. 1990b) or somaclonal variation. It is, therefore, important to find suitable primers which can produce a sum of both parents banding patterns in somatic hybrids (Xu et al. 1993a). Furthermore, it is good for a single primer to generate more than one polymorphic amplification product from each parent (Baird et al. 1992).

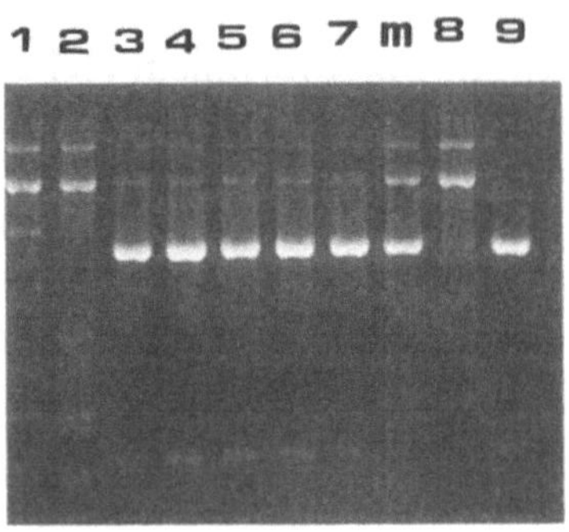

Fig. 10.1. RAPD profiles generated by Operon primer A10. *Lanes 1, 2* Regenerants from one fusion parent *S.tuberosum* PDH40; *lanes 3–7* somatic hybrids; *lanes 8–9* fusion parents. (Xu et al. 1993a)

10.3.1.2
RFLPs

This procedure involves the extraction of total genomic DNA, restriction enzyme digestion of the DNA, Southern blotting, probe isolation and labelling plus detection of probes on Southern blots. These techniques have been discussed elsewhere in the book, mainly in Chapters 1 and 2. The theory behind RFLP analysis has been extensively discussed in Chapter 7.

A complete linkage group analysis of somatic hybrids of plant species of which a molecular map is available, such as some *Lycopersicon* and *Solanum* species, can be performed with linkage group-specific RFLP markers. Nearly saturated linkage maps of potato and tomato have been produced, which contain more than 1000 markers on all 12 chromosomes of the complement (Gebhardt et al. 1991; Tanksley et al. 1992). Utilizing the RFLP map, Williams et al. (1990a) used a selection of linkage group specific RFLP markers to distinguish the chromosomes of *S. brevidens* and *S. tuberosum* in interspecific somatic hybrids. In their study, the symmetric somatic hybrids of *S. tuberosum* and *S. brevidens* had all of the chromosomes of both parents. An internal deletion was detected on *S. brevidens* chromosome 8 (Williams et al. 1990a). In our study, the same approach was employed to analyze the genomic composition of asymmetric hybrids between *S. tuberosum* and irradiated *S. brevidens* (Xu and Pehu 1993). The use of RFLP probes for analysis of hybrids made it possible to account for the presence or absence of individual chromosomes and chromosome arms.

Analysis of the Results

RFLP analysis of somatic hybrids using nuclear probes can be used to answer three questions:

- Are the putative hybrids true nuclear hybrids ?
- What is the parental nuclear genome dosage in the hybrids ?
- Which parental chromosomes are present in the hybrids ?

If the aim is to verify nuclear hybridity, almost any low-copy number nuclear sequence could be used. If the focus is on studying the nuclear genome dosage in the hybrids, laser densitometer analysis can be used to determine the relative intensities of the diagnostic parental bands (Pehu et al. 1989).

In asymmetric hybridization the aim is to combine only a part of the donor genome with the genome of the recipient. The fragmentation of the donor genome is most commonly effected by gamma-irradiation. In these cases, it is of great interest to know which chromosomes are retained in the hybrids. For such a detailed analysis a set of linkage group specific RFLP markers need to be available. For example, we used genomic clones (TG clones) representing all of the 12 linkage groups of potato to study the chromosome composition of somatic hybrids between *S. tuberosum* and *S. brevidens* (Xu and Pehu 1993). The first step in the analysis is to survey polymorphisms between the parental species. For example, HindIII digestion of potato DNA revealed a species-specific polymorphism for the TG43 sequence, located on chromosome 10: a single fragment was observed in DNA isolated from *S. brevidens*, while another single fragment was observed in DNA isolated from *S. tuberosum* following probing with TG43. The enzyme/probe combination having differences between the parental species is then used to study the somatic hybrids. If both diagnostic parental fragments or patterns are present, the hybrid contains at least one chromosome of both species; if it lacks the bands of one of the parental lines, the hybrid carries that particular chromosome of one parental species only (Xu and Pehu 1993). In addition to presence/absence data, linkage group specific RFLPs can also be used to study sequence rearrangements in the chromosomes of the hybrids (Wijbrandi et al. 1990; Xu and Pehu 1993).

RFLP-analysis of somatic hybrids can also result in banding patterns which are not present in either of the parental species; however, in our species combination this was quite rare. The most likely reason is chromosomal rearrangements generated by somaclonal variation.

To identify clear species-specific RFLP patterns several enzyme/probe combinations should be surveyed.

For major crop species such as rice (*Oryza sativa*), tomato (*Lycopersicon esculentum*) and wheat (*Triticum aestivum*), RFLP maps are already available (Graner and Wenzel 1992). So far, linkage group-specific RFLP analysis has been carried out on hybrids between *Lycopersicon esculentum* and irradiated. *peruvianum*, and *L. pennellii* (Wijbrandi et al. 1990; Melzer and O'Connell 1990, 1992) and *Solanum tuberosum* and irradiated *S. brevidens* (Xu and Pehu 1993).

10.3.1.3 *Species-Specific Repetitive DNA*

Repetitive DNA has been shown to evolve rapidly and to have a high degree of sequence variation between species (Lapitan 1992). Southern hybridization of species-specific repetitive DNA sequences with digested genomic DNA provides a quick and reliable way to identify nuclear hybridity of somatic hybrids (Saul and Potrykus 1984). Simple procedures such as squash-blots and dot-blots to identify nuclear hybridity of somatic hybrids using the species-specific sequences have been developed (Schweizer et al. 1988; Pehu et al. 1990; Xu et al. 1993b). In dot-blot hybridizations on genomic DNA from hybrids, species-specific repetitive DNA can be used to quantify the relative contribution of each parent's genome in the somatic hybrids (Imamura et al. 1987; Piastuch and Bates 1990; Wolters et al. 1991; Derks et al. 1992). By in situ hybridization of species-specific probes to hybrid chromosome spreads, it is possible to identify parental chromosomes, chromosome fragments, and translocations (Piastuch and Bates 1990; Itoh et al. 1991). Recently, a new technique named genomic in situ hybridization (GISH), where total genomic DNA is used as a probe, is used to determine the parental origin of somatic hybrids (Parokonny et al. 1992; See Chap. 12).

The procedure involves several stages: extraction of total DNA, restiction enzyme digestion, cloning and transformation of bacteria, colony blotting etc. All of these techniques have been previously described elsewhere in the book and therefore details will not be given here.

Isolation of Species-Specific Sequences

Extract total DNA from 2 g fresh weight of frozen leaf tissue from glasshouse grown plants by the method of Dellaporta et al. (1983). Digest total DNA from both parental species using HaeIII or similar 4-cutter and clone into an appropriate site in the plasmid, e.g. SmaI of pUC plasmids. Use the ligation products to transform *E. coli* and select recombinant clones on plates containing ampicillin (100 μg/ml) and 2% X-Gal and transfer a copy of them to two identical nitrocellulose/nylon filters.

Reagents

- Prehybridization/hybridization solution: 50% formamide/5x Denhardt's/5x SSPE/0.1% SDS/100 µg/ml denatured salmon sperm
- Hybridization washes: 1x SSC/0.1% SDS
- Autoradiography film
- Solutions for developing autoradiography film
- Nitrocellulose/nylon membrane
- Denaturation solution: 1.5 M NaCl/0.5 M NaOH
- Neutralization solution: 1.5 M NaCl/0.5 M Tris-HCl pH 7.2/ 0.001 M EDTA
- Nick translation kit
- α-32 dATP

Equipment

- 15 °C water bath
- Hybridization oven
- Autoradiographic cassettes
- Slot-blot apparatus

Procedure

1. Hybridize one set of the nitrocellulose/nylon filters containing the recombinant clones to total DNA (and purified chloroplast DNA if available) of one of the parental species and the other set with total DNA of the other species previously nick-translated to a specific activity of 53×10^6 cpm/µg DNA.

2. Carry out the prehybridization and hybridization in 50% formamide (5x Denhardt's, 5x SSPE, 0.1% SDS, 100 µg/ml denatured salmon sperm DNA) at 42 °C for 4 h and at 37 °C overnight, respectively.

3. Wash the filters in 1x SSC, 0.1% SDS at 62 °C for 1 h.

4. Select putative species-specific clones, which give a strong signal when probed with homologous DNA but no signal when probed with heterologous DNA.

5. To verify the species specificity, bind 3–4 µg of the plasmid DNA carrying the putative specific sequence onto a nitrocellulose/nylon membrane using a slot-blot apparatus according to the manufacturers instructions. Place the filter on a pad of 3MM paper soaked in denaturation solution for 5 min followed by 1 min of neutralization solution. Air-dry and fix the DNA to the membrane according to the manufacturer's recommendations.

6. Probe the slot-blot filter sequentially with nick-translated total DNA of each fusion parent. Carry out the prehybridization and hybridization in 50% formamide (5x Denhardt's, 5x SSPE, 0.1% SDS, 100 µg/ml denatured salmon sperm DNA) at 42 °C for 4 h and at 37 °C overnight, respectively. Wash the filters in 1x SSC, 0.1% SDS at 62 °C for 1 h. Select putative species-specific clones, which give a strong signal when probed with homologous DNA but no signal when probed with heterologous DNA.
7. If the sequence gives a strong signal when probed with the parental total DNA of the species it is derived from and no or a faint signal with the total DNA of the other species, select it for further analysis of hybrids.

The isolated species-specific probes can be further characterized regarding their copy number, nucleotide sequence and physical localization in the complement using in situ hybridization (Imamura et al. 1987; Pehu et al. 1990; Piastuch and Bates 1990; Itoh et al. 1991).

Quantitative Analysis of the Nuclear Composition by Dot Blotting

This method is a slight modification of Wolters et al. (1991).

Reagents

- SDW
- Nylon/nitrocellulose membranes
- Denaturation solution: 1.5 M NaCl/0.5 M NaOH
- Neutralization solution: 1.5 M NaCl/0.5 M Tris-HCl pH 7.2/ 0.001 M EDTA
- Random primed labelling kit
- α–32 dATP
- Autoradiography film

Equipment

- Spectrophotometer
- Dot-blot apparatus
- Micropipettes (1–200 µl, 200 µl-1 ml vol)
- 1.5 ml Eppendorftubes
- 37 °C water bath
- Tri-carb liquid scintillation analyzer

Procedure

1. Measure the parental and hybrid DNA concentrations to be analyzed by spectrophotometer.

2. Transfer 12 concentrations, ranging from 0 ng DNA/dot to 400 ng DNA/dot, of DNA of the parents and DNA of somatic hybrids (200 ng/dot, three replicates) to a nitrocellulose/nylon membrane using a dot-blot appartus. Place the filter on a pad of 3-MM paper soaked in denaturation solution for 5 min followed by 1 min of neutralization solution. Air-dry and fix the DNA to the membrane according to the manufacturer's recommendations. Prepare two identical dot-blots.
3. Label the species-specific clone with [α-^{32}P]dATP using the Random Primed DNA Labeling Kit.
4. Probe one of the filters with the specific probe of one species and the other with the specific probe of the other species.
5. Cut the dot from the membrane and measure radioactivity with a Tri-carb liquid scintillation analyzer after exposure to an X-ray film, correct the reading for background.
6. Make a calibration plot of the radioactivity per dot in relation to the amount of DNA from the two species, separately.
7. Estimate the amount of both parental DNAs in the hybrids per dot with the calibration plots for both species-specific probes, and determine the percentage of parental nuclear DNAs in the fusion products.

Notes and troubleshooting

- The size of the libraries to search for a clone is somewhat dependent on the phylogenetic distance of the parental species. After review of the available literature on isolation of species-specific sequences and based on our own experience, we recommend that a library of approximately 1000 recombinant clones for each species be used.
- The species specificity is a function of the stringency of hybridization and washing conditions. Based on our experience, it is recommended that when characterizing the specificity of the probes a series of washes allowing 30%, 15% and 5% mismatch be used.
- Analysis of the nuclear composition by dot-blotting can also be carried out by exposing the hybridized filters to autoradiographic film and analyzing the black exposed areas by photodensitometry.
- The physical localization of the species-specific sequences is of importance in relation to the degree of accuracy required to identify the genome composition of the hybrids. If the sequences are used

for a quick "squash-blot"-type analysis of putative, symmetric hybrids any repetitive, species-specific sequence, irrespective of its chromosomal localization, could be used (Pehu et al. 1990). If, however, the sequences are used to estimate the relative amount of each parental DNA in the hybrids, the accuracy of the estimation, using this method, depends on the copy number and distribution of the repetitive sequences used to represent each genome. In order to obtain this information, the chromosomal localization should be clarified using in situ hybridization (Imamura et al. 1987; Piastuch and Bates 1990).

10.3.2
Organellar Genomes

This includes both chloroplast (cpDNA) and mitochondrial DNA (mtDNA). The organellar DNA of somatic hybrids can be compared with that of the fusion partners either by hybridizing suitable organellar DNA probes to the total cellular DNA of the putative hybrids or by examining the restriction profiles of purified (enriched) organellar DNA (Fig. 10.2). The results obtained are then used to classify the chloroplasts or mitochondria in each fusion product as identical to those of one or both of the fusion partners or as novel (Earle et al. 1992). Southern blot hybridization analysis of digested total DNA probed with cloned cpDNA or mtDNA fragments is a suitable method for characterization of organellar genomes of somatic hybrids/cybrids because of the small amount of plant material needed (Hanson 1984; Galun and Aviv 1986). However, the possibility of the different organellar DNAs sharing homology and thus giving non-specific signals needs to be kept in mind (Kemble et al. 1983; Schuster and Brennicke 1988).

10.3.2.1
cpDNA

There are two major steps in cpDNA isolation:

- Isolation of intact chloroplasts
- cpDNA extraction from the chloroplast preparation.

Full methods are given in Chapter 5. For a good yield of good quality cpDNA, optimization of the first step is most critical. Within

Fig. 10.2. Restriction pattern of chloroplast DNA isolated from *Brassica oleracea (lane B.o), B. campestris (lane B.c)* and five somatic *B. oleracea* (+) *B. campestris* hybrids (*lanes 1–5*) digested with BamHI. Lambda DNA digested with EcoRI/HindIII was used as molecular weight standard (*lane M*). (Courtesy K. Glimelius and M. Landgren)

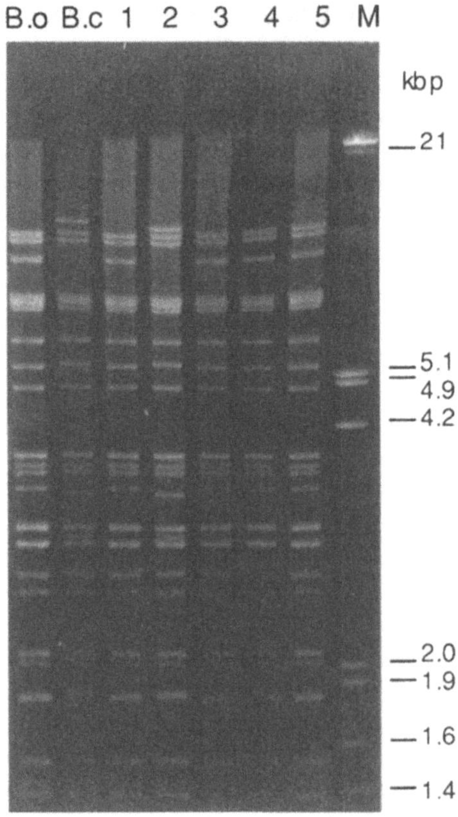

this first step important stages are the homogenization and gradient centrifugation. If the Waring blender homogenization generates a lot of foam and broken chloroplasts, try to release the chloroplasts by cutting the tissue in thin strips with a razor blade. If the collection of the chloroplasts in the gradient centrifugation is not satisfactory, try different sucrose concentrations or the use of Percoll gradients (for details, see Mills and Joy 1980).

If there is high background in the gel after digestion of the chloroplast DNA, methylation-sensitive restriction enzymes (for example: SmaI and PstI) could be tried. This is because they digest preferably non-methylated cp- and mtDNA. Nuclear DNA with abundant methylated restriction sites is hardly digested by these enzymes.

Analysis of cpDNA of the hybrids can also be carried out by Southern hybridization of total genomic DNA with cpDNA-specific probes. Differences in the cpDNA RFLP patterns with certain enzyme/probe combinations were easy to identify in the species we have used, *S. tuberosum* and *S. brevidens* (Xu and Pehu 1994).

10.3.2.2
mtDNA

As in the case of cpDNA, there are two methods for analyzing mtDNA in somatic hybrids:

- Extraction of mtDNA from plant tissue its restriction and comparison of the restriction patterns (Galun and Aviv 1986).
- Southern hybridization of total genomic DNA with specific mtDNA probes.

Isolation of intact mitochondria to otain a sufficient yield of mtDNA has proven quite difficult compared to the same approach applied to the isolation of cpDNA. Therefore, if the aim is to study the fate of parental mitochondria following fusion, we recommend that the second method is used to characterize the hybrids.

Extraction, restriction and Southern transfer of total cellular DNA and Southern analysis are the same as described in other sections of the book. To identify different RFLP patterns between mtDNAs of the fusion partners RFLP patterns produced by different mtDNA probes in combination with different restriction endonucleases (such as BamHI, EcoRI, HindIII and PstI) should be tested.

We have successfully used mtDNA probes nad5 and petunia orf25 to analyze mtDNA genomes of somatic hybrids between *Solanum tuberosum* and *S. brevidens* (Xu et al. 1993c). Tomato somatic hybrids have also been analyzed by several mtDNA probes (San et al. 1990; Wachocki et al. 1991).

References

Austin S, Baer MA, Helgeson JP (1985) Transfer of resistance to potato leaf roll virus from *Solanum brevidens* into *Solanum tuberosum* by somatic fusion. Plant Sci 39:75–82

Baird E, Cooper-Bland S, Eaugh R, DeMaine M, Powell W (1992) Molecular characterisation of inter- and intra-specific somatic hybrids of potato using randomly amplified polymorphic DNA (RAPD) markers. Mol Gen Genet 233:469–475

Bates GW (1992) Molecular analysis of nuclear genes in somatic hybrids. Physiol Plant 85:308–314

Binding H, Jain SM, Finger J, Mordhorst G, Nehls R, Gressel J (1982) Somatic hybridization of an atrazine-resistant biotype of *Solanum nigrum* with *Solanum tuberosum*. Theor Appl Genet 63:273–277
Bromova M, Knopf UC (1991) High frequency recovery of intergeneric fusion products of tobacco (*Nicotiana tabacum*) and potato (*Solanum tuberosum*) by electro-mechanical selection. Plant Sci 74:127–133
Chaput MH, Sihachakr D, Ducreux G, Marie D, Barghi N (1990) Somatic hybrid plants produced by electrofusion between dihaploid potatoes:BF15 (H1), Aminca (H6) and Cardinal (H3). Plant Cell Rep 9:411–414
Debnath SC, Wenzel Z (1987) Selection of somatic fusion products in potato by hybrid vigor. Potato Res 30:371–307
Deimling S, Zitzlsperger J, Wenzel G (1988) Somatic fusion for breeding of tetraploid potatoes. Plant Breed 101:181–189
Dellaporta SL, Wood J, Hicks JB (1983) A simple DNA minipreparation, version II. Plant Mol Biol Rep 1:19–21
Derks FHM, Hakkert JC, Verbeek WHJ, Colijn-Hooymans CM (1992) Genome composition of asymmetric hybrids in relation to the phylogenetic distance between the parents. Nucleus-chloroplast interaction. Theor Appl Genet 84:930–940
Earle ED, Temple M, Walters TW (1992) Organelle assortment and mitochondrial DNA rearrangements in *Brassica* somatic hybrids and cybrids. Physiol Plant 85:325–333
Edwards K, Johnstone C, Thompson C (1991) A simple and rapid method for the preparation of plant genomic DNA for PCR analysis. Nucleic Acids Res 19:1349
Fahleson J, Dixelius J, Sundberg E, Glimelius K (1988) Correlation between flow cytometric determination of nuclear DNA content and chromosome number in somatic hybrids within Brassicaceae. Plant Cell Rep 7:74–77
Famelaer I, Gleba YY, Sidorov VA, Kaleda VA, Parakonny AS, Borshuk NV, Negrutiu I, Jacobs M (1989) Intrageneric asymmetric hybrids between *Nicotiana plumbaginifolia* and *Nicotiana sylvestris* obtained by gamma-fusion. Plant Sci 61:105–117
Fish N, Karp A, Jones MGK (1987) Improved isolation of dihaploid *Solanum tuberosum* protoplasts and the production of somatic hybrids between dihaploid and *S. brevidens*. In Vitro 23:575–580
Fish N, Karp A, Jones MGK (1988) Production of somatic hybrids by electrofusion in *Solanum*. Theor Appl Genet 76:260–266
Galbraith DW (1984) Selection of somatic hybrid cells by fluorescence-activated cell sorting. Cell culture and somatic cell genetics of plants, vol 1, Chap 50. Academic Press, New York
Galun E, Aviv D (1986) Organelle transfer. In: Weissbach A, Weissbach H (eds) Plant molecular biology. Methods in enzymology, vol 118. Academic Press, Orlando, pp595–611
Gebhardt C, Ritter E, Debener T, Schachtschabel U, Walkemeier B, Kaufmann H, Thompson RD, Bonierbale MW, Ganal MW, Tanksley SD, Salamini F (1991) RFLP maps of potato and their alignment with the homologous tomato genome. Theor Appl Genet 83:49–57

Gleba YY, Hoffmann F (1978) Hybrid cell lines *Arabidopsis thaliana* + *Brassica campestris*:no evidence for specific chromosome elimination. Mol Gen Genet 165:257–264

Gleba YY, Momot VP, Cherep NN, Skarzynskaya MV (1982) Intertribal hybrid cell lines of *Atropa belladonna* × *Nicotiana chinensis* obtained by cloning individual protoplast fusion products. Theor Appl Genet 62:75–79

Gleba YY, Hinnisdaels S, Sidorov VA, Kaleda VA, Parokonny AS, Boryshuk NV, Cherep NN, Negruiu I, Jacobs M (1988) Intergeneric asymmetric hybrids between *Nicotiana plumbaginifolia* and *Atropa belladonna* obtained by "gamma-fusion". Theor Appl Genet 76:760–766

Gleddie S, Keller WA, Setterfield G (1986) Production and characterization of somatic hybrids between *Solanum melongena* and *Solanum sisymbriifolium* Lam. Theor Appl Genet 71:613–621

Glimelius K, Djupsjöbacka M, Fellner-Feldegg H (1986) Selection and enrichment of plant protoplast heterokaryons of Brassicaeae by flow sorting. Plant Sci 45:133–141

Graner A, Wenzel G (1992) Towards an understanding of the genome – new molecular markers increase the efficiency of plant breeding. Agro-food-Industry, Hi-Tech, March/April 18–23

Hanson MR (1984) Stability, variation, and recombination in plant mitochondrial genome via tissue culture and somatic hybridization. Oxf Surv of Plant Mol Cell Biol 1:33–52

Imamura J, Saul MW, Portrykus I (1987) X-ray irradiation promoted asymmetric somatic hybridization and molecular analysis of the products. Theor Appl Genet 74:445–450

Innis MA, Gelfand DA, Sninsky JJ, White TJ (1990) PCR Protocols Academic Press, San Diego

Itoh K, Iwabuchi M, Shimamoto K (1991) In situ hybridization with species-specific DNA probes gives evidence for asymmetric nature of *Brassica* hybrids obtained by X-ray fusion. Theor Appl Genet 81:356–362

Jain SM (1993) Recent advances in plant genetic engineering. Curr Sci 64:715–714

Jain SM, Shahin EA, Sun S (1988) Interspecific protoplast fusion for the transfer of atrazine resistance from *Solanum nigrum* to tomato (*Lycopersicon esculentum* L.). Plant Cell Tissue Organ Cult 12:189–192

Jain SM, Oker-Blom C, Pehu E, Newton RJ (1992) Genetic engineering:an additional tool for plant improvement. Agric Sci Finl 1:323–337

Kao KN (1977) Chromosomal behavior in somatic hybrids of soybean + *Nicotiana glacua*. Mol Gen Genet 150:225–230

Kemble RJ (1987) A rapid, single leaf nucleic acid assay for determining the cytoplasmic organelle complement of rapeseed and related *Brassica* species. Theor Appl Genet 73:364–370

Kemble RJ, Mans RJ, Gabay-Laughnan S, Laughnan JR (1983) Sequences homologous to episomal mitochondrial DNAs in the maize nuclear genome. Nature 304:744–747

Khandjian EW (1987) Optimized hybridization of DNA blotted and fixed to nitrocellulose and nylon membranes. Bio/Technology 5:165

Kobayashi S, Ohgawara T, Fujiwara K, Oiyama I (1991) Analysis of cytoplasmic genome in somatic hybrids between navel orange (*Citrus sinensis* Osb.) and Murcott tangor. Theor Appl Genet 82:6–10

Komari T, Saito Y, Nakakido F, Kumashiro T (1989) Efficient selection of somatic hybrids in *Nicotiana tabacum* L. using a combination of drug resistance markers introduced by transformation. Theor Appl Genet 77:547–552
Lapitan NLV (1992) Organization and evolution of higher plant nuclear genome. Genome 35:171–181.
Lindsey K (1992) Genetic manipulation of crop plants. J Biotechnol 26:1–28
Masson J, Lancelin D, Bellini C, Lecerf M, Guerche P, Pelletier G (1989) Selection of somatic hybrids between diploid clones of potato (*Solanum tuberosum* L.) transformed by direct gene transfer. Theor Appl Genet 78:153–159
Melzer JM, O'Connell MA (1990) Molecular analysis of the extent of asymmetry in two asymmetric somatic hybrids of tomato. Theor Appl Genet 79:193–200
Melzer JM, O'Connell MA (1992) Effect of radiation dose on the production of and the extent of asymmetry in tomato asymmetric somatic hybrids. Theor Appl Genet 83:337–344
Mills WR, Joy KW (1980) A rapid method for isolation of purified physiologically active chloroplasts used to study the intracellular distribution of amino acida in pea leaves. Planta 148:75–83
Naton B, Ecke M, Hampp R (1992) production of fertile hybrids by electrofusion of vacuolated and evacuolated tobacco mesophyll protoplasts. Plant Sci 85:197–208
Neuhaus G, Spangenberg G, Mittelsten Scheid O, Schwiger HG (1987) Transgenic rapeseed plants obtained by the microinjection of DNA into microspore-derived embryoids. Theor Appl Genet 75:30–36
Neuhaus-Url G, Neuhaus G (1993) The use of the nonradioactive digoxigenin chemiluminescent technology for plant genomic Southern blot hybridization:a comparison with radioactivity. Transgen Res 2:115–120
Parokonny AS, Kenton AY, Gleba YY, Bennett MD (1992) Genome reorganization in *Nicotiana* asymmetric somatic hybrids analysed by in situ hybridization. Plant J 2:863–874
Pehu E, Karp A, Moore K, Steele S, Dunckley R, Jones MGK (1989) Molecular, cytogenetic and morphological characterization of somatic hybrids of dihaploid *Solanum tuberosum* and diploid *S. brevidens*. Theor Appl Genet 78:696–704
Pehu E, Thomas M, Poutala T, Karp A, Jones MGK (1990) Species-specific sequences in the genus *Solanum*: identification, characterization and application to study somatic hybrids of *S. brevidens* and *S. tuberosum*. Theor Appl Genet 80:693–698
Piastuch WC, Bates GW (1990) Chromosomal analysis of *Nicotiana* asymmetric somatic hybrids by dot blotting and in situ hybridization. Mol Gen Genet 222:97–103
Puite KJ, Schaart JG (1993) Nuclear genomic composition of asymmetric fusion products between irradiated transgenic *Solanum brevidens* and *S. tuberosum*: limited elimination of donor chromosomes and polyploidization of the recipient genome. Theor Appl Genet 86:237–244
San HL, Vedel F, Sihachakr D, Remy R (1990) Morphological and molecular characterization of fertile tetraploid somatic hybrids by protoplast electrofusion and PEG-induced fusion between *Lycopersicon esculentum* Mill. and *Lycopersicon peruvianum* Mill. Mol Gen Genet 221:17–26

Saul MW, Potrykus I (1984) Species-specific repetitive DNA used to identify interspecific somatic hybrids. Plant Cell Rep 3:65–67

Schieder O (1978) Somatic hybrids of *Datura innoxia* Mill. + *Datura discolor* Bernh. and of *Datura innoxia* Mill. + *Datura stramonium* L. var. tatula L. I. Selection and characterisation. Mol Gen Genet 162:113–119

Schuster W, Brennicke A (1988) Interorganellar sequence transfer: plant mitochondrial DNA is nuclear, is plastid, is mitochondrial. Plant Sci 54:1–10

Schweizer G, Ganal M, Ninnemann H, Hemleben V (1988) Species-specific DNA sequences for identification of somatic hybrids between *Lycopersicon esculentum* and *Solanum acaule*. Theor Appl Genet 75:679–684

Sundberg E, Glimelius K (1986) A method for production of interspecific hybrids within Brassiceae via somatic hybridization, using resynthesis of *Brassica napus* as a model. Plant Sci 43:155–162

Tanksley SD, Ganal MW, Prince JP, deVicente MC, Bonierbale MW, Broun P, Fulton TM, Giovannoni JJ, Grandillo S, Matin GB, Messesguer R, Miller JC, Miller L, Paterson AH, Pineda O, Roder M, Wing RA, Wu W, Young ND (1992) High density molecular linkage maps of the tomato and potato genomes. Genetics 132:1141–1160

Thomas MR, Johnson LB, White FF (1990) Selection of interspecific somatic hybrids of *Medicago* by using *Agrobacterium*-transformed tissue. Plant Sci 69:189–198

Waara S, Tegeiström H, Wallin A, Eriksson T (1989) Somatic hybridization between anther-derived dihaploid clones of potato (*Solanum tuberosum* L.) and the identification of hybrid plants by isozyme. Theor Appl Genet 77:49–56

Waara S, Wallin A, Eriksson T (1991) Production and analysis of intraspecific somatic hybrids of potato (*Solanum tuberosum* L.) Plant Sci 75:107–115

Wachocki SE, Bonnema AB, O'Connell MA (1991) Comparison of the organization of the mitochondrial genome in tomato somatic hybrids and cybrids. Theor Appl Genet 81:420–427

Wijbrandi J, Zabel P, Koornnef M (1990) Restriction fragment length polymorphism analysis of somatic hybrids between *Lycopersicon esculentum* and irradiated *L. peruvianum*: evidence for limited donor genome elimination and extensive chromosome rearrangements. Mol Gen Genet 222:270–277

Williams CE, Hunt GJ, Helgeson JP (1990a) Fertile somatic hybrids of *Solanum* species: RFLP analysis of a hybrid and its sexual progeny from crosses with potato. Theor Appl Genet 80:545–551

Williams JGK, Kubelik AR, Livak J, Antoni Rafalski J, Tingey SV (1990b) DNA polymorphisms amplified by arbitrary primers are useful as genetic markers. Nucleic Acids Res 18:6531–6535

Wolters AMA, Schoenmarkers HCH, van der Meulen-Muisers JJM, van der Knaap E, Derks FHM, Koornneef M, Zelcer A (1991) Limited DNA elimination from the irradiated potato parent in fusion products of albino *Lycopersicon esculentum* and *Solanum tuberosum*. Theor Appl Genet 83:225–232

Xu YS, Pehu E (1993) RFLP analysis of asymmetric hybrids of *Solanum tuberosum* and irradiated *S. brevidens*. Theor Appl Genet 86:754–760

Xu YS, Pehu E (1994) Analysis of chloroplast and mitochondrial DNA in asymmetric somatic hybrids between *Solanum tuberosum* and irradiated *S. brevidens*. Transgen Res 3:256–259

Xu YS, Clark M, Pehu E (1993) Use of RAPD markers to screen somatic hybrids between *Solanum tuberosum* and *S. brevidens*. Plant Cell Rep 12:107–109

Xu YS, Murto M, Dunckley R, Jones MGK, Pehu E (1993b) Production of asymmetric hybrids of *Solanum tuberosum* and irradiated *S. brevidens*. Theor Appl Genet 85:729–734

Xu YS, Jones MGK, Karp A, Pehu E (1993c) Analysis of mitochondrial DNA of somatic hybrids between *Solanum tuberosum* and *S. brevidens* using digoxigenin labeled probes. Theor Appl Genet 85:1017–1022

Chapter 11

Cytological Characterization of Transformed Plants: Mapping of Low-Copy and Repetitive DNA Sequences by Fluorescent In Situ Hybridization (FISH)

I.J. Leitch, A.Y. Kenton, A.S. Parokonny, and M.D. Bennett

11.1 Introduction

The physical localization of DNA sequences to plant chromosomes by in situ hybridization (ISH) is becoming increasingly important in many areas of plant molecular biology. ISH has provided new information on the position and order of genes and DNA sequences on chromosomes (e.g. Plate 1d, f and g; Leitch and Heslop-Harrison 1992, 1993; Leitch et al. 1994; Bennett et al. 1995; Cox et al. 1993; Lehfer et al. 1993). This will assist chromosome walking studies aiming to isolate genes of interest, and has enabled comparisons to be made between genetic and physical maps (e.g. Gustafson et al. 1990; Pedersen et al. 1995 and Laurie et al. 1993). Such data are increasingly being used to link the DNA sequence with higher orders of genome structure, gene activities and genetic recombination (e.g. Leitch et al. 1992, 1993; Lapitan 1992; Moore et al. 1993) and to understand the mechanisms of chromosome evolution (Parokonny et al. 1992; Kenton et al. 1993; Leitch et al. 1996). Moreover, ISH has been used to confirm plant transformation and to identify the site(s) of DNA insertion (Plate 1h; Kenton et al. 1995; Ambros et al. 1986; Mouras et al. 1987; Mouras and Negrutiu 1989). Increased knowledge of how the function and expression of transformation sequences are influenced by their location in the host's chromatin domain will enable a more targeted approach to genetic transformation experiments.

While repetitive DNA sequences can now be readily localized by ISH (Plate 1d, f and g), there are still relatively few reports of successful low-copy sequence detection in plants (e.g. Ambros et al. 1986; Lehfer et al. 1993; Leitch and Heslop-Harrison 1993). This stems from a number of problems inherent in plant material (e.g. Gustafson et al. 1990). However, by paying particular attention to certain steps in the schedule the detection of low copy sequences is possible (Plate 1h). The current sensitivity of detection in plants is in the region of 10 kbp (Leitch and Heslop-Harrison 1993).

11.2 Optimizing the ISH Protocol for Detecting Low-Copy Sequences

The main problem associated with detecting low-copy sequences is the weak ISH signal. It is often difficult to distinguish between "genuine" ISH signal (due to specifically hybridized probe), and background signal (due to non-specific binding of the probe or detection reagents). It is therefore essential to take steps to maximize signal strength (e.g. amplification of signal, image enhancement systems) and reduce background to an absolute minimum. To achieve these aims, and so optimize the chances of success, the following points should be considered.

11.2.1 Chromosome Preparations

Chromosome preparations should be free of cytoplasm and cell wall debris; both have been shown to bind probe and detection reagents non-specifically and so contribute to background signal (Ambros et al. 1986). Chromic acid-cleaned slides for chromosome preparations are also recommended, as dirt and grease on slides can also contribute to background signal. Methods for making plant chromosome preparations are discussed in Section 12.3, together with a protocol for preparing chromic-acid cleaned slides.

Sensitivity of detection may be increased by using polytene chromosome preparations. Polytene chromosomes result from repeated cycles of endoreduplication producing many identical copies of each chromatid. The presence of numerous copies means that a low-copy sequence is represented many times. Polytene chromosomes have been used to map low-copy sequences in *Drosophila* species (e.g. Whiting et al. 1989). In plants, repetitive sequences have been mapped to polytene chromosomes using ISH (e.g. Avanzi et al. 1972; Durante et al. 1977), but these chromosomes have not yet been used to map low-copy sequences.

11.2.2
High Metaphase Index

The successful detection of low copy sequences in plants is usually low, so chances are improved by increasing the number of metaphase cells per slide. Methods to accumulate metaphases in plants are described in Sect. 12.3.1.

11.2.3
Probe Types

Many probes used for ISH are prepared from plasmid clones which contain the DNA sequence of interest plus vector DNA. To reduce background only the sequence of interest should be used as a probe. Vector sequences can bind non-specifically to target DNA (chromosomal DNA), thereby increasing background. Cosmids and yeast artificial chromosomes (YACs), which can contain up to 1 Mb of DNA, are increasingly being used to map low-copy sequences in animals (e.g. Baldini et al. 1992), but their potential has not yet been realized in plants. Owing to the large insert size in cosmids and YACs, the probe also contains many repetitive DNA sequences. Hybridization of these to the target DNA can be suppressed by including blocking DNA in the hybridization mixture (e.g. Lichter et al. 1988).

11.2.4
Choice of Labels

Both radioactive and non-radioactive labels have been used for ISH. Initially, radioactive labels were found to be most sensitive. However, the long detection times required (sometimes days or weeks), and the poor signal resolution means that non-radioactive labels are now usually preferred.

Non-radioactive labels can be divided into two classes – the direct and the indirect. The direct labels are the fluorochrome-conjugated nucleotides (e.g. fluorescein-11-dUTP) which can be visualized immediately following the post-hybridization washes. Indirect labels, of which the most common are biotin-11-dUTP and digoxigenin-11-dUTP, require further detection steps following post-hybridization washing in order to be visualized. Direct labels are quick and easy to use, and generally result in lower background, but they tend to be less sensitive (Wiegant et al. 1991). Signal amplification is possible only for some fluorochromes (Bauman et al. 1981). Indirect labels are therefore recommended for detecting low-copy sequences. Both biotin and digoxigenin have been successfully used as described below (Sect. 11.6).

11.2.5 Choice of Labelling Method

DNA for use as a probe can be labelled by a variety of methods using either enzymes [e.g. nick translation, random priming, end-labelling, polymerase chain reaction (PCR)] or chemical modifications (e.g. mercuration, sulphonation), and reviewed by Leitch et al. (1994). Enzymatic labelling methods have been used most widely and are reported to produce cleaner probes (especially important for detecting low copy sequences). For probes greater than 1 kb nick translation is recommended, whereas random priming is preferred for probes 100 bp-1 kb. For oligonucleotide probes (less than 100 bp) end labelling is necessary. Section 11.5 gives a protocol for labelling DNA with fluorochromes, biotin or digoxigenin by nick translation. Alternatively, kits for enzymatic labelling are available from Amersham International, Boehringer Mannheim or BRL Life Technologies.

PCR is a simple and quick method to prepare probes up to 4 kbp if suitable primers are available (e.g. Lo et al. 1988; Albertson et al. 1991). By incorporating labelled nucleotides into the reaction, the sequence of interest is specifically amplified and labelled. Vector sequences are not amplified in the reaction, so do not have to be removed.

The optimum size of a probe for ISH is estimated to be 300–600 bp. Commercially available enzymes for nick translation (DNA polymerase/DNase) and random priming (Klenow DNA polymerase) are optimized to produce probes of this length. However, PCR generates probes the length of the insert. If the sequence is greater than 1 kbp, then there may be problems of probe penetration into the material. To avoid this, probes greater than 1 kbp should be sheared or autoclaved to give smaller fragments.

11.2.6
Choice of Detection System

Most indirect labels are visualized using a primary antibody (e.g. anti-digoxigenin) conjugated to a detection system. Secondary antibodies can then be used for further amplification. For probes labelled with biotin, an alternative detection system incorporating (strept) avidin can be used. The avidin-biotin system may be more sensitive than antibody-based detection systems owing to the relatively greater affinity of biotin for avidin and the greater potential for signal amplification.

Three types of detection system are available; fluorochromes, enzymes and colloidal gold. Usually fluorochromes are preferred owing to their excellent signal resolution, and the opportunity for multiple labelling strategies (Section 11.2.8; Trask 1991).

11.2.7
Imaging Low-Copy Signal

The sensitivity of ISH signal detection can be increased by recording the image electronically using a camera sensitive to low intensity light (Viegas-Piquignot et al. 1989). The most sensitive camera system to date is the cooled charge-coupled device (CCD) camera, capable of recording signals that are invisible to the eye. Wiegant et al. (1991) estimated that detection sensitivity was increased 30-fold using a CCD camera.

11.2.8
Multiple Labelling and Reprobing

Simultaneous detection of more than one probe is not recommended to detect very low-copy sequences in plants, as the additional detection reagents may increase background signal. If, however, background is not a problem, the additional information obtained can be invaluable for plant chromosome mapping programmes as mapping speed is cnhanced and sequences can be ordered in relation to each other (e.g. Plate 1f, g; Leitch et al. 1991; Mukai et al. 1993; Heslop-Harrison and Schwarzacher 1993). In addition, if a probe of known location is used together with a low-copy probe this can help determine if the low-copy "signal" is real. If background signal is a problem, using two or more probes then sequential reprobing of the material is possible (Plate 1d, e; Heslop-Harrison et al. 1993; Sect. 11.6, Procedure, step **27**).

Section 11.6 describes a protocol for simultaneously detecting a biotin- and digoxigenin-labelled probe. An alternative detection system for digoxigenin in a double labelling experiment is described by Kenton et al. (1993).

11.3 Visualizing Fluorescent ISH Signal

Fluorescence microscopy is based on the principle that light of a certain wavelength (excitation wavelength) excites electrons in the fluorochrome making them jump into an outer electron shell. The excited electrons are unstable and on returning to the stable state energy is released which is emitted as fluorescence (emission wavelength). The excitation light is usually a high-pressure mercury or xenon arc lamp that emits high-intensity radiation at the required wavelengths of light. Filters are used to select appropriate wavelengths of light for exciting and visualizing a particular fluorochrome. The most commonly used fluorochromes are DAPI (fluoresces blue on excitation with UV light), FITC and fluorescein (fluoresces green on excitation with blue light) and rhodamine and Texas red (fluoresce red on excitation with green light).

In multiple labelling experiments, the different fluorochromes can be photographed by multiple exposures of the same image using different filter combinations (Plate 1f, g). If the alignment of each filter combination is the same and the surfaces of the emission filter are absolutely parallel, than there is no problem and the exposure times can be optimized for each fluorochrome. Problems arise when the different filters are not aligned, because the superimposed pictures from the multiple exposures will be offset. Dual and triple filter blocks which are now available overcome this problem, enabling two or three fluorochromes, respectively, to be visualized and photographed simultaneously. This may be important when the relative positions of different probes need to be accurately determined.

11.4
Chromosome Preparation

Methods for preparing chromosomes suitable for ISH are described in Chap. 12.3. All slides should be screened rigorously to ensure that only those with well spread metaphases, free from cell wall debris and cytoplasm are used.

11.5
Probe Preparation

Different methods to label DNA depending on the size of the DNA are discussed in Section 11.2.5. This section describes the labelling of cloned DNA sequences by nick translation.

11.5.1
Purifying DNA by Gel Extraction

The DNA insert must first be released from the vector by digesting at least 3 µg of the plasmid with an appropriate restriction enzyme. The reaction products are separated by agarose gel electrophoresis. The band of the correct size is cut out of the gel and can be purified by a variety of methods. These have already been covered in Chapter 1.

11.5.2
Labelling DNA by Nick Translation

DNA fragments isolated by gel extraction can be labelled by nick translation. The method is the same as described in Chapter 12.4.2, except leave out the blocking DNA from genome B in Step **4**.

It is important to know that label incorporation has been successful. Checking biotin incorporation is described in Section 12.4.3. Digoxigenin incorporation can be checked in the same way with the following modifications:

a) At step **7** substitute streptavidin with antidigoxigenin-alkaline phosphatase (Boehringer Mannheim), diluted 1: 5000 in 0.05 M Tris-HCl, pH 7.5.
b) Omit steps **9** and **10**.

Incorporation of fluorochrome-labelled nucleotides is checked by pipetting a small (1–2 μl) drop of probe onto a microscope slide. Using the correct filters (Table 11.1) in the epifluorescence microscope the whole drop should fluoresce uniformly at the appropriate wavelength.

11.6 Protocol for In Situ Hybridization

This section describes the steps of ISH using two probes, one labelled with biotin and the other with digoxigenin. If a biotin-labelled probe is

Table 11.1. Examples of filter blocks used to visualize fluorescent ISH signal[a]

<table>
<tr><th>Fluorochrome</th><th>Leica Ploemopak filter block</th><th>Nikon filter block</th><th>Zeiss filter block</th><th colspan="2">Chroma Dual filter block</th><th>Chroma Triple filter block</th><th>Colour of fluorescence</th></tr>
<tr><td>DAPI
Excitation
max = 355 nm
Emission
max = 450 nm</td><td>A or D</td><td>UV-1A</td><td>01 or 02</td><td rowspan="2">51 000</td><td></td><td rowspan="3">61 002</td><td>Blue</td></tr>
<tr><td>Fluorescein, FITC
Excitation
max = 495 nm
Emission
max = 515 nm</td><td>I3
(L3
selective)</td><td>B-2A
(B-2H
selective)</td><td>09
(10
selective)</td><td rowspan="2">51 006</td><td>Green</td></tr>
<tr><td>Texas red
Excitation
max = 595 nm
Emission
max = 695 nm</td><td>TX</td><td>G-1B</td><td>00</td><td></td><td>Red</td></tr>
<tr><td>Rhodamine
Excitation = 550 nm
Emission
max = 575 nm</td><td>N2.1</td><td>G-2A</td><td>15</td><td colspan="2"></td><td></td><td>Red</td></tr>
<tr><td>Propidium iodide
Excitation
max = 340, 530 nm
Emission
max = 615 nm</td><td>N2.1</td><td>G-2A</td><td>14</td><td colspan="2"></td><td></td><td>Red</td></tr>
</table>

[a]All filters are available from Micro Instruments Ltd, Long Hanborough, Oxon, UK.

used on its own, then add an appropriate volume of water to the hybridization mix to give the correct final volume (step **6**, Sect. 11.6.3 Procedure), use only avidin-Texas red in BSA block for step **17** and biotinylated anti-avidin in goat serum block for step **20**. If using only a digoxigenin-labelled probe, then adjust volume of hybridization mix as above, use only anti-digoxigenin-fluorescein in BSA block for step **17**, sheep anti-rabbit FITC in goat serum block for step **20**, and omit step **22**.

Reagents

All reagents should be made up in fresh sterile distilled water (SDW) unless otherwise indicated.

General Buffers

- 20x SSC, pH 7.0: 3 M NaCl, 0.3 M Na citrate adjusted to pH 7.0 with HCl
- 100x TE, pH 8.0: 1 M Tris-HCl, pH 8, 0.1 M EDTA, pH 8.0
- BT buffer: 0.1 M $NaHCO_3$, 0.05% (v/v) Tween 20, (a 10x stock, i.e. 1 M $NaHCO_3$, 0.5% Tween 20, can be prepared and stored at room temperature)
- PBS, pH 7.4: 120 mM NaCl, 7 mM Na_2HPO_4, 3 mM NaH_2PO_4, 2.7 mM KCl

Reagents for Pretreatments

- RNase A. Stock solution is prepared by dissolving 10 mg/ml of RNase A in 10 mM Tris-HCl, pH 7.5 and 0.15 M NaCl. Boil for 15 min. Allow to cool then store at –20 °C in aliquots. For use (50 µg/ml) dilute stock 1: 200 in 2x SSC.
- 70, 90 and 100% ethanol

Reagents for Denaturation and Hybridization

- Formamide. Good quality (e.g. Fison's electrophoresis grade) should be used. Store in aliquots at –20 °C.
- 50% (w/v) dextran sulphate. Prepare by dissolving in SDW (heating to 70 °C can speed up the process). Filter sterilize using a Sartorius Minisart filter (0.22 µm + glass fibre prefilter; Sartorius SM 17823K) because the liquid is very viscous. Store filtered solutions at –20 °C in aliquots.

- 20x SSC (see above)
- Probe DNA. One low-copy probe labelled with biotin (100 ng/μl) and a second probe labelled with digoxigenin (100 ng/μl). For probes labelled as described in Section 11.5, increase the volume to 30 μl by adding 10 μl 1x TE. This gives a probe concentration of 100 ng/μl.
- 10% (w/v) sodium dodecyl sulphate (SDS). Prepare by dissolving in SDW, then filter (0.22 μm) sterilize.
- Denaturing solution. 70% (v/v) formamide in 2x SSC. Place in plastic Coplin jar and heat to 70 °C in a water bath.
  Safety note. The denaturing solution contains formamide so place water bath in fume hood. Alternatively, wear an organic vapour respirator (e.g. Merck 5651) which provides protection against 10x the occupational exposure limit (OEL).
- Freezer-cold ethanol series: 50, 70, 90 and 100% ethanol stored at –20 °C. Transfer to Coplin jars standing on ice immediately before denaturation (Sect. 11.6.3, step **7**).

Reagents for Post-Hybridization Washes

- 2x SSC (see above) preheated to 42 °C
- Formamide wash. 50% (v/v) formamide in 2x SSC, preheated to 42 °C.
  Safety note. The post-hybridization wash contains formamide, so place shaking water bath in a fume hood. Alternatively, wear a respirator fitted with a grade A1 filter which provides protection against 10x the occupational exposure limit.
- BT buffer (see above) at room temperature

Reagents for Detection

- Bovine serum albumin (BSA) block. 5% (w/v) BSA (Sigma; Fraction V, globulin-free) in BT buffer. This must be freshly made every time. Filter (0. 22 μm) sterilize before use.
- Detection reagent mix (prepare immediately before use). Use 5 μg/ml anti-digoxigenin-fluorescein (Boehringer Mannheim) and 5 μg/ml avidin-Texas red (Vector Laboratories). Dilute in BSA block. Filter (0.22 μm) sterilize.
- BT buffer (see above) pre-warmed to 40 °C.

- Goat serum block. 5% (v/v) normal goat serum (Vector Laboratories) in BT buffer. Filter (0.22 µm) sterilize.
- Amplification mix (prepare immediately before use). Use 25 µg/ml biotinylated anti-avidin D (Vector Laboratories) and 10 µg/ml rabbit anti-sheep FITC (DAKO). Dilute in goat serum block. Filter (0.22 µm) sterilize.
- Avidin-Texas red (prepare immediately before use). Use 5 µg/ml avidin-Texas Red (Vector Laboratories) diluted in BSA block. Filter (0.22 µm) sterilize.

Reagents for Counterstaining and Mounting

- Propidium iodide (PI). Prepare stock solution of 1 mg/ml in SDW. Store in aliquots at –20 °C. Immediately before use, dilute 1 µl of stock with 2 ml of PBS, pH 7.4 (see above). Unused, diluted solution should be discarded, because it does not keep.
  Safety note: PI is a mutagen, treat with caution!
- DAPI (4′, 6-diamidino-2-phenylindole) stock. Prepare stock solution of 100 µg/ml. Store in aliquots at –20 °C.
- Vectashield/DAPI mountant. Add 10 µl of DAPI stock to 10 ml of Vectashield (Vector Laboratories). Store in aliquots at –20 °C.

Reagents for Reprobing

- Histo-Clear (National Diagnostics, New Jersey)

Equipment

General

- Plastic cover slips for all incubations of 50 µl or less. These are made by cutting autoclavable waste disposal bags to ca. 22 mm^2.
- Plastic dishes and slide holders for all washing steps (Raymond A Lamb E/100)
- Fine (No.5), clean forceps
- Humid chamber for incubations. Place tissues at bottom of a plastic sandwich box and soak with 2x SSC. Put drinking straws on tissues to support slides. Put on lid and preheat to the appropriate temperature required for incubation.
- 37 °C incubator
- Rotary shaker

Equipment for Denaturation

- 70 °C water bath in a fume hood
- Ice bucket to keep ethanol series cold

Equipment for Post-Hybridization Washes

- Shaking water bath at 42 °C in a fume hood

Equipment for Visualization of ISH Signal

- Epifluorescence photomicroscope and filter sets for DAPI, FITC and Texas red (Table 11.1)
- Objective lenses: Use fluorescence objectives with high numerical apertures (e.g. Zeiss Plan Neofluor, Leica NPL Fluotar). Fluorescence intensity is proportional to numerical aperture (see also Sect. 12.2.7).
- Glass cover slips. 24×40 mm or 24×50 mm
- Rubber solution. e.g. Dunlop
- Photographic film. For projection slides use Kodak Ektachrome Professional (daylight) 800/1600 ASA, for prints use Fujicolor HG 400 ASA.

Procedure Slides should be handled with care throughout the ISH protocol to avoid scratches. Unless specifically stated, slides should not be allowed to dry out. Washing steps are carried out in racks placed in plastic dishes using enough solution to cover all the slides. All steps are carried out at room temperature (20–25 °C) unless otherwise stated.

Pretreatments

1. Thaw slides and air-dry.

2. Pipette 50 µl of RNase A onto each slide and cover with a plastic coverslip. Place slides in humid chamber and incubate at 37 °C for 2 h (note 1).

3. Carefully remove coverslips with fine forceps and immediately place slides in slide rack in 2x SSC. Wash slides in 2x SSC for 2×5 min (note 2).

4. Dehydrate slides by placing them sequentially in 70, 90 and 100% ethanol for 2×5 min each.

5. Dry slides and store overnight in a vacuum desiccator at 4 °C (note 3).

Denaturation and Hybridization

6. Prepare hybridization mix (20 μl per slide; note 4) : For 10 slides (final concentration in brackets) add 100 μl formamide (50%), 40 μl dextran sulphate (10%), 20 μl 20x SSC (2x), 10 μl of low copy biotin-labelled probe (5 ng/μl) and 7.5 μl of digoxigenin-labelled probe (3.5 ng/μl; note 5), 10 μl SDS (0.5%; note 6) and 12.5 μl SDW. Store on icc.

7. Handle slides in pairs. First denature in denaturing solution for 2 min at 70–72 °C then dehydrate through freezer-cold ethanol series, 2 min in each solution and air-dry. Check the temperature of denaturing solution before adding each pair of slides.

8. When all slides have been denatured and dehydrated, vortex hybridization mix and denature in water bath at 76 °C for 15 min. Store on ice for 5 min.

9. Place 20 μl of hybridization mix on each slide. Apply coverslip taking care to remove any air bubbles. Place slides in humid chamber (pre-warmed to 37 °C) and incubate at 37 °C overnight (12 to 15 h; note 7).

Post-Hybridization Washes

10. Place pre-warmed 2x SSC and formamide wash solutions into plastic dishes and allow the temperatures to equilibrate to 42 °C.

11. Remove slides from humid chamber. Carefully remove coverslips and place slides immediately into pre-warmed 2x SSC (42 °C). Shake gently for 5 min.

12. Place slides into pre-warmed (42 °C) formamide wash. Shake gently for 5 min (note 8).

13. Transfer slides to fresh, pre-warmed (42 °C) 2x SSC. Shake gently for 5 min.

14. Move slides to fresh 2x SSC at room temperature. Shake gently for a further 5 min.

15. Place slides in BT buffer for 10 min.

Detection of Biotin- and Digoxigenin-Labelled Probes

16. Drain each slide individually, add 50 µl of BSA block and apply cover slip (note 9). When BSA block has been added to all slides, incubate for 5 min.

17. For each slide, carefully remove coverslip and drain. Add 50 µl of detection reagent mix and apply a clean coverslip. When detection reagent mix has been added to all slides, transfer to humid chamber and incubate at 37 °C for 1 h.

18. Remove coverslips and place slides immediately in pre-warmed (40 °C) BT buffer. Wash, shaking gently, for 2×5 min.

19. Repeat step **16** but replace BSA block with goat serum block (note 9).

20. Repeat step **17** but replace detection reagent mix with amplification mix.

21. Wash slides as for step 18.

22. Repeat steps **16–18** but replace detection reagent mix with avidin-Texas red in step 17 (note 10).

Counterstaining and Mounting (note 11)

23. *Propidium iodide (PI) counterstaining (note 12).* Remove each slide from BT buffer and drain well. Add 50 µl diluted PI and apply a coverslip. When PI has been added to all slides, incubate in humid chamber, in the dark for 10 min. Remove coverslips and place slides immediately in PBS buffer. Wash, shaking gently for 5 min.

24. *DAPI counterstaining and mounting.* Remove each slide from buffer and drain well. Add 15 µl of Vectashield/DAPI mountant and apply glass coverslip. Remove excess mountant from slide by placing between filter paper and pressing firmly. Seal edge of cover slip with rubber solution.

Viewing and Photography

25. The ISH signal is visualized using an epifluorescence photomicroscope (notes 13, 14). Slides can be viewed immediately but the signal stabilizes after storing for a few days in the dark at 4 °C (notes 15, 16).

26. Photograph areas of interest (note 17) using either slide or print film (note 18).

Reprobing Slides (note 19)

27. Remove rubber solution and clean slides with Histo-Clear to remove oil. Wash in BT buffer for 3×2 min.

28. Place slides in 2x SSC for 5 min then proceed as described from step **4.**

Notes

- *Note* 1. RNase A treatment removes endogenous RNA, which might otherwise contribute to background signal.
- *Note* 2. If there is a lot of cytoplasm associated with the chromosome preparations, then a pepsin treatment is recommended at this step. Pepsin increases probe and reagent accessibility by digesting proteins. The reaction steps are as follows. Place slides in 0.01 M HCl for 2 min. Add 50 µl 5 µg/ml pepsin (porcine stomach mucosa, activity: 3200 to 4500 U/mg, Sigma) in 0.01 M HCl and apply a plastic coverslip. Incubate for 10 min at 37 °C. Stop reaction by washing in SDW for 2 min and then in 2x SSC for 2×5 min. Dehydrate and store slides as described in steps 4 and 5 (Sect. 11.6). It may be necessary to adjust the time and concentration of pepsin treatment for different material.
- *Note* 3. Storage of slides overnight results in better preservation of chromosome morphology.
- *Note* 4. The prepared hybridization mix can be stored at –20 °C for up to 6 months.
- *Note* 5. Typically probe concentrations of 1 to 5 ng/µl are used. Increasing the concentration tends to result in higher background rather than stronger ISH signal. For highly repeated DNA se-

quences, less probe can be used and the remaining volume of the hybridization mix made up with SDW. For probes which cannot be successfully purified by gel extraction use 20 μl probe (final concentration = 10 ng/μl) to compensate for the presence of vector sequences in the probe.

- *Note* 6. SDS is a wetting agent assisting probe penetration. Typically, a concentration of 0.05–1% is used.
- *Note* 7. Make sure the tissues in the humid chamber are well soaked; if they dry out, then so will the slides!
- *Note* 8. The stringency of an ISH reaction determines the approximate percentage of nucleotides correctly matched between the probe and target hybrid. Stringency can be increased by raising the temperature and percent formamide, or decreased by increasing the concentration of 2x SSC. The washing conditions described in step **12** allow sequences with 80–85% homology to remain hybridized, assuming the conditions in situ are the same as they are in solution (Meinkoth and Wahl 1984). Stringency can be increased to 85–90% by increasing the percent formamide in the formamide wash (step **12**) to 60%. A lower stringency of 75–80% is most easily achieved by performing post-hybridization washes (steps **10–13**) at 37 °C instead of 42 °C.
- *Note* 9. BSA and goat serum block reduce background signal by blocking sites, e.g. in the cytoplasm, which might otherwise bind detection reagents.
- *Note* 10. For unique sequences, a further amplification of the ISH signal may be necessary, although background will also increase with successive rounds of amplification. To amplify biotin-labelled probes, repeat steps **16–22**, replacing detection reagent mix with 5 μg/ μl avidin-Texas red and amplification mix with 25 μg/μl biotinylated anti-avidin.
- *Note* 11. The choice of counterstain depends on the fluorochromes used for detecting the ISH signal. When multiple labelling with both fluorescein and Texas red (as described in Section 11.6), only counterstains which fluoresce blue, such as DAPI, are suitable, because there is no overlap of the excitation and emission wavelengths between the counterstain and the fluorescent ISH signal.

- *Note* 12. PI is suitable as a counterstain when green fluorochromes such as fluorescein and FITC are used on their own. PI fluoresces orange at the same wavelengths used to visualize fluorescein and FITC (green) fluorescence. PI fluoresces red under wavelengths of light used to excite Texas red and rhodamine fluorochromes, and therefore obscures the ISH signal. Care must be taken not to overstain cells with PI, as this can obscure weak ISH signal.

- *Note* 13. The epifluorescence microscope should ideally be in a darkened room and set up correctly to maximize fluorescence intensity. For example, ensure that the mercury vapour lamp is correctly aligned to produce even illumination across the slide. Replace lamp when the intensity begins to diminish, otherwise signal brightness will also diminish. The average life of a lamp varies depending on the type (e.g. average lamp life for HBO50 = 100 h, HBO100W/2 DC = 200 h, HBO200W/4 = 200 h). The field iris in the epifluorescent illuminator should be adjusted so that just the field of view of the objective being used is illuminated. This eliminates extraneous fluorescing material and so optimizes contrast. This will also reduce photobleaching of adjacent cells.

- *Note* 14. Suppliers usually offer a matched combination of filters for specific fluorochromes. Some typical filter sets are listed in Table 11.1. Older and/or cheaper filter sets incorporate glass emission filters which are long pass filters (i.e. they allow light longer than a certain wavelength to pass through). However, if the Texas red of rhodamine signal is particularly bright the signal may "bleed through" and so is visible under the FITC filter. This can be avoided by using filter sets that incorporate interference filters that let only a narrow, specific band of light to pass through. The dual and triple band filter sets always incorporate narrow band interference filters. The fluorescent intensity of a particular fluorochrome using a dual or triple band filter set is reduced in comparison with a filter set designed for a specific fluorochrome. However, this is usually outweighed by the convenience of a single exposure and the avoidance of image shift associated with multiple exposures.

- *Note* 15. Slides can be stored up to a year at 4 °C but background fluorescence may increase with time. To re-examine old slides remove coverslips carefully and wash as described in step **27**. Reapply Vectashield/DAPI mountant (step **24**).

- *Note* 16. Patchiness of ISH signal can occur on all slides. However, the lower success of ISH with low-copy probes can compound this problem. Consequently, each slide must be scanned very thoroughly and each metaphase examined to assess fully the success of the experiment.

- *Note* 17. Fluorescence fades on viewing, although this is greatly reduced by mounting material in antifade reagents such as Vectashield. It is therefore important to record the image photographically as soon as possible. Short, high energy wavelengths, e.g. near UV/violet, used to excite DAPI, bleach the fluorochromes more quickly than the longer, weaker energy wavelengths used to visualize Texas red and rhodamine. It is recommended to photograph red fluorescent signals first, followed by the green fluorescence. Photograph blue fluorescence images last of all. Photographing fluorescence ISH can be difficult, because often the fluorescent signal fills only a small part of the frame and is set against a dark background. If the microscope has an automatic camera attachment that uses light from the entire photographic field to calculate the exposure, it averages out the fluorescence and the surrounding dark areas and consequently over-exposes the photograph. This can be overcome by using a spot measuring facility or darkfield setting on the camera, if available, or controlling the exposure time manually. Alternatively, rate the film at least two stops below the predicted exposure setting. Exposure times for ISH signals that appear bright are around 10 to 30 s. For weaker signals, such as from low-copy probes, up to 5 min exposure may be necessary, DAPI images, which are usually much brighter than the ISH signals, require exposures of only a few seconds.

- *Note* 18. The recommended films represent a good contrast between sensitivity in recording the signal and resolution.

- *Note* 19. After photography, slides can be reprobed (e.g. plate 1d, e) with additional probes to obtain further information and/or confirm the identity of chromosomes using a well-characterized probe.

Controls For any ISH experiment, it is essential to include suitable controls to determine the specificity of probe hybridization. Controls can also be useful for identifying the source of problems where an experiment has been unsuccessful. Different types of controls are used to evaluate particular aspects of the experiment.

- *Positive controls.* These are used to demonstrate that denaturation, hybridization and detection reagents have all worked properly. When the positive control has worked but the test slide has failed this points to a problem with the test probe rather than the ISH reaction per se. Examples of positive controls include using probes whose target sequences are known to be present in many copies in the material such as ribosomal DNA probes.
- *Negative controls.*
  – To determine the level of non-specific labelling, conduct experiments either without any probe or with an unrelated probe known to be absent from the material, or with the DNA sequence of interest unlabelled.
  – To determine the specificity of probe hybridization perform the experiment at different stringencies (note 8). As stringency increases, the number of hybridization sites should decrease, leaving only the specific hybridization sites, corresponding to the sequence of interest, at high stringency.

Troubleshooting

Given the length and number of steps of ISH inevitably numerous problems can arise. Table 11.2 lists common problems, their possible causes and suggested remedies.

Acknowledgements. We are grateful to Dr. A.R. Leitch (Queen Mary and Westfield College, London, UK), and B.C. Cowen and H. Shearman (Microinstruments Ltd, Oxford, UK) for their helpful comments on the manuscript.

Table 11.2. Troubleshooting – possible causes and remedies

Problem	Possible cause of problem	Possible remedy
• Loss of material	– Dirt or grease on the slides used for chromosome preparation	– Slides for chromosome preparations should be thoroughly cleaned (e.g. chromic acid cleaned). Alternatively slides may be coated with solutions designed to "stick" chromosome material (e.g. Vectabond see Chap. 12.3)
• Weak or no signal	– Probe not labelled, so detection reagents cannot work	– Always check probe to ensure the label has been incorporated (Sect. 12.3.1)
– Chromosomes with good morphology	– If there is no signal on either the test or the positive control slide with a highly repeated probe, then the detection reagents may not be working	– Make up fresh detection reagents and do not store diluted antibodies or avidin/anti-avidin solutions. Wherever possible use the same detection reagents for dot-blot testing of the probe as for ISH to check that they work
	– Probe not fully denatured into single strands so unable to hybridize to target sequences on chromosomes	– Check that the water bath temperature is 76°C. After heating and transferring the hybridization mix to ice (step **8**), leave for at least 5 min to prevent "snap-back of the single-stranded DNA into duplexes
	– Chromosomeal DNA not fully denatured into single strands so probe cannot gain access to the DNA sequence	– Check temperature of denaturing solution. It may be necessary to remove DNA-associated proteins, especially if the sequence of interest is located near or in heterochromatin. This can be done by treating material with pepsin (see note 2). Alternatively, repeat ISH procedure on the same slide increasing denaturation time by 1 min (step **7**). Redenaturation may improve probe penetration
	– Stringency (note 8) too high	– If probe is not identical to target sequence then reduce stringency to enable non-identical but closely related sequences to hybridize (note 8)

Table 11.2. (*Contd.*)

Problem	Possible cause of problem	Possible remedy
	– Excessive counter-staining with propidium iodide can obscure ISH signal expecially if it is weak	– Destain by removing coverslip as described in step **27**, and wash in BT buffer for 2 × 15 min. Remount as described in step **24**
• Weak or no signal – Chromosomes weakly counter-stained with DAPI and ragged-looking	– Chromosomes over-denatured resulting in excessive loss of DNA from material. Exact denaturing conditions depend on the target sequence and the extent to which it is protected by DNA-associated proteins. Different sequences may require slightly different pretreatments for optimal denaturation	– Chromosomes can be fixed to reduce DNA loss during denaturation. For example, fixing slides in 3:1 ethanol: acetic acid for 10 min prior to dehydration (step **4**) may help. Alternatively, they can be denatured for a shorter time (step **7**, e.g. 1 min rather than 2 min). The conditions may need to be determined empirically for each new material used
	– Slides too old (> 1 month) and/or incorrectly stored, chromosomal DNA in old slides can become degraded and is readily lost during the ISH procedure	– Slides should be stored desiccated at –20 °C for no longer than 1 month
• Weak or no signal – Bright, green auto fluorescence (seen under FITC filter) around chromosomes	– Too much cytoplasm is present inhibiting penetration of probe and detection reagents into chromosomes. Cytoplasm can also prevent satisfactory chromosome denaturation	– Pepsin treatment of chromosomes can help where there is lots of cytoplasm (note 2). However, use cytoplasm-free preparations where possible
• Signal present but patchy	– Patchiness can result if the reagents have not been well mixed or if air bubbles are not removed from underneath the coverslips prior to incubations	– Vortex all reagents before use, especially the hybridization mix because dextran sulphate is very viscous and readily sinks to the bottom of tube. Apply coverslips carefully to minimize introduction of bubbles. If they do appear, gently lift and lower cover slip several times to pull bubbles to edge

Table 11.2. (*Contd.*)

Problem	Possible cause of problem	Possible remedy
• Background signal over all the slide	– Non-specific binding of the probe to material	– Increase the number and duration of all washing steps but take care not to damage or lose material with overzealous washing
	– Non-specific binding of detection reagents to material	– Always use freshly diluted detection reagents and filter sterilize before use. Decreasing the concentration of detection reagents may also help to reduce background
	– If target sequence is expressed transcribed RNA will be homologous to the target sequence and will bind the probe	– RNase A treatment (step **2**) is essential for removing this source of background
• Background signal restricted to chromosomes	– Stringency (note 8) too low	– Increase stringency to eliminate probe hybridization to DNA sequences which have some homology to the target DNA
• Glare of ISH signal and counterstain	– Too much glycerol remaining on slide after mounting	– Always squeeze out as much mountant as possible from underneath coverslip (step **24**). Glycerol in mountant gives rise to glare
	– Slides which have been stored for several months often show increased glare and background fluorescence	– Treat slides as described in note 15

References

Albertson DG, Sherrington P, Vaudin M (1991) Mapping nonisotopically labeled DNA probes to human chromosome bands by confocal microscopy. Genomics 10:143–150

Ambros PF, Matzke MA, Matzke AJM (1986) Detection of a 17 kb unique sequence (T-DNA) in plant chromosomes by in situ hybridization. Chromosoma 94:11–18

Avanzi S, Durante M, Cionini PG, D'Amato FD (1972) Cytological localization of ribosomal cistrons in polytene chromosomes of *Phaseolus coccineus*. Chromosoma 39:191–203

Baldini A, Ross M, Nizetic D, Vatcheva R, Lindsay EA, Lehrach H, Siniscalco M (1992) Chromosomal assignment of human YAC clones by fluorescent in situ hybridization: use of single-yeast-colony PCR and multiple labelling. Genomics 14:181–184

Bauman JGJ, Wiegant J, Van Dujin P (1981) Cytochemical hybridization of fluorochrome labeled RNA. III. Increased sensitivity by the use of anti-fluorescein antibodies. Histochemistry 73:181–193

Bennett ST, Leitch IJ, Bennett MD (1995) Chromosome identification and mapping in the grass *Zingeria biebersteiniana* (2n = 4) using fluorochromes. Chromosome Res 3:101–108

Cox AV, Bennett ST, Parokonny AS, Kenton A, Callimassia MA, Bennett MD (1993) Comparison of plant telomere locations using a PCR-generated synthetic probe. Ann Bot 72:239–247

Durante M, Cionini PG, Avanzi S, Cremonini R, D'Amato FD (1977) Cytological localization of the genes for the four classes of ribosomal RNA (25S, 18S, 5.8S, and 5S) in polytene chromosomes of *Phaseolus coccineus*. Chromosoma 60:269–282

Gustafson JP, Butler E, McIntyre CL (1990) Physical mapping of a low-copy DNA sequence in rye (*Secale cereale* L.). Proc Natl Acad Sci USA 87:1899–1902

Heslop-Harrison JS, Schwarzacher T (1993) Molecular cytogenetics – biology and applications in plant breeding. In: Sumner AT, Chandley AC (eds) Chromosomes Today, vol 11. Chapman & Hall, London, pp 191–198

Heslop-Harrison JS, Harrison GE, Leitch IJ (1993) Reprobing of DNA:DNA in situ hybridization preparations. Trends Genet 8:372–373

Kenton A, Parokonny AS, Gleba YY, Bennett MD (1993) Characterization of the *Nicotiana tabacum* L. genome by molecular cytogenetics. Mol Gen Genet 240:159–169

Kenton A, Khashoggi A, Parokonny A, Bennett MD, Lichtenstein C (1995) Chromosomal location of endogenous geminivirus-related DNA sequences in *Nicotiana tabacum* L. Chromosome Res 3:346–350

Lapitan NLV (1992) Organization and evolution of higher plant nuclear genomes. Genome 35:171–181

Laurie DA, Pratchett N, Devos KM, Leitch IJ, Gale MD (1993) The distribution of RFLP markers on chromosome 2(2H) of barley in relation to the physical and genetic location of 5S rDNA. Theor Appl Genet 87:177–183

Lehfer H, Busch W, Martin R, Herrmann RG (1993) Localization of the B-hordein locus on barley chromosomes using fluorescence in situ hybridization. Chromosoma 102:428–432

Leitch IJ, Heslop-Harrison JS (1991) Physical mapping of plant DNA sequences by simultaneous in situ hybridization of two differently labelled fluorescent probes. Genome 34:329–333

Leitch IJ, Heslop-Harrison JS (1992) Physical mapping of the 18S-5.8S-26S rRNA genes in barley by in situ hybridization. Genome 35:1013–1018

Leitch IJ, Heslop-Harrison JS (1993) Physical mapping of four sites of 5S rDNA sequences and one site of the α-amylase-2 gene in barley (*Hordeum vulgare*). Genome 36:517–523

Leitch AR, Mosgoller W, Shi M, Heslop-Harrison JS (1992) Different patterns of rDNA organization at interphase in nuclei of wheat and rye. J Cell Sci 101:751–757

Leitch AR, Schwarzacher T, Wang ML, Leitch IJ, Šurlan-Momirovich G, Moore G, Heslop-Harrison JS (1993) Molecular cytogenetic analysis of repeated sequences in a long term wheat suspension culture. Plant Cell Tissue Organ Cult 33:287–296

Leitch AR, Schwarzacher T, Jackson D, Leitch IJ (1994) In situ hybridization: a practical approach. RMS microscopy handbook, vol 27. Bios Scientific Publishers, Oxford

Leitch AR, Schwarzacher T, Leitch IJ (1994) The use of fluorochromes in the cytogenetics of the small-grained cereals (Triticeae). Histochemical Journal 26:471–479

Leitch IJ, Parokonny AS, Bennett MD (1996) New insights into chromosome evolution in plants from molecular cytogenetics. Chromosomes Today 12 (Proceedings of the 12th International chromosome conference, Madrid) Parker JS (ed) Chapman and Hall, London (in press)

Lichter P, Cremer T, Borden J, Manuelidis L, Ward DC (1988) Deliniation of individual human chromosomes in metaphase and interphase cells by in situ suppression hybridization using recombinant DNA libraries. Hum Genet 80:224–234

Lo Y-MD, Mehal WZ, Fleming KA (1988) Rapid production of vector-free biotinylated probes using the polymerase chain reaction. Nucleic Acids Res 16:8719

Meinkoth J, Wahl G (1984) Hybridization of nucleic acids immobilized on solid supports. Ann Biochem 138:267–284

Moore G, Abbo S, Cheung W, Foote T, Gale M, Koebner R, Leitch A, Leitch I, Money T, Stancombe P, Yano M, Flavell R (1993) Key features of cereal genome organization as revealed by the use of cytosine methylation-sensitive restriction endonucleases. Genomics 15:472–482

Mouras A, Negrutiu I (1989) Localization of the T-DNA on marker chromosomes in transformed tobacco cells by in situ hybridization. Theor Appl Genet 78:715–720

Mouras A, Saul MW, Essad S, Potrykus I (1987) Localization by in situ hybridization of a low copy chimaeric resistance gene introduced into plants by direct gene transfer. Mol Gen Genet 207:204–209

Mukai Y, Nakahara Y, Yamamoto M (1993) Simultaneous discrimination of the three genomes in hexaploid wheat by multicolor fluorescence in situ hybridization using total genomic and highly repeated DNA probes. Genome 36:489–494

Parokonny AS, Kenton AY, Gleba YY, Bennett MD (1992) Genome reorganization in *Nicotiana* asymmetric somatic hybrids analysed by in situ hybridization. Plant J 2:863–874

Pedersen C, Giese H, Linde-Laursen I (1995) Towards an integration of the physical and the genetic chromosome maps of barley by in situ hybridization. Hereditas 123:77–88

Sambrook J, Fritsch EF, Maniatis T (1989) Molecular cloning: a laboratory manual. Cold Spring Harbor Laboratory Press, Cold Spring Harbor

Trask BJ (1991) Fluorescent in situ hybridization: applications in cytogenetics and gene mapping. Trends Genet 7:149–154

Viegas-Piquignot E, Dutrillaux B, Magdelenat H, Coppey-Moisan M (1989) Mapping of single-copy DNA sequences on human chromosomes by in situ hybridization with biotinylated probes: enhancement of detection sensitivity by intensified-fluorescence digital-imaging microscopy. Proc Natl Acad Sci USA 86:582–586

Whiting JH Jr, Pliley MD, Farmer JL, Jeffery DE (1989) In situ hybridization analysis of chromosomal homologies in *Drosophila melanogaster* and *Drosophila virilis*. Genetics 122:99–109

Wiegant J, Ried T, Nederlof PM, van der Ploeg M, Tanke HJ, Raap AK (1991) In situ hybridization with fluoresceinated DNA. Nucleic Acids Res 19:3237–3241

Chapter 12

Cytological Characterization of Somatic Hybrids: Detection of Genome Origin by Genomic In Situ Hybridization (GISH)

A.Y. KENTON, A.S. PAROKONNY, I.J. LEITCH, and M.D. BENNETT

12.1 Introduction

Genomic in situ hybridization (GISH) allows genomes differing in molecular composition to be discriminated visually in cell nuclei. Total genomic DNA from a given genotype (e.g. *AA*) is used to probe for chromosomes of A in hybrids (e.g. *AB*; Plate 1 b, c), allopolyploids (e.g. *AABB;* Bennett et al. 1992; Kenton et al. 1993b; Leggett and Markhand 1995), or lines of *B* introgressed with DNA from *A* (Mukai et al. 1993a; Parokonny and Kenton 1995). The hybridization target is chromosomal DNA in spreads (Plate 1 b, c), sections (Leitch et al. 1990) or whole, isolated nuclei (van Dekken et al. 1989). The last two examples are useful for studying genome domains in hybrid nuclei, a major application of GISH (Schwarzacher et al. 1989; Leitch et al. 1990).

Completeness of the GISH reaction depends mainly on the degree of hybridization between dispersed, highly repeated DNA sequences common to probe and target (chromosomal) DNA (Anamthawat-Jonsson et al. 1990; Anamthawat-Jonsson and Heslop-Harrison 1993a). GISH can thus be used to give a visual impression of the extent of sequence divergence among related species (e.g. from AA to A′ A′; Anamthawat-Jonsson et al. 1990; Kenton et al. 1993a). Although, initially, its use in plants was confined to intergeneric hybrids or other wide crosses, (Le et al. 1989; Schwarzacher et al. 1989), GISH has more recently been shown to discriminate between DNA from closely related species (Parokonny et al. 1992a; Kenton et al. 1993a; Anamthawat-Jonsson et al. 1993b). GISH has been used in plant breeding to determine the cytogenetic constitution of hybrids and the fate of their component genomes at meiosis (Le and Armstrong 1991; Bailey et al. 1993; Kenton et al. 1993b; Parokonny et al. 1994). It has also characterized addition, substitution and translocation lines in both sexual (Heslop-Harrision et al. 1990; Schwarzacher et al. 1992; King et al. 1993; Chen et al. 1995) and somatic hybrids (Piastuch and Bates 1990; Parokonny et al. 1992b; Plate 1 c). The smallest translocated segment so far detected by GISH is believed to contain around 50 Mbp of DNA (Mukai et al. 1993a; Parokonny and Kenton 1995). Sequential hybridizations with specific gene probes and GISH (Kenton et al. 1993b, 1995) are applicable to plant genetic engineering, for allocating transgene integration sites to a parental genome.

12.2 Optimizing GISH for Distinguishing Between Genomes

The resolution of GISH depends on many physical and chemical factors, including probe size, target accessibility and hybridization stringency. Only when these conditions have been optimized can meaningful conclusions about sequence homology be made. This is especially true for applications which aim to determine relationships between species (e.g. for biosystematic studies).

12.2.1 Choice of Labelling Method

A review of the labeling methods available is given by Leitch et al. (1994) and in Section 11.2.5. Of these, random priming (Mukai et al. 1993b) and nick translation have been successfully used for GISH. Nick translation is generally preferred, as it results in a probe size (300–600 bp) optimal for tissue penetration. This is especially important when the original size of the DNA is relatively large.

12.2.2 Total Genomic DNA as a Probe for In Situ Hybridization

Prior to nick translation, total genomic DNA must be mechanically sheared to a size suitable as a substrate for DNase I/DNA polymerase I activity (Sect. 12.4). Fragments of this length (around 10–12 kb; Schwarzacher et al. 1989), can be generated by hard vortexing (Parokonny et al. 1992a), repeated freezing and thawing, passing the DNA through a fine gauge needle (Mukai et al. 1993b), or sonication. Alternatively, a longer period of DNase I digestion or increased DNase I concentration may be used for nick translation (Manuelidis 1985). The size of the sheared DNA can be checked by gel electrophoresis in 1% agarose against a size marker. Total genomic DNA that is slightly degraded, and therefore unsuitable for digestion with restriction enzymes, is readily usable for GISH probes.

12.2.3
Probe Penetration/Target Accessibility

This presents one of the major difficulties of working with plant chromosomes. Cell walls and metabolites can present particular problems (Sect. 12.3.4). DNA in cytoplasmic organelles may bind strongly to the probe, increasing background signal, especially in meiocytes. The only certain remedy is to remove all possible cytoplasm. Denaturation protocols that also remove cytoplasm (e.g. acetic acid hydrolysis, (Sect. 12.5, note 2) or pepsin treatment detailed in Chapter 11 (Sect. 11.5, note 2) may be helpful. Non-specific binding to the probe by cell proteins and polysaccharides must also be minimized, especially when avidin is being used in the detection steps (Sect. 12.5). Protein binding sites can be blocked with immunoglobulins prior to detection (Sect. 12.4, note 6). Non-specific hybridization to chromosomes is blocked by adding a large excess of unlabelled, sonicated, unrelated DNA (e.g. salmon sperm DNA) to the hybridization mix. The small fragments bind readily to short sequences of universal homology in the target.

12.2.4
Stringency

Stringency describes the conditions (in terms of salt concentration and temperature) under which stable duplexes are permitted to form between single-stranded DNAs of a given length and G/C ratio (Britten et al. 1974). Sequences with a progressively greater degree of mismatch re-anneal only under comparatively less stringent conditions (higher salt concentration and lower temperature). Formamide allows the same stringency to be achieved at relatively lower temperatures, thus minimizing damage to chromosomes. The stringent washes given here (0.3 M NaCl, 50% formamide, 42 °C) allow stable duplexes to persist only between single-stranded DNAs sharing 80–85% of sequence homology (Meinkoth and Wahl 1984). Stringency can be modified by

(1) varying the salt concentration between 0.3 M and 15 mM NaCl ($2 \times$ SSC to $0.1 \times$ SSC).
(2) varying the formamide concentration between 10 and 60%
(3) varying the temperature of hybridization and post-hybridization washes

or a combination of (1)–(3).

These conditions, combined with the use of blocking DNA (see below) can be used to "fine-tune" the degree of sequence homology detectable between probe and target (Schwarzacher et al. 1992; Leitch et al. 1994; Plate 1 a,b).

12.2.5 Blocking DNA

Short, highly repeated DNA sequences re-anneal faster than longer, unique sequences (Britten et al. 1974). Thus, binding sites of probe *A* are rapidly saturated by any highly repeated sequences shared between genomes *A* and *B*, reducing their differentiation by GISH. Hybridization specificity for *A* is increased by including in the hybridization mix a large excess of unlabelled, total genomic DNA from *B*, sheared to ca. 250 bp fragments by autoclaving or sonication (Plate 1a,b; Schwarzacher et al. 1989; Anamthawat-Jonsson et al. 1990; Kenton et al. 1993a; Bailey et al. 1993). Repetitive sequences in the unlabelled (blocking) DNA saturate sites of common sequence homology, leaving only *A*-specific sites available for hybridization. The amount of blocking DNA required to maximize genome differentiation is determined empirically. Published examples have used blocking DNA in 10–60 $\times$ excess of the probe concentration (Schwarzacher et al. 1992).

12.2.6 Choice of Detection System

Fluorescence-based detection systems (described in this chapter and in Chap. 11) have the advantage of optimal resolution and contrast against a dark background. This makes them especially suitable for multiple detections or re-probings. However, they cost more than brightfield detection and need special equipment. Excellent results have been obtained for GISH using brightfield microscopy and enzyme-mediated detection systems (e.g. Mukai and Gill 1991). The latter rely on cleavage of a chromogenic substrate by either alkaline phosphatase or horseradish peroxidase, and subsequent reduction, respectively, of either nitroblue tetrazolium (NBT) to give a purple precipitate (outlined in Sect. 12.4), or diaminobenzidine (DAB) to give a brown precipitate (Mukai and Gill 1991). The brown DAB signal is counterstained blue with Giemsa. Carmine may be used as a pink counterstain for NBT, but is water-soluble and therefore incompatible with aqueous mountants. Xylene-based solutions also remove the colour from NBT preparations. Immersion oil is the best mounting medium for NBT and DAB.

12.2.7 Viewing and Recording the Results

The principles of fluorescence microscopy are outlined in Section 11.3. Optimal combinations of excitor and barrier filters make an important contribution to the intensity and contrast of the fluorescent image. The correct choice of appropriate objectives is also critical. Flat-field objectives of high numerical aperture (NA), such as Zeiss Plan Neofluar, are normally used, but cheaper phase contrast or fluorite objectives (e.g. Leica Plan Phaco or EF) may also be suitable at lower magnifications (e.g. x40). The intensity of the fluorescent image is proportional to $(NA)^4$. Several manufacturers therefore offer special immersion fluorescence objectives at low and medium power magnifications to enable the NA to be increased. A range of suitable objectives is available from Micro Instruments Ltd, Long Hanborough, Oxford, UK. A high speed, professional (daylight) film that maintains an acceptable grain size (e.g. Kodak Ektachrome 800/1600 ASA or Fuji 1000 ASA), should be used to record fluorescent images, using relatively short exposures (e.g. 30 s for GISH). The above films have the advantage that they can be processed at any one of four different film sensitivities.

12.3 Cytological Preparations

Because of local variation in GISH efficiency among cells within preparations, the mitotic index must be maximized by correct choice of material and chromosome pre-treatment. For fluorescence, absolute numbers of cells per slide may need to be lower than for stained preparations, due to interference by fluorescent materials in the same microscope field. Optimally, the specimen should be free of cell walls and cytoplasm, and slides should be free of dust or other light-scattering debris.

Plant Material

Nuclei from whole plants should be obtained from actively dividing meristems with a minimum of accumulated metabolites, e.g. root tips from in vitro culture or from young plants in full growth. Collect young root tips from potted plants that have been watered 24 h earlier, or from cuttings in hydroponic culture. Protocols for obtaining mitosis and meiosis from plant tissues are given below. Dividing nuclei can also be harvested directly from protoplasts (van Dekken et al. 1989; Ciupercescu 1991).

Reagents

- Pretreating agent of choice (Table 12.1)
- 96% ethanol
- Glacial acetic acid
- Chromium trioxide in 80% sulphuric acid (Merck)
- Vectabond (Vector Laboratories SP-1800)
- Acetone
- 45% acetic acid
- Enzyme mixture 2%(w/v) cellulase from *Trichoderma viride* (Cellulysin, Calbiochem 219466), 20% (v/v) pectinase (Sigma 40% solution in glycerol, P 4802) in citrate buffer, pH 4.8. Store in aliquots at –20°C
- Citrate buffer, pH 4.8: 0.1 M citric acid, 0.1 M trisodium citrate
- 2% (w/v) aceto-orcein; prepare a 4.4% (w/v) stock solution of synthetic orcein (Gurr) in glacial acetic acid. Reflux at least 30 min. For 2% solution, mix 9 ml stock, 11 ml distilled water and 1 ml 1 M HCl. Filter before use.

Equipment

- Glass tubes, 5-ml capacity
- Fine forceps
- Metal racks for 25 slides (Raymond A. Lamb E/102)
- 120 mm^2 glass staining jars (Raymond A. Lamb E/105)
- Microscope slides
- Coverslips (18 or 20 mm^2, see note 4)

Table 12.1. Pretreatments for the accumulation of metaphases

Treatment	Concentration[a]	Time (h)	Temp. (°C)
8-Hydroxyquinoline (Merck AnalaR) HARMFUL	290 mg/l in distilled water; dissolve at 60 °C	2.5–4.5	18
Colchicine (Sigma 9754) TOXIC	5 mg/l	3.5–4.0	18
"	"	1.5–2.0	Room temp.
1.4-Dichlorobenzene Aldrich 32, 933–9 HARMFUL	Saturated solution in water	20–24	4
1-Bromonaphthalene Merck HARMFUL	2–3 drops in 5 ml tap water, well shaken	20–24	4
Iced tap water	–	Up to 24	1

[a]All stock solutions are stored at 4 °C.

- Razor blade
- Line-free cloth, e.g. Selvyt (Agar Scientific)
- Dissecting microscope with x10 objective
- Microscope with phase contrast objectives (x10, x40)
- Fine scalpel with extra-small blade (Swann-Morton SM65)
- "Squasher" made from a flattened mounting needle
- Pair of hypodermic needles 25G, 25 mm (these can be mounted on supports with Superglue for easier handling)
- Drawn-out pasteur pipettes
- Humid chamber: line a lidded plastic sandwich box with paper towels moistened with distilled water (for method Sect. 12.3.3) or 45% acetic acid (for method Sect. 12.3.4). Place drinking straws, cut to size, over the towels as slide supports
- Strips of filter paper or chromatography paper
- Diamond pencil
- Dewar of liquid nitrogen or block of dry ice
- Tongs or long forceps
- Plastic boxes for slide storage (e.g. Merck 406/0287)

12.3.1
Pretreatment and Fixation (notes 1, 2 and 7)

Procedure

1. Cut root tips ca. 1 cm long from germinated seeds or young plants with a pair of fine forceps.
2. Transfer roots to one of the pretreating solutions in Table 12.1 (note 1).
   After the specified incubation time, transfer roots to freshly prepared fixative (3 : 1 96% ethanol: glacial acetic acid; note 2).
3. Leave at 4 °C for 30 min-1 h.
4. Transfer roots to fresh fixative, store at 4 °C for 24 h-2 weeks.
5. For longer storage, transfer the material to fresh fixative or 70% ethanol and store at –20 °C.

Super-Cleaning Slides with Chromic Acid (note 3)

1. Place slides in chromium trioxide in 80% sulphuric acid for at least 3 h at room temperature.
2. Wash in running water, 5 min.
3. Rinse thoroughly in distilled water and air-dry.
4. Transfer slides to 100% ethanol. Remove and dry as required.

Vectabond Treatment to Increase Slide Retentiveness

1. Set up a row of five staining jars containing, sequentially: 400 ml of acetone; 350 ml acetone mixed with 25 ml Vectabond; and three changes of distilled water.
2. Place slides in a metal rack. A rack of 25 slides will just be covered by 375 ml of solution in a standard 120 mm^2 glass staining jar. Do not place the slides back to back, as this hinders evaporation at the drying stage. Do not use plastic vessels or racks.
3. Holding the rack with tongs, dip into each solution, allowing 5 min in Vectabond. Agitate well for the water washes. Washing is improved if the rack can be placed in running distilled water. Diluted Vectabond cannot be stored, so prepare slides in batches of 500.

4. Drain slides very well, shaking off all excess liquid, and dry at room temperature or at 37 °C.
5. Store slides in a cardboard box at room temperature.
   Safety note. Vectabond can cause serious damage to human tissues. Wear gloves and eye protection. Preferably, carry out the whole process in a fume hood.

Treatment of Coverslips (note 4)

1. Scrape coverslips on both sides with a razor blade.
2. Immerse in ethanol for a few seconds.
3. Drain and polish with a lint-free cloth.

12.3.2 Chromosome Preparations from Root-Tip Squashes

All steps may be done using a dissecting microscope.

Procedure

1. Excise tips from one to three roots, depending on size.
2. Place tips on a Vectabonded slide. Cut off any root cap.
3. For small root tips, add 10 µl 45% acetic acid and macerate tips with "squasher" until meristematic cells are released. Split large root tips longitudinally and scrape out meristems with a mini-scalpel into 10–20 µl of 45% acetic acid. Then macerate meristems as above.
4. Remove and discard debris with hypodermic needles or fine forceps. (Do not allow the cells to dry out).
5. Stir the cell suspension with a hypodermic needle and adjust the volume to 10–15 µl with 45% acetic acid.
6. Place a prepared coverslip over the suspension.
7. Suck out any excess liquid with filter paper strips.
8. Pass the slide quickly through a flame once or twice. Do not overheat – the slide should not be too hot to touch.
9. Place filter paper over the coverslip and apply gentle pressure.

10. Look at the cells with phase contrast optics and repeat steps 8–9 until cells are optimally spread.

11. Note the co-ordinates of well-spread metaphases, and mark the position of the coverslip with a diamond pencil.

12. Freeze each preparation rapidly by dipping the slide in liquid nitrogen (use tongs), or placing it, coverslip upwards, on a briquette of dry ice for at least 5 min.

13. Quickly flip off coverslip with a scalpel and plunge the slide immediately into 96% ethanol for at least 5 s.

14. Air dry. Place slides in sealed plastic slide boxes with a sachet of silica gel and store at –20 °C.

12.3.3
Chromosome Preparations from Enzymic Digests (note 5)

Procedure

1. Excise fixed root tips with a scalpel.

2. Remove any hard root cap.

3. Place one to three tips on each Vectabonded slide, in a drop of citrate buffer.

4. Suck out buffer with a strip of filter paper, replace with a drop of enzyme mixture. Do not add a coverslip.

5. Place slides in a humid chamber at 37 °C for up to 2 h.

6. *Carefully* remove the enzyme mixture with a drawn out Pasteur pipette or micropipette tip (the roots will by now be very soft), watching the process under a dissecting microscope.

7. Add a drop of 45% acetic acid. The roots should disintegrate. If necessary, gently tease them apart.

8. Place a prepared coverslip over the drop.

9. Look at the cells using a low power (x10) phase contrast objective. Tap gently on the cover slip, if necessary, until the roots disintegrate into single cells.

10. Continue as Section 12.3.2, procedure steps **9** through**14**.

12.3.4
Meiotic Chromosome Preparations (notes 6–8)

Procedure

1. Extract one anther from a developing bud and press out the contents into 2% aceto-orcein to check stage of division.
2. Press out pollen mother cells (PMCs) from meiotic anthers into a drop of 45% acetic acid on a Vectabonded slide.
3. Make a suspension of PMCs with a "squasher". Remove debris with a pair of fine forceps, and place a prepared coverslip over the cells.
4. Examine with phase contrast, making sure that PMCs are well separated. If not, very light tapping on the coverslip with a needle will usually separate clumps. Suck out any surplus liquid with filter paper.
5. Place slides in a humid chamber containing 45% acetic acid, and leave at room temperature for 30–60 min.
6. Remove slides individually. Apply the edge of a strip of filter paper to the coverslip, drawing out any excess liquid. Then lay filter paper over the coverslip and press gently until cell walls break, examining constantly with phase contrast.

   Safety note. Acetic acid produces a harmful vapour. Open the box in the fume hood or in a very well-ventilated area.
7. When walls have broken, re-flood the slide with 45% acetic acid until protoplasts float free.
8. Press on the coverslip by degrees with filter paper until cells are optimally flattened.
9. Continue as Section 12.3.2, procedure steps **11** through **14.**

Notes

- *Note* 1. Pre-treatments improve chromosome morphology, synchronize mitosis and accumulate metaphases. Low temperature increases chromosome contraction, and improves the morphology of large chromosomes (e.g. *Allium cepa*, *Vicia faba*). Small chromosomes (e.g. *Phaseolus*) often respond better to no pre-treatment or shorter pre-treatments at room temperature. Experiment with different pre-treatments when examining a species for the first time, sampling root tips at intervals to obtain the optimum incubation time. Alternative pre-treatments for protoplast suspensions include hydroxyurea (Wang et al. 1986) and amiprophos-methyl (Pan et al. 1993).

- *Note* 2. Fixation should prevent DNA loss, while maintaining optimal chromosome morphology and permeability. Paraformaldehyde, a cross-linking fixative, minimizes DNA loss but may reduce permeability and destroy DAPI banding. Fixatives containing acetic acid soften cell walls, clear cytoplasm, enhance DAPI bands and are good for squash preparations. Fixation in 3:1 ethanol: acetic acid gives best results for root tip chromosomes after 24 h of fixation at 4 °C. Longer periods (> 2 weeks) may result in DNA loss or unacceptable fixation of cellular proteins. Material stored in 70% ethanol may require enzymatic digestion or soaking in 45% acetic acid prior to squashing.

- *Note* 3. To prevent loss of material from slides throughout the GISH procedure, slides are treated to make them more retentive, either by coating them (e.g. with silane, poly-L-lysine, or Denhardt's solution; Huang et al. 1983; Mouras 1991; Leitch et al. 1994) or by modifying the glass (e.g. with Vectabond). Slide coatings adhere better to slides that have been thoroughly cleaned. Vectabonded slides do not require special cleaning but this may be done optionally if slides appear greasy.

- *Note* 4. Use 20 mm^2 coverslips for making preparations and 22 mm^2 coverslips for mounting hybridized slides. This ensures that the edge of the coverslip does not obliterate cells at the periphery.

- *Note* 5. Alternative protocols, using suspensions, are given by Ambros et al. (1986) and Schwarzacher and Leitch (1993).

- *Note* 6. Pollen mother cells (PMCs) have several advantages over somatic tissue for in situ hybridization:
  - divisions are numerous and higly synchronized;
  - chromatin is condensed to a different degree at different stages, often enabling problems associated with DNA density and compaction to be overcome;
  - a wider range of cytological parameters can be studied.

- *Note* 7. Meiocyte walls are autofluorescent and may prevent successful probe penetration. Walls may also be removed enzymatically (Dickinson and Sheldon 1984).

- *Note* 8. Fixation in 3:1 ethanol: acetic acid facilitates extrusion of protoplasts from PMC walls, but results in "sticky" meiotic chromosomes. Proteases, sometimes used to reduce cytoplasm, may further increase stickiness.

12.4 Probe Preparation

Reagents

- 10x nick translation buffer (NTB); 0.5 M Tris-HCl, pH 7.5, 50 mM $MgCl_2$, 0.5 mg/ml nuclease-free BSA (Sigma B 1752)
- 10x dNTP mix; 0.5 mM dATP, 0.5 mM dGTP, 0.5 mM dCTP, 0.1 mM dTTP (prepared from 100 mM stock solutions of dNTPs, Boehringer Mannheim 1051440, 1051458, 1051466, 1051482)
- 0.1 M dithiothreitol (DTT; Sigma D 5545)
- 1 mM biotin-11-dUTP; dilute 20 mM stock solution (Sigma B 8280) to 1 mM with sterile 50 mM Tris-HCl, pH 7.5)
- 10x enzyme mix (0.4 U/μl DNA polymerase I + 40 pg/μl DNase I); Life Technologies, 5108162SA
- 0.5 M EDTA (pH 7.5)
- Sterile distilled water (SDW)
- Sonicated salmon sperm DNA, 10 mg/ml (Stratagene)
- 3 M sodium acetate
- Freezer-cold 96% ethanol
- 10x buffer 1: 1 M Tris-HCl, pH 7.5, 1 M NaCl, 20 mM $MgCl_2$, 0.5% Trition-X-100. Prepare the buffer without Triton initially. Autoclave, cool to room temperature and add Triton. Working solution = 1x.
- Buffer 2: 3% bovine serum albumin (BSA, note 1) in 1x buffer 1, filter-sterilize
- 10x buffer 3: 1 M Tris-HCl (pH 9.5), 1 M NaCl, 0.5 M $MgCl_2$. Allow any precipitate to settle before decanting the buffer for use. Working solution = 1x
- Alkaline phosphatase buffer, pH 7.3* (note 2): 3 M NaCl, 1 mM $MgCl_2$, 0.1 mM $ZnCl_2$. Make up to 10 ml with SDW. Filter-sterilize, add 1.86 mg triethanolamine. Dissolve well, store at 4 °C.
- Biotinylated alkaline phoshatase*: Add 250 μl alkaline phosphatase buffer to 0.25 g lyophilized biotinylated alkaline phosphatase (Sigma P 8024). Store at 4 °C.

- Streptavidin buffer: sterile 50 mm Tris-HCl, pH 7.5, 0.2 mg/ml sodium azide
  Safety note. Sodium azide (a preservative) is *Toxic* and should be used with caution.
- Streptavidin*, add 0.5 ml streptavidin buffer to 0.5 mg of lyophilized streptavidin (Sigma S 4762), store at 4 °C
- NBT/BCIP* stable mix (75 mg/ml NBT, 50 mg/ml BCIP; Life Technologies, 193–5984SA)
- 100x TE (pH 8.0): 1 M Tris-HCl, pH 8, 0.1 M EDTA, pH 8, working solution = 1x

Equipment

- Nitrocellulose membrane (Schleicher & Schuell)
- Vacuum oven at 80 °C
- Plastic tweezers
- Small plastic containers, e.g weighing boats, for 1 ml incubations
- Plastic washing dishes (e.g. small sandwich boxes)
- Vortexer
- Rotary shaker

12.4.1 Isolation of Genomic DNA

Procedure

Standard protocols can be used to isolate total genomic DNA for GISH (e.g. the CTAB method of Doyle and Doyle (1990). Protocols are provided in Section 1.1. Resuspend the DNA in 1x TE at a convenient concentration (e.g. 0.5 μg/μl).

1. Shear DNA of genome A (probe) to 12 kb or less by vortexing hard for 30–60 s (see Sect. 12.2.2 for alternatives). Check size in a 1% agarose gel.

2. Autoclave DNA from genome B (blocking DNA), in aliquots of 150 μg, for 5 min at 15 lb/in^2 (ca.2.65 kg/cm^2).

12.4.2 Biotin Labelling of Genomic DNA by Nick Translation (notes 3, 4)

Procedure

The reaction is prepared in a total volume of 150 μl.

1. Place in an Eppendorf tube on ice:
   x μl DNA from genome A (to contain 3 μg)
   y μl SDW (y = 150-subtotal)

15 μl 10x NTB
3 μl DTT (note 5)
12 μl 10x dNTP mix
3 μl biotin-11-dUTP (note 6)
15 μl 10x enzyme mix

2. Incubate at 16 °C for 1.5 h.

3. Add 3 μl 0.5 M EDTA and mix

4. Add 9 μl salmon sperm DNA and (optionally) 15 μg blocking DNA from species B. Mix very thoroughly.

5. Add 15 μl 3 M sodium acetate. Mix and add 125 μl freezer-cold ethanol.

6. Mix by inverting the tube, store for 2 h at –20 °C.

7. Centrifuge at 12 000 *g* for 10–20 min.

8. Carefully pour off supernatant and blot mouth of tube.

9. Add 450 μl freezer-cold 70% ethanol.

10. Centriguge at 12 000 *g*, 10 min.

11. Pour off ethanol, drain pellet and air-dry.

12. Resuspend in 21 μl of 1x TE and leave to dissolve at 37 °C for at least 30 min.

12.4.3
Testing Biotin Incorporation by Spot Hybridization

Procedure To avoid problems later, probes should always be analyzed for biotin incorporation before proceeding with GISH.

1. To prepare a test strips, spot 1 μl of biotinylated probe (step 12, Sect. 12.4.2) onto a small piece of nitrocellulose.

2. Air-dry and bake under vacuum for 30 min at 80 °C. Several samples can be spotted on the same test strip.

3. Hydrate test strip in 1x buffer 1 for 1 min

4. Transfer to 10 ml buffer 2, pre-warmed to 42 °C, in a small weighing boat. Incubate 20 min.

5. Remove test strips, blot gently. Optionally, bake under vacuum at 80 °C for 15 min.

6. Re-hydrate test strips in 10 ml of buffer 2 at room temperature for 10 min. Perform all subsequent steps with gentle agitation (e.g. on a rotary shaker). Never let the strips dry out.

7. In an Eppendorf tube, mix 1 ml of lx buffer 1 and 2 µl of streptavidin. Drain test strips, add streptavidin solution. Incubate, agitating, for 10 min.

8. Wash in at least 20 ml of 1x buffer 1, 3 x 3 min, on a rotary shaker at low speed.

9. In an Eppendorf tube, mix 1 ml of buffer 1 and 1 µl of biotinylated alkaline phosphatase. Place in a clean boat, add the washed test strips and incubate 10 min.

10. Wash 2 × 3 min in a clean dish containing at least 20 ml of 1x buffer 1.

11. As step **10**, replacing 1x buffer 1 with 1x buffer 3.

12. Transfer test strips to a plastic bag, sealed on three sides, containing 5 ml BCIP/NBT mix

13. Seal the fourth side and allow reaction to develop in the dark; 1µl samples from biotinylated genomic DNA should produce an intense blue spot within 5–10 min.

Notes

- *Note* 1. For blocking non-specific antibody binding to filters or chromosome preparations, we obtain good results with a general-purpose BSA (Advanced Protein Products). BSA solutions are filter-sterilized immediately before use, to prevent contamination with micro-organisms.

- *Note* 2. Products marked with * are also included as part of the DNA Detection Kit (Life Technologies, 18239–012).

- *Note* 3. The protocol given is for labelling 3 µg total genomic DNA, resulting in a final concentration of 15 µg/ml total DNA in the hybridization mixture, to be used for 10–12 slides. The concentration of labelled probe depends on the incorporation level achieved (ca. 25% for nick translation, i.e. about 4 µg/ml).

- *Note* 4. A proprietary kit, e.g. Bionick labelling system (Life Technologies, 18247–015), may be used to biotinylate 3 μg of DNA, following the manufacturer's protocol.
- *Note* 5. Recommended concentrations of DTT vary in different nick translation protocols.
- *Note* 6. The above protocol is for biotinylated probes. Other conjugates can be substituted for biotin-11-dUTP, e.g. 3 μl 1 mM digoxygenin-11-dUTP (Boehringer Mannheim 1093 088) or 1 mM fluorescein-11-dUTP (Amersham FluoroGreen, RPN 2121) or 1 mM rhodamine-4-dUTP (Amersham FluoroRed, RPN 2122); or 1.5 μl FluoroRed + 1.5 μl FluoroGreen. Probes labelled by other methods require modified detection systems (Sect. 11.5.2).

12.5 Genomic In Situ Hybridization

Reagents

- 20 μl (3 μg) of biotinylated DNA (Sect. 12.4.2, step **12**)
- 1x TE (pH 8)
- Denaturing solution (70% v/v formamide in 2x SSC)
- 50, 70 and 100% ethanol on ice
- Formamide (Fison's electrophoresis grade, F/1570/08)
- 50% dextran sulphate, filtered through a Sartorius Minisart Plus filter (note 1)
- 20x SSC (3 M NaCl, 0.3 M trisodium citrate); autoclave
- Formamide wash: 50% (v/v) formamide in 2x SSC
- 10x BT buffer (1 M sodium bicarbonate, 0.5% (v/v) Tween 20), autoclave before adding Tween, working solution = 1x
- BSA block (5% BSA in 1x BT buffer, filter sterilize)
- F-avidin: 1 mg fluorescein avidin DN (Vector Laboratories, A-3101), reconstituted in 1 ml of SDW
- Anti-avidin: 0.5 mg biotinylated anti-avidin D (Vector Laboratories, BA 0300) reconstituted in 1 ml SDW
- Goat serum block: 5% normal goat serum (Vector Laboratories, S-1000) in 1x BT, filter-sterilize
- Phosphate buffered saline, pH 7.4 (PBS); 120 mM NaCl, 7 mM Na_2HPO_4, 3 mM NaH_2PO_4, 2.7 mM KCl, autoclave
- Propidium iodide (PI) stock solution, 1 mg/ml in PBS
  Safety note. PI is a *MUTAGEN* – handle with caution!

- Vectashield antifade (Vector Laboratories H-1000) containing 1 μg/ml DAPI (Sect. 11.6)

Equipment

- Plastic Coplin jars (Raymond A. Lamb E/96)
- Water bath at 70–72 °C (preferably in fume hood)
- Small water bath at 76 °C
- Plastic "coverslips" (approx. 22 mm^2 pieces cut from an autoclave bag)
- Fine, clean forceps
- Humid chamber(s) as Sect. 12.3, substituting 2x SSC for distilled water or acetic acid
- Incubator at 37 °C
- Shaking water bath, preferably in fume hood
- Rotary shaker
- Prepared 22 mm^2 glass coverslips (as Sect. 12.3)
- Rubber solution (Dunlop)
- Epifluorescence photomicroscope with appropriate filters for FITC, DAPI and PI (see Table 11.1)

Procedure

Denaturation and Hybridization (note 2)

1. Prepare the hybridization mix. Increase the volume of the biotinylated probe to 40 μl with 1x TE.
2. Votex well and add 100 μl formamide.
3. Add 40 μl of 50% dextran sulphate.
4. Mix well by vortexing and store on ice.
5. Remove slides from freezer, thaw and air dry (note 3). Re-fix by immersing for 30 min in 3:1 ethanol:glacial acetic acid and 2×10 min in 96% ethanol. Air-dry.
6. Heat denaturing solution to 70–72 °C in a plastic Coplin Jar standing in a water both. (**Safety note.** For all steps involving formamide (*TOXIC*) either use a fume hood or wear a respirator (Sect. 11.6)).
7. Denature slides in pairs for exactly 2 min at 70–72 °C.
8. Dehydrate each pair through 50, 70 and 100% ethanol, and air-dry.
9. Denature hybridization mix in a waterbath at 76 °C for 15 min. Place on ice and add 20 μl of 20x SSC. Mix thoroughly, centrifuge

briefly (5 s), store on ice for 2–3 min. Aim to complete slide denaturation slightly ahead of probe.

10. Stir probe and pipette 15–20 µl onto each slide. Cover with a plastic square, avoiding air bubbles.

11. Place slides in a humid chamber at 37 °C, 12–15 h.

Post-Hybridization Washing (note 3)

All washing steps are done with gentle agitation (ca. 120 strokes/min) in a shaking bath or rotary shaker.

12. Pre-heat two dishes of 2x SSC (one containing a slide rack) and one dish of formamide wash to 42 °C in a shaking water bath.

13. Peel plastic from each slide and place immediately in the slide rack in 2x SSC at 42 °C. Agitate for 5 min.

14. Using forceps, transfer rack containing slides to formamide wash at 42 °C for exactly 10 min.

15. Transfer to 2x SSC at 42 °C for 10 min.

16. Transfer to 2x SSC at room temperature for 5 min.

17. Transfer to 1x BT at room temperature for 10 min.

Detection of Biotinylated Probe

18. Drain each slide individually and pipette 50 µl of BSA block onto the cells. Cover slides with plastic coverslips and incubate at room temperature for 5 min.

19. Dilute F-avidin to 10 µg/ml in BSA block. Peel plastic off each slide in turn, drain and add 50 µl diluted F-avidin solution. Cover with clean plastic coverslip. Incubate slides in a humid chamber at 37 °C for 1 h. Meanwhile, heat two dishes of 1x BT (one containing a slide rack) to 40 °C in a shaking water bath.

20. Peel plastic from each slide and place immediately in warmed 1x BT. When all slides are in BT, wash 2×5 min.

21. As step **18,** substituting goat serum block for BSA block.

22. Dilute anti-avidin to 25 µg/ml in goat serum block. Peel plastic from each slide in turn, drain and add 50 µl diluted anti-avidin.

Cover with clean plastic cover slips and incubate in a humid chamber at 37 °C for 1 h.

23. Wash 2 × 5 min in 1x BT at 40 °C (as step **20**).

24. Repeat the incubation in F-avidin (as step **19**).

25. Wash 2 × 5 min 1x BT at 40 °C (as step **20**).

Counterstaining and Mounting (note 4)

26. Dilute PI to 0.5 μg/ml in PBS.

27. Drain slides individually and immediately add 50 μl diluted PI. Cover with plastic coverslips and place in a humid chamber, in the dark, for 10 min (note 4).

29. Wash off each coverslip in turn by squirting slide gently with PBS from a clean wash bottle.

30. Drain each slide and mount immediately with a prepared glass coverslip in 12–15 μl of Vectashield /DAPI

31. Blot slides well, ring with rubber solution (note 5).

Re-Probing Slides for Sequential ISH

The above protocol can be applied after first hybridizing the slides with a probe for a specific DNA sequence (e.g. a gene or repetitive DNA, Chap. 11 and Plate 1 e, f). Wash slides (Sect. 11.6), then proceed to Section 12.5.

Notes

- *Note* 1. The Minisart Plus filter contains a binder-free glass fibre pre-filter and 0.22-μm sterilizing filter. This greatly facilitates the filtration of dextran sulphate.
- *Note* 2. Always check the temperature of the denaturation mix before proceeding. When working with a species for the first time, experiment with different times and temperatures of denaturation. Alternative protocols are:
  - 45% acetic acid at 60 °C for 15–30 min (Hagemann et al. 1992),
  - exonuclease digestion (van Dekken et al. 1988),
  - simultaneous denaturation of probe and target DNA at 90 °C (Leitch et al. 1990).

- *Note* 3. RNase treatment may be used prior to denaturation, to remove endogenous RNA which might otherwise bind the probe. Binding of RNA is, however, unlikely to contribute significantly to background when whole genomic DNA is used as a probe. A protocol for RNase treatment is included in Section 11.6.
- *Note* 4. Incorporating PI into the antifade mountant results in an intense red fluorescence which may mask FITC signal. A separate staining step is therefore preferable.
- *Note* 5. When using glycerol-based solutions, e.g. Vectashield ensure that as much liquid as possible has been removed from beneath the coverslip. Too much glycerol causes "glare" and poor signal resolution,

Controls It is essential to run controls to set a baseline for the amount of non-specific hybridization, and to help trace any problems with the protocol.

- *Positive control.* Use total genomic DNA to probe chromosomes of the same species or (preferably) genotype.
- *Negative control.* Use total genomic DNA to probe chromosomes of an unrelated species (e.g. a different genus of the same family).
- *Internal control.* Mix cells of an unrelated species, whose chromosomes are easy to recognize, on the same slide as the test species.

roubleshooting See Table 11.2.

12.6 Microscopy And Photography

Optimization of microscopy and photography are discussed in Section 12.2.7 (see also Sect. 11.6, notes 11–15). Excellent hints on equipment and procedures are also provided by larger manufacturers of microscopes.

Acknowledgements. We are grateful to B.C. Cowen and H. Shearman (Micro Instruments Ltd, Oxford, UK) and Dr. A.R. Leitch (Queen Mary and Westfield College, London, UK) for helpful comments on the manuscript.

References

Ambros PF, Matzke MA, Matzke AJM (1986) Detection of a 17 kb unique sequence (T-DNA) in plant chromosomes by in situ hybridization. Chromosoma 94:11–18

Anamthawat-Jonsson K, Schwarzacher T, Leitch AR, Bennett MD, Heslop-Harrison JS (1990) Discrimination between closely related Triticeae species using genomic DNA as a probe. Theor Appl Genet 79:721–728

Anamthawat-Jonsson K, Schwarzacher T, Heslop-Harrison JS (1993a) Behavior of parental genomes in the hybrid *Hordeum vulgare* x *H. bulbosum*. J Hered 84: 78–82

Anamthawat-Jonsson K, Heslop-Harrison JS (1993b) Isolation and characterization of genome-specific DNA sequences in Triticeae species. Mol Gen Genet 240:151–158

Bailey JP, Bennett ST, Bennett MD, Stace CA (1993) Genomic in situ hybridization identifies parental chromosomes in the wild grass hybrid × *Festulpia hubbardii*. Heredity 71:413–420

Bennett ST, Kenton AY, Bennett MD (1992) Genomic in situ hybridization reveals the allopolyploid nature of *Milium montianum* (Gramineae). Chromosoma 101:420–424

Britten RJ, Graham DE, Neufeld BR (1974) Analysis of repeating DNA sequences by reassociation. In: Grossman L, Moldave K (eds) Methods in enzymology, XXIX. Nucleic acids and protein synthesis, part E. Academic Press, New York, pp 363–418

Chen Q, Conner RL, Laroche A (1995) Identification of the parental chromosomes of the wheat-alien amphiploid Agrotana by genomic in situ hybridization. Genome 38:1163–1169

Ciupercescu DD (1991) Karyotyping with protoplast procedures. In: Negrutiu I, Gharti-Chhetri G (eds) A laboratory guide for cellular and molecular plant biology. Birkhauser, Berlin, pp 289–295

Dickinson HG, Sheldon JM (1984) A radial system of microtubules extending between the nuclear envelope and the plasma membrane during early male haplophase in flowering plants. Planta 161:86–90

Doyle JJ, Doyle JL (1990) Isolation of plant DNA from fresh tissue. Focus 12:13–15

Hagemann S, Scheer B, Schweizer D (1993) Repetitive DNA sequences in the genome of *Anemone blanda*: identification of tandem arrays and of dispersed repeats. Chromosoma 102:312–324

Heslop-Harrison JS, Leitch AR, Schwarzacher T, Anamthawat-Jonsson K (1990) Detection and characterization of 1B/1R translocations in hexaploid wheat. Heredity 65:385–392

Huang WM, Gibson SJ, Facer P, Gu J, Polak JM (1983) Improved section adhesion for immunohistochemistry using high molecular weight polymers of L-lysine as a slide coating. Histochemistry 77:275–279

Kenton A, Parokonny AS, Bennett ST, Bennett MD (1993a) Does genome organization influence speciation? A reappraisal of karyotype studies in evolutionary biology. In: Lees DR, Edwards D (eds) Evolutionary patterns and processes, Chap. 12. Linnean Society, London, pp 189–206

Kenton A, Parokonny AS, Gleba YY, Bennett MD (1993b) Characterization of the *Nicotiana tabacum* genome by molecular cytogenetics. Mol Gen Genet 240:159–169

Kenton A, Khashoggi A, Parokonny A, Bennett MD, Lichtenstein C (1995) Chromosomal location of endogenous geminivirus-related DNA sequences in *Nicotiana tabacum* L. Chromosome Res 3:346–350

King IP, Purdie KA, Rezanoor HN, Koebner RMD, Miller TE, Reader SM, Nicholson P (1993) Characterization of *Thinopyrum bessarabicum* chromosome segments in wheat using random amplified polymorphic DNAs (RAPDs) and genomic in situ hypbridization. Theor Appl Genet 86:895–900

Le HT, Armstrong KC (1991) In situ hybridization as a rapid means to assess meiotic pairing and detection of alien DNA transfers in interphase cells of wide crosses involving wheat and rye. Mol Gen Genet 225:33–37

Le HT, Armstrong KC, Miki B (1989) Genomic DNA as a probe. Plant Mol Biol Rep 7: 150–158

Leggett JM, Markhand GS (1995) The genomic structure of *Avena* revealed by GISH. In: Brandham PE, Bennett MD (eds.) *Kew Chromosome Conference IV* Royal Botanic Gardens, Kew UK, pp 133–139

Leitch AR, Mosgoller W, Schwarzacher T, Bennett MD, Heslop-Harrison JS (1990) Genomic in situ hybridization to sectioned nuclei shows chromosome domains in grass hybrids. J Cell Sci 95:335–341

Leitch AR, Schwarzacher T, Jackson D, Leitch IJ (1994) In situ hybridization: a practical guide. RMS Microscopy Handbooks, no 27. Bios Scientific Publishers Ltd, Oxford

Manuelidis L (1985) In situ detection of DNA sequences using biotinylated probes. Focus 7:4–8

Meinkoth J, Wahl G (1984) Hybridization of nucleic acids immobilized on solid supports. Anal Biochem 138:267–284

Mouras A (1991) Gene targeting in plant metaphase chromosomes by in situ hybridization with tritiated probe DNA. In: Negrutiu I, Gharti-Chhetri G (eds) A laboratory guide for cellular and molecular plant biology. Birkhauser, Berlin, pp 296–307

Mukai Y, Gill BS (1991) Detection of barley chromatin added to wheat by genomic in situ hybridization. Genome 34:448–452

Mukai Y, Friebe B, Hatchett JH, Yamamoto M, Gill BS (1993a) Molecular cytogenetic analysis of radiation-induced wheat-rye terminal and intercalary chromosomal translocations and the detection of rye chromatin specifying resistance to Hessian fly. Chromosoma 102:88–95

Mukai Y, Nakahara Y, Yamamoto M (1993b) Simultaneous discrimination of the three genomes in hexaploid wheat by multicolour fluorescence in situ hybridization using total genomic and highly repeated DNA probes. Genome 36:489–494

Pan WH, Houben A, Schlegel R (1993) Highly effective cell synchronization in plant roots by hydroxyurea and amiprophos-methyl or colchicine. Genome 36:387–390

Parokonny AS, Kenton AY (1995) Comparative physical mapping and evolution of the *Nicotiana tabacum* L. karyotype. In: Brandham PE, Bennett MD (eds.) *Kew Chromosome Conference IV* Royal Botanic Gardens, Kew UK, 301–320

Parokonny AS, Kenton AY, Meredith L, Owens SJ, Bennett MD (1992a) Genomic divergence of allopatric sibling species studied by molecular cytogenetics of their F1 hybrids. Plant J 2:695–704

Parokonny AS, Kenton A, Gleba YY, Bennett MD (1992b) Genome reorganization in *Nicotiana* asymmetric somatic hybrids analyzed by in situ hybridization. Plant J 2:863–874

Parokonny AS, Kenton A, Gleba YY, Bennett MD (1994) The fate of recombinant chromosomes and genome interaction in *Nicotiana* asymmetric somatic hybrids and their sexual progeny. Theor Appl Genet 89:488–497

Piastuch WC, Bates GW (1990) Chromosomal analysis of *Nicotiana* asymmetric somatic hybrids by dot blotting and in situ hybridization. Mol Gen Genet 222:97–103

Schwarzacher T, Leitch AR (1993) Enzymatic treatment of plant material to spread chromosomes for in situ hybridization. In: Isaac PG (ed) Methods in molecular biology. XX. Protocols for nucleic acid analysis by nonradioactive probes. Humana Press, Totowa, New Jersey

Schwarzacher T, Leitch AR, Bennett MD, Heslop-Harrison JS (1989) In situ localization of parental genomes in a wide hybrid. Ann Bot 64:315–324

Schwarzacher T, Anamthawat-Jonsson K, Harrison GE, Islam AKMR, Jia JZ, King IP, Leitch AR, Miller TE, Reader SM, Rogers WJ, Shi M, Heslop-Harrison JS (1992) Genomic in situ hybridization to identify alien chromosomes and chromosome segments in wheat. Theor Appl Genet 84:778–786

Van Dekken H, Pinkel D, Mullikin J, Gray JW (1988) Enzymatic production of single-stranded DNA as a target for fluorescence in situ hybridization. Chromosoma 97:1–5

Van Dekken H, Pinkel D, Mullikin J, Trask B, van den Engh G, Gray J (1989) Three-dimensional analysis of human chromosome domains in human-hamster hybrid interphase nuclei. J Cell Sci 94:299–306

Wang AS, Phillips RL, Mi CC (1986) Cell cycle parameters and accumulation of metaphase cells in maize suspension cultures. Plant Sci 46:53–61

Platel 1a–h. (a) and **(b)** *Gibasis consobrina* × *G. karwinskyana* (Commelinaceae) GISH probe = *G. consobrina* DNA **a** No blocking DNA **b** 10× excess of blocking DNA from *G. karwinskyana*. Post-hybridization washes were 37 °C for **a** and 42 °C for **b**. **c, d, e** and **f** Asymmetric somatic hybrid of *Nicotiana sylvestris* (+) *N. plumbaginifolia* (Solanaceae).**c** GISH probe = *N. sylvestris* DNA. *Arrowhead* shows intergenomic chromosome translocation. **d** Meiocyte probed with repetitive DNA sequence (HRS 60.1) **e** The same meiocyte as in **(d)** sequentially reprobed with *N. sylvestris* genomic DNA. **f** Meiocyte simultaneously probed with biotinylated HRS 60.1 (detected with fluorescein, green fluorescence) and dioxigenin-labelled telomere-specific repeated sequence (detected with rhodamine, red fluorescence). Results recorded by triple exposure (rhodamine, fluorescein and DAPI filters). **g** *Zingeria biebersteiniana* (2n = 4; Gramineae); mitotic metaphase simultaneously probed with digoxigenin-labelled 5S ribosomal DNA (detected by green fluorescence) and rhodamine-dUTP-labelled 18S-5.8S-25S ribosomal DNA (detected by red fluorescence). Left DAPI fluorescence; centre double exposure (rhodamine and fluorescein filters), Right triple exposure (rhodamine, fluorescein and DAPI filters), Bennett et al. 1995. **h** *Nicotiana tabacum* (Solanaceae) meiocyte probed with virus-related DNA sequence (arrowhead), 900 bp long, c.20 copies per genome Kenton et al. 1995. In **a** to **e** and **h** probes were labelled with biotin, detected with fluorescein and counterstained with PI; **f** and **g** were counterstained with DAPI. Bar 10 μm

Acknowledgement. We greatefully acknowledge the collaboration of Dr. C. Lichtenstein and A. Khashoggi (Centre for Biotechnology, Imperial College, London), which led to the physical mapping of the virus-related DNA sequence in **h**.

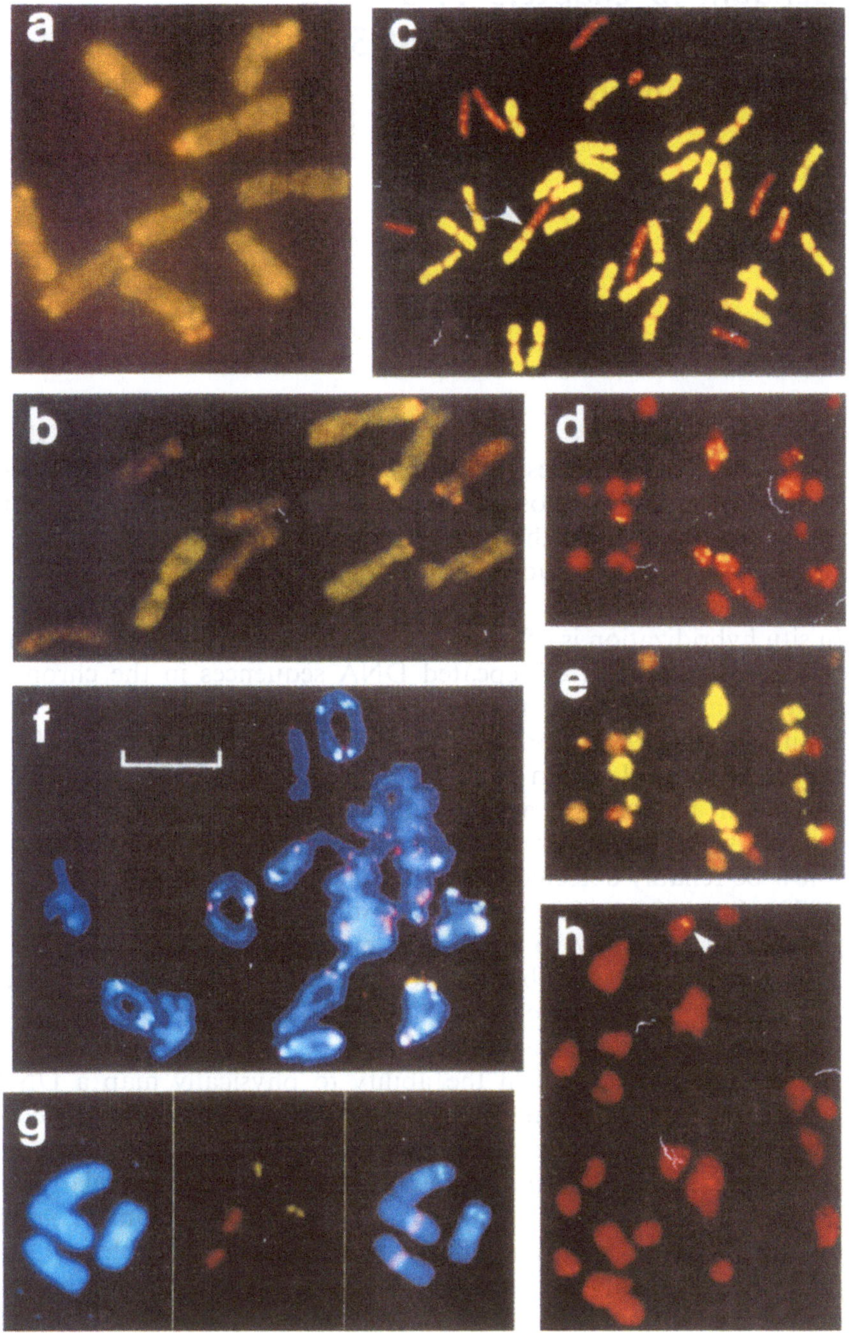
a
c
b
d
e
f
h
g

Addendum to Chapters 11 and 12 In Situ Hybridization in Plant Species with Small Chromosomes

N. Lapitan

A.1 Introduction

In situ hybridization is a technique for physically localizing DNA sequences on chromosomes. It involves hybridizing a DNA or RNA probe to denatured chromosomes fixed on a glass slide. If the probe is labelled non-isotopically, the sites of hybridization are detected either colorimetrically or with fluorescence dyes and then visualized under a microscope.

In situ hybridization is often used for studying the organization and distribution of families of repeated DNA sequences in the chromosomes of animal and plant species (for example, Gall and Pardue 1969; Hutchinson et al. 1981; Manuelidis et al. 1982; Rayburn and Gill 1986; Ganal et al. 1991; Lapitan 1992). Recent developments in the technique and in microscopy have resulted in improved resolution and sensitivity, so that hybridization sites of single- or low-copy sequences can now be reliably detected (Lichter et al. 1991; Trask 1991; Ried et al. 1992). In some species, such as *Drosophila* and *humans*, the technique is used not only to physically map cloned genes and DNA markers on chromosomes, but also to determine the physical distances between DNA markers on chromosomes (Kafatos et al. 1991; Trask et al. 1991; van den Engh et al. 1992).

A factor that is critical in the ability to physically map a DNA sequence on a chromosome is the length or condensation state of chromosomes to which a probe is hybridized. In small and very condensed chromosomes, the spatial relationship of a sequence relative to other parts of a chromosome, such as centromere and telomeres, may not be possible to determine. This is particularly a problem if the centromeres of small chromosomes are not obvious under the microscope. Furthermore, in small chromosomes, hybridization signals of repeated DNA sequences may occupy an entire chromosome or chromosome arm. The resolution that can be achieved in gene mapping by in situ hybridization depends on the condensation of chromosomes. In human metaphase chromosomes, probes can be mapped by in situ hybridization with a resolution >3 million base pairs (Mbp; Trask 1991). In comparison, in interphase cell nuclei, the order of sequences separated by as little as 50 kilo base pairs (kb) can be determined (Lawrence et al. 1990; van den Engh et al. 1992).

To improve the resolution of in situ hybridization, elongated chromosomes obtained by different methods have been used. These include the use of prometaphase mitotic chromosomes (Lapitan et al. 1989), surface-spread pachytene chromosomes (Moens and Pearlman 1989, 1990; Albini and Schwarzacher 1992), interphase cell nuclei (Lawrence et al. 1988; Trask et al. 1991), free chromatin (Heng et al. 1992), and naked DNA stretched by allowing DNA in solution to flow on a glass slide (Parra and Windle 1993). High resolution can be achieved with the last three methods. Interphase and chromatin mapping can resolve DNA sequences as close as 50 and 20 kb, respectively (Lawrence et al. 1988; Trask et al. 1991). DNA sequences as close as 5 kb have been resolved by in situ hybridization on stretched, naked DNA (Parra and Windle 1993). However, a disadvantage with these methods is that they do not show the position of DNA sequences on chromosomes. Furthermore, the usefulness of these materials for in situ hybridization in plants has not been demonstrated. In comparison, the use of prometaphase and surface-spread pachytene chromosomes provides information on chromosome arm location of DNA sequences and their spatial relationship to centromeres and telomeres. However, these methods give considerably much lower levels of resolution compared to the use of interphase nuclei, chromatin, or stretched DNA. Therefore, the choice of material to use for in situ hybridization depends on the purpose of the study.

In this chapter, a procedure for obtaining elongated prometaphase chromosomes from plants using a DNA replication-inhibiting compound is described. A protocol for preparing cytoplasm-free chromosome squashes, is also included. Biotin-labelling of DNA probes, in situ hybridization and colorimetric and fluorescence detection are described in detail in the previous two chapters (11 and 12).

A.2
Obtaining Prometaphase Chromosomes Using Aphidicolin

Mitotic chromosomes of plants are usually obtained with the use of mitostatic agents such as colchicine, hydroxyquinoline and alpha-bromonaphthalene. All of these function by arresting dividing cells at the metaphase stage, where chromosomes have already reached a highly condensed state. In order to obtain elongated mitotic chromosomes, I developed a pretreatment technique for obtaining prometaphase chromosomes using aphidicolin.

Aphidicolin is a chemical which inhibits eukaryotic DNA replication by binding to DNA polymerase alpha (Bryant and Dunham 1988). In this pretreatment procedure, aphidicolin is applied to root tip cells in order to block the cells at the S-phase thereby synchronizing them. Aphidicolin is then removed and cells are allowed to proceed through the cell cycle up to the prometaphase stage where they are fixed.

This procedure has been applied for cytogenetic analysis of tomato (Lapitan et al. 1989) and potato (Pehu, pers. comm.), both of which have small genomes (9.5×10^8 bp/haploid genome) (Arumuganathan and Earle 1991) and small chromosomes. Table A.1 shows a comparison of average sizes of tomato chromosomes obtained using alpha-bromonaphthalene (Ramanna and Prakken 1967) and aphidicolin (Lapitan et al. 1989). Chromosomes obtained with aphidicolin are approximately twice as long as those obtained with alpha-bromonaphthalene.

A.2.1
Aphidicolin Treatment for Synchronizing Tomato Root Tip Cells

Reagents

- Aphidicolin (Sigma 0781)
- Dimethyl sulphoxide (DMSO)
- Sodium hypochlorite
- Fixative: 3 parts ethanol, 1 part acetic acid

Equipment

- Petri dish (100 × 15 mm)
- Filter papers (9-cm circles)

Table A.1. Comparison of size of tomato mitotic chromosomes obtained with alpha-bromonaphthalene and aphidicolin

Chromosome no.	Alpha-bromonaphthalene[a] (in μm)	Aphidicolin[b] (in μm)
1	4.58	8.2
2	3.11	6.6
3	3.05	6.7
4	2.69	6.8
5	2.63	5.7
6	2.61	6.0
7	2.97	5.5
8	2.95	5.5
9	2.95	5.7
10	2.83	5.3
11	2.71	5.1
12	2.14	4.6
Mean	2.94	5.98

[a]Data taken from Ramanna and Prakken (1967). Lengths given are averages from 30 measurements for each chromosome.
[b]Data taken from Lapitan et al. (1989). Lengths given are averages from 20 measurements for each chromosome.

Procedure

1. Prepare the aphidicolin stock solution by dissolving 1 mg of aphidicolin in 1 ml DMSO. This stock solution is stored at 4 °C and thawed just before use.

2. Germinate seeds by soaking in 50% sodium hypochlorite for 15 min. Rinse thoroughly. Place seeds in petri dish with wet filter paper.

3. After three to four days (or when roots are approximately 0.5 cm long), transfer seedlings to a petri dish containing a filter paper moistened with 10 μg/ml aphidicolin. This working solution is prepared by adding 60 μl of the stock solution (1 mg/ml) to 6 ml distilled water. Mix and pour this solution into a petri dish containing a filter paper.

4. Leave the seedlings in aphidicolin overnight (approximately 14–16 h).

5. The next morning, remove aphidicolin by thoroughly rinsing the seedlings with distilled water.

6. Transfer the seedlings to another petri dish containing a filter paper moistened with distilled water.

7. After 4 h, cut the root tips and transfer to a freshly made solution of fixative. At this time, most of the cells should be in the stages of prometaphase to early metaphase. As seedlings are allowed to grow for a longer period of time after removal of aphidicolin, cells with more condensed chromosomes will be recovered.

Notes and troubleshooting

- This length of time (4 h) for allowing seedlings to grow further before fixation has been specifically worked out for *L.esculentum* (VF 36). This time may vary with different species. For pepper, for example, 6 h gives the best results. To determine the optimum amount of time to allow cells to proceed through the cell cycle before fixing them, first fix root tips from several seedlings at different time intervals (e.g. 2,4,6 and 8 h) after removal of aphidicolin. Prepare chromosome squashes from five to six seedlings from each treatment and determine the treatment that gives the highest numbers of cells at the prometaphase and early metaphase stages.

- The age of root tips used for aphidicolin treatment is critical. Root tips containing actively dividing cells must be used. When older tomato root tips are used for this procedure (i.e. more than 5-day-old seedlings), no mitotic cells are obtained. Similarly, root tips from stressed seedlings, such as those left in petri dishes with dry filter paper, will not give good chromosome preparations.

A.2.2
Chromosome Squashes

Reagents

- 1% acetocarmine
- 45% acetic acid
- Root tips fixed for 1 to 3 days

Equipment

- Glass slides
- Cover slips
- Scalpel with replaceable surgical blades
- Needle
- Double-edged razor blade, cut in half

Procedure

1. Soak root tips in 1% acetocarmine for 15 to 20 min to stain cells and soften the tissue.

2. Take out one root tip from the stain and blot on a paper for a few seconds to remove most of the stain. Place the root tip on a glass slide. Using a scalpel with a very sharp blade, cut off the root cap (outer 0.1 mm) and discard. Then cut a very thin cross section (about 0.1 mm), and apply one drop of 45% acetic acid.

3. Place half of a double-edged razor blade to one side of the root tip section so that one corner of the cover slip will lie on top of the razor blade (see Fig. A.1). Put a cover slip on top of the root tip section.

4. Gently tap the centre of the cover slip with a needle until the tissue section disperses.

5. Carefully remove the blade while holding down the opposite edge of the cover slip with your fingers to prevent the cover slip from moving.

6. Heat the glass slide on an alcohol lamp or gas burner for 3 to 4 s. Press the glass slide under a blotting paper to flatten the chromosome squash.

7. View under 20x or 40x objective. Store slides containing at least five well-spread cells free of cytoplasmic material in a –80 °C freezer. Use slides for in situ hybridization within a week, since the quality of chromosome morphology decreases after this time.

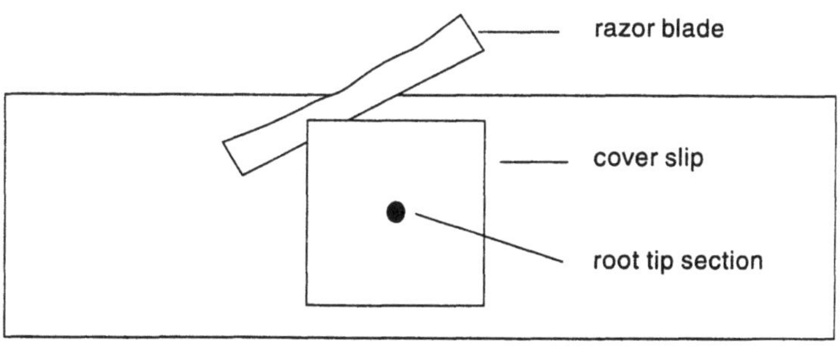

Fig. A.1. Diagram illustrating preparation of chromosome spreads free of cell wall and cytoplasm from root tips by tapping on a tissue section under a glass slide

Notes and troubleshooting

- If well-spread chromosome preparations are not obtained, perhaps your root tips sections are too thick. Keep practicing to obtain sections as thin as possible.

A.3 In Situ Hybridization

Protocols for in situ hybridization are extensively documented in the previous two chapters. Fluorescence staining is recommended over colorimetric detection for analyzing in situ hybridization signals of species with small chromosomes because total DNA can be more clearly visualized.

References

Albini SM, Schwarzacher T (1992) In situ localization of two repetitive DNA sequences to surface-spread pachytene chromosomes of rye. Genome 35:551–559

Arumuganathan K, Earle ED (1991) Estimation of nuclear DNA content of plants by flow cytometry. Plant Mol Biol Rep 9:229–241

Bryant JA, Dunham VL (1988) The biochemistry of DNA replication. In: Bryant JA, Dunham VL (eds) DNA replication in plants. CRC Press, Boca Raton, pp 17–54

Gall JG, Pardue ML (1969) Formation and detection of DNA-DNA hybrid molecules in cytological preparations. Proc Nat Acad Sci USA 63:378–383

Ganal MW, Lapitan NLV, Tanksley SD (1991) Macrostructure of the tomato telomeres (*Lycopersicion esculentum*). Plant cell 3:87–94

Heng HHQ, Squire J, Tsui L (1992) High resolution mapping of mammalian genes by in situ hybridization to free chromatin. Proc Nat Acad Sci USA 89: 9509–9513

Hutchinson J, Flavell RB, Jones J (1981) Physical mapping of plant chromosomes by in situ hybridization. In: Setlow J, Hollaender A (eds) Genetic engineering, vol 3. Plenum, New York, pp 207–222

Kafatos FC, Louis C, Savakis C, Glover DM, Ashburner M, Link AJ, Siden-Kiamos I, Saunders RDC (1991) Integrated maps of the *Drosophila* genome: progress and prospects. Trends Genet 7:155–161

Lapitan NLV (1992) Organization and evolution of higher plant nuclear genomes. Genome 35:171–181

Lapitan NLV, Ganal MW, Tanksley SD (1989) Somatic chromosome karyotype of tomato based on in situ hybridization of the TGRI satellite repeat. Genome 32:992–998

Lawrence JB, Villnave C, Singer RH (1988) Sensitive, high-resolution chromatin and chromosome mapping in situ: presence and orientation of two closely integrated copies of EBV in a lymphoma line. Cell 52:51–61

Lawrence JB, Villnave C, Singer RH, McNeil JA (1990) Interphase and metaphase resolution of different distances within the human dystrophin gene. Science 249:928–932
Lichter P, Boyle AL, Cremer T, Ward DC (1991) Analysis of genes and chromosomes by non-isotopic in situ hybridization. Gen Anal Tech Appl 8:24–35
Manuelidis L, Langer-Safer PR, Ware DC (1982) High-resolution mapping of satellite DNA using biotin-labelled DNA probes. J Cell Biol 95:619–625
Moens PB, Pearlman RE (1989) Satellite DNA I in chromatin loops of rat pachytene chromosomes and in spermatids. Chromosoma 98:287–294
Moens PB, Pearlman RE (1990) High resolution in situ DNA sequence mapping with surface-spread pachytene chromosomes. Cytogenet Cell Genet 53:219–220
Parra I, Windle B (1993) High resolution visual mapping of stretched DNA by fluorescent hybridization. Nature Genet 5:17–21
Ramanna MS, Prakken R (1967) Structure of and homology between pachytene and somatic metaphase chromosomes of the tomato. Genetica 38:115–133
Rayburn AL, Gill BS (1986) Molecular identification of the D-genome chromosomes of wheat. J Hered 77:253–255
Ried T, Baldini A, Rand TC, Ward DC (1992) Simultaneous visualization of seven different DNA probes by in situ hybridization using combinatorial fluorescence and digital imaging microscopy. Proc Nat Acad Sci USA 89:1388–1392
Trask B (1991) Fluorescence in situ hybridization. Trends Genet 7:149–154
Trask BJ, Massa H, Kenwrick S, Gitschier J (1991) Mapping of human chromosome Xq28 by two-color fluorescence in situ hybridization of DNA sequences to interphase cell nuclei. Am J Hum Genet 48:1–15
van den Engh G, Sachs R, Trask BJ (1992) Estimating genomic distance from DNA sequence location in cell nuclei by a random walk model. Science 257:1410–1412

Appendix Nuclear Genome Sizes of Important Plant Species

C.-N. Liu

Appendix. Nuclear Genome Sizes of Important Plant Species[a]

Scientific name	Common name	Family	kb/1C[b]
Arabidopsis thaliana	Arabidopsis	Cruciferae	1.5×10^5
Avena sativa	Oats	Gramineae	1.1×10^7
Beta vulgaris ssp. *saccharifera*	Sugar beet	Chenopodiaceae	7.6×10^5
Brassica napus	Rapeseed	Cruciferae	1.2×10^6
Brassica oleracea ssp. *botrytis*	Cauliflower	Cruciferae	6.4×10^5
Brassica oleracea ssp. *italica*	Broccoli	Cruciferae	6.0×10^5
Citrus sinensis	Orange	Rutaceae	3.8×10^5
Daucus carota	Carrot	Umbelliferae	4.7×10^5
Glycine max (2n = 4x)	Soybean	Leguminosae	1.1×10^6
Gossypium hirsutum (2n = 4x)	Cotton	Malvaceae	2.2×10^6
Helianthus annuus	Sunflower	Compositae	2.9×10^6
Hordeum vulgare	Barley	Gramineae	4.9×10^6
Ipomoea batatas (2n = 6x)	Sweet potato	Convolvulaceae	1.6×10^6
Lycopersicon esculentum	Tomato	Solanaceae	1.0×10^6
Nicotiana tabacum (2n = 4x)	Tobacco	Solanaceae	4.4×10^6
Oryza sativa ssp. *Indica*	Rice	Gramineae	4.3×10^5
Petunia hybrida	Petunia	Solanaceae	1.3×10^6
Phaseolus vulgaris	Common bean	Leguminosae	6.4×10^5
Solanum tuberosum (2n = 4x)	Potato	Solanaceae	1.7×10^6
Sorghum bicolor	Sorghum	Gramineae	7.6×10^5
Triticum aestivum (2n = 6x)	Wheat	Gramineae	1.6×10^7
Zea mays	Corn	Gramineae	2.5×10^6

[a]Data are based on K. Arumuganathan, E.D. Earle (1991) Nuclear DNA content of some important plant species. Plant Mol Biol Rep 9: 208–218.
[b]DNA content of the unreplicated haploid chromosome complement.
Supplied by C-N Liu, author Chapter 2.4.

Subject Index

Please note that this index should be used in conjunction with the comprehensive chapter listings at the front of the book. Many of the chapter sub-headings are self-explanatory and it is felt, do not require further duplication.

Only major text references to organisms have been included in the index.

MIX
Papier aus verantwortungsvollen Quellen
Paper from responsible sources
FSC® C105338

If you have any concerns about our products,
you can contact us on
ProductSafety@springernature.com

In case Publisher is established outside the EU,
the EU authorized representative is:
Springer Nature Customer Service Center GmbH
Europaplatz 3, 69115 Heidelberg, Germany

Printed by Libri Plureos GmbH
in Hamburg, Germany